VOLUME FIVE HUNDRED AND EIGHTY THREE

METHODS IN ENZYMOLOGY

Enzymology at the Membrane Interface: Interfacial Enzymology and Protein-Membrane Binding

METHODS IN ENZYMOLOGY

VOLUME FIVE HUNDRED AND EIGHTY THREE

METHODS IN ENZYMOLOGY

Enzymology at the Membrane Interface: Interfacial Enzymology and Protein-Membrane Binding

Edited by

MICHAEL H. GELB
University of Washington, Seattle, WA, United States

ELSEVIER

ACADEMIC PRESS
An imprint of Elsevier
elsevier.com

Academic Press is an imprint of Elsevier
50 Hampshire Street, 5th Floor, Cambridge, MA 02139, United States
525 B Street, Suite 1800, San Diego, CA 92101–4495, United States
The Boulevard, Langford Lane, Kidlington, Oxford OX5 1GB, United Kingdom
125 London Wall, London, EC2Y 5AS, United Kingdom

First edition 2017

Notices
Knowledge and best practice in this field are constantly changing. As new research and experience broaden our understanding, changes in research methods, professional practices, or medical treatment may become necessary.

Practitioners and researchers must always rely on their own experience and knowledge in evaluating and using any information, methods, compounds, or experiments described herein. In using such information or methods they should be mindful of their own safety and the safety of others, including parties for whom they have a professional responsibility.

To the fullest extent of the law, neither the Publisher nor the authors, contributors, or editors, assume any liability for any injury and/or damage to persons or property as a matter of products liability, negligence or otherwise, or from any use or operation of any methods, products, instructions, or ideas contained in the material herein.

ISBN: 978-0-12-809419-8
ISSN: 0076-6879

Publisher: Zoe Kruze
Acquisition Editor: Zoe Kruze
Editorial Project Manager: Helene Kabes
Production Project Manager: Magesh Kumar Mahalingam
Cover Designer: Greg Harris

CONTENTS

CONTRIBUTORS

S. Amara
CNRS, Aix Marseille Université, Enzymologie Interfaciale et de Physiologie de la Lipolyse, Marseille, France

K. Baek
Kyung Hee University, Yongin, South Korea

S.E. Barbour
University of Georgia at Athens, Athens, GA, United States

S. Bezzine
University of Sfax, Laboratoire de Biochimie et de Génie Enzymatique des Lipases, Sfax, Tunisia

A. Bénarouche
CNRS, Aix Marseille Université, Enzymologie Interfaciale et de Physiologie de la Lipolyse, Marseille, France

C. Bourlieu
INRA CIRAD, UMR1208 IATE, Montpellier, France

V. Brglez
Institut de Pharmacologie Moléculaire et Cellulaire, Université de Nice Sophia Antipolis and Centre National de la Recherche Scientifique (CNRS), Valbonne, France

J.E. Burke
University of Victoria, Victoria, BC, Canada

C. Cambillau
Aix-Marseille Université, CNRS, Architecture et Fonction des Macromolécules Biologiques, Marseille, France

F. Carrière
CNRS, Aix Marseille Université, Enzymologie Interfaciale et de Physiologie de la Lipolyse, Marseille, France

J.F. Cavalier
CNRS, Aix Marseille Université, Enzymologie Interfaciale et de Physiologie de la Lipolyse, Marseille, France

W. Cho
University of Illinois at Chicago, Chicago, IL, United States; Kyung Hee University, Yongin, South Korea

G.L. Dornan
University of Victoria, Victoria, BC, Canada

G. Du
University of Texas Health Science Center at Houston, Houston, TX, United States

S. Feng
European Molecular Biology Laboratory (EMBL), Heidelberg, Germany; NCCR Chemical Biology, University of Geneva, Geneva, Switzerland

M.A. Frohman
Center for Developmental Genetics, Stony Brook University School of Medicine, Stony Brook, NY, United States

M.H. Gelb
University of Washington, Seattle, WA, United States

F. Ghomashchi
University of Washington, Seattle, WA, United States

J. Gomez-Cambronero
Wright State University, Boonshoft School of Medicine, Dayton, OH, United States

E.E. Ha
Center for Developmental Genetics, Stony Brook University School of Medicine, Stony Brook, NY, United States

K.M. Henkels
Wright State University, Boonshoft School of Medicine, Dayton, OH, United States

Y. Hu
University of Illinois at Chicago, Chicago, IL, United States

L. Jeammet
Institut de Pharmacologie Moléculaire et Cellulaire, Université de Nice Sophia Antipolis and Centre National de la Recherche Scientifique (CNRS), Valbonne, France

M.L. Jenkins
University of Victoria, Victoria, BC, Canada

H. Kim
University of Illinois at Chicago, Chicago, IL, United States

G. Lambeau
Institut de Pharmacologie Moléculaire et Cellulaire, Université de Nice Sophia Antipolis and Centre National de la Recherche Scientifique (CNRS), Valbonne, France

C.C. Leslie
National Jewish Health, Denver; University of Colorado Denver, Aurora, CO, United States

S.-L. Liu
University of Illinois at Chicago, Chicago, IL, United States

S. Longhi
Aix-Marseille Université, CNRS, Architecture et Fonction des Macromolécules Biologiques, Marseille, France

E. Mateos-Diaz
CNRS, Aix Marseille Université, Enzymologie Interfaciale et de Physiologie de la Lipolyse, Marseille, France

Y. Miki
Lipid Metabolism Project, Tokyo Metropolitan Institute of Medical Science, Tokyo, Japan

A.J. Morris
The Gill Heart Institute, College of Medicine, Lexington Veterans Affairs Medical Center, University of Kentucky, Lexington, KY, United States

M. Murakami
Lipid Metabolism Project, Tokyo Metropolitan Institute of Medical Science; AMED-CREST, Japan Agency for Medical Research and Development, Tokyo, Japan

R. Murase
Lipid Metabolism Project, Tokyo Metropolitan Institute of Medical Science, Tokyo, Japan

C. Payré
Institut de Pharmacologie Moléculaire et Cellulaire, Université de Nice Sophia Antipolis and Centre National de la Recherche Scientifique (CNRS), Valbonne, France

F. Philip
Center for Developmental Genetics, Stony Brook University School of Medicine, Stony Brook, NY, United States

V. Point
CNRS, Aix Marseille Université, Enzymologie Interfaciale et de Physiologie de la Lipolyse, Marseille, France

D.M. Raben
Johns Hopkins University, Baltimore, MD, United States

S. Ramanadham
Comprehensive Diabetes Center, University of Alabama at Birmingham, Birmingham, AL, United States

A. Roussel
Aix-Marseille Université, CNRS, Architecture et Fonction des Macromolécules Biologiques, Marseille, France

L. Sams
CNRS, Aix Marseille Université, Enzymologie Interfaciale et de Physiologie de la Lipolyse, Marseille, France

H. Sato
Lipid Metabolism Project, Tokyo Metropolitan Institute of Medical Science, Tokyo, Japan

M. Schifferer
European Molecular Biology Laboratory (EMBL), Heidelberg, Germany

C. Schultz
European Molecular Biology Laboratory (EMBL), Heidelberg, Germany

M.A. Seeliger
Center for Developmental Genetics, Stony Brook University School of Medicine, Stony Brook, NY, United States

R. Sheng
University of Illinois at Chicago, Chicago, IL, United States

F. Stein
European Molecular Biology Laboratory (EMBL), Heidelberg, Germany

Y. Taketomi
Lipid Metabolism Project, Tokyo Metropolitan Institute of Medical Science, Tokyo, Japan

S.A. Tatulian
College of Sciences, University of Central Florida, Orlando, FL, United States

L.W.R. Tay
University of Texas Health Science Center at Houston, Houston, TX, United States

B. Tu-Sekine
Johns Hopkins University, Baltimore, MD, United States

O. Vadas
Pharmaceutical Sciences Section, University of Geneva, Geneva, Switzerland

V. Vié
Université Rennes 1, Institut de Physique de Rennes, UMR CNRS 6251, Rennes cedex, France

Z. Wang
University of Texas Health Science Center at Houston, Houston, TX, United States

K. Yamamoto
Lipid Metabolism Project, Tokyo Metropolitan Institute of Medical Science; PRIME, Japan Agency for Medical Research and Development, Tokyo; Faculty of Bioscience and Bioindustry, Tokushima University, Tokushima, Japan

Y. Yoon
Konkuk University, Seoul, South Korea

B. Yun
National Jewish Health, Denver, CO, United States

PREFACE

Interfacial enzymology is concerned with enzymes that must access their substrates from a boundary between two phases (an interface), the most common type being the lipid–water interface of biological membranes (Berg & Jain, 2002). Interfacial enzymes are distinct from noninterfacial enzymes (Gelb, Jain, Hanel, & Berg, 1995). The latter includes enzymes in the water layer that act on substrates in the water layer. One example is hexokinase, which is a water-soluble enzyme that acts on the highly water-soluble substrate glucose. Noninterfacial also include enzymes in the membrane that act on water-soluble substrates. One example is the phosphodiesterase in the visual transduction cycle that hydrolyzes cyclic GMP. These noninterfacial enzymes have in common that their active site should access substrate in the water layer even though one of the enzymes is an integral membrane protein. Presumably the active site is well exposed to the water layer even if the enzyme is bound to membranes. Noninterfacial enzymes are sensitive to the concentration of substrate in the aqueous phase (moles of substrate per volume of aqueous phase). Rate equations describing noninterfacial enzymes are composed of rate constants and aqueous phase concentrations of interacting partners (enzyme and substrate). By contrast, interfacial enzymes must access their substrates from the interface, and the rate equations that describe their action depend on rate constants and the moles of substrate per volume of interface. Among interfacial enzymes, we also include those that access their substrates from the membrane core, and in these cases the rate equation contains terms for the concentration of substrate in the membrane core.

For enzymes that act on highly water-insoluble substrates, for example, phospholipids with long fatty acyl chains or transmembrane segments of proteins, the concentration of substrate in the aqueous phase is vanishingly small, so much so that the enzyme is forced to access its substrate from the membrane phase. These are the interfacial enzymes. Strictly speaking, it is not possible to conclude that an enzyme is interfacial based on solubility arguments alone. Proof comes from the observation of processive behavior. Consider the enzyme secreted phospholipase A_2. This is a water-soluble enzyme that absorbs onto the surface of phospholipid vesicles and thus exists in a water-soluble state (E) and an interfacial state (E*). It has been experimentally demonstrated under some conditions that enzyme prebound to vesicles of one type of phospholipid is not able to act on vesicles of a different phospholipid added later (under conditions where there is no intervesicle transfer of phospholipids), yet if enzyme is added to a premixture of both vesicles, it acts on both types of phospholipids (Gelb et al., 1995). Enzyme bound to the first vesicle is able to remain bound to the vesicle and catalyze the hydrolysis of several phospholipids before leaving the interface. If the enzyme was acting on the trace amount of phospholipid in the water phase, this type of processive process would clearly not be possible. These results also show that binding of enzyme to the interface (E to E*) is not the same step as loading the active site of the enzyme with substrate (Michaelis complex formation, E* to E*S). If these steps were the same, release of product from the enzyme's active site would necessarily result in release of enzyme from the membrane interface to the water layer and processive behavior would not be possible. Processive behavior of interfacial enzymes has been called "scooting" by Mahendra Jain and coworkers, whereas nonprocessive behavior is called "hopping" (Jain & Berg, 1989).

An interesting dilemma results with an enzyme that acts on a substrate that has significant solubility in both the membrane and aqueous phases. One example is platelet-activating factor acetylhydrolase that cleaves the ester of a phospholipid containing a short acetyl group instead of a long-chain fatty acyl group. The enzyme is water soluble but found *in vivo* bound tightly to lipoproteins in blood. Does the enzyme access its substrate only in the aqueous phase (for example, the membrane-bound enzyme has its active site exposed mainly to the water layer) or does the enzyme access its substrate only in the lipoprotein (for example, its active site is not well exposed to the aqueous phase)? The question is nontrivial to answer because the substrate readily partitions between the aqueous and membrane phases. Thus, at equilibrium, the concentration of substrate in the membrane phase is equal to the concentration of substrate in the aqueous multiplied by the partition equilibrium constant. Also during equilibrium, variation of the concentration of substrate in one phase leads to a proportional variation of substrate in the other phase. Any steady-state rate equation that is written in terms of moles of substrate in the aqueous phase divided by the volume of the aqueous phase (conventional concentration) can be rewritten in terms of the moles of substrate in the membrane phase divided by the volume of membrane phase (interfacial concentration) times a constant. The two equations are mathematically equivalent, and it is thus impossible to design any steady-state kinetic experiment to determine which equation applies, i.e., whether the enzyme is interfacial or not. The problem has been solved for platelet-activating factor acetylhydrolase by studying the kinetics during the presteady-state phase in which the substrate has not yet equilibrated between membrane and aqueous phases (Min et al., 1999). This study showed that the enzyme operates on platelet-activating factor in the aqueous phase and is thus not an interfacial enzyme. One may wonder whether a membrane-bound signaling receptor that is gated by a ligand that can exist in the membrane and aqueous phases interacts with ligand in the membrane phase (interfacial receptor) or the aqueous phase (noninterfacial receptor). This question has never been answered.

For interfacial enzymes that display fast turnover numbers, the catalytic cycle may be limited by the rate of exchange of substrate between substrate aggregates containing bound enzyme. For example, some secreted phospholipases A_2 display a turnover number of $\sim 100\ s^{-1}$. If the enzyme is bound to a small unilamellar vesicle containing say 500 phospholipids, substrate would become exhausted in just a few seconds. To continue, enzyme must move to a new vesicle or there must be intervesicle exchange of phospholipids, and these processes may become rate-limiting for steady-state turnover. In this case the steady-state parameters do not reflect the true catalytic properties of the enzyme but rather those of substrate and enzyme intervesicle dynamics. This problem is especially pertinent to phospholipid–detergent mixed micelles that contain only a few phospholipids per particle (Dennis, Cao, Hsu, Magrioti, & Kokotos, 2011).

For enzymes such as secreted phospholipase A_2, there are two components to their substrate specificity. The first issue is what are the structure–function relationships that determine the affinity of the enzyme for the membrane surface (E to E*). Once bound to the interface, then comes the issue of the structural requirements of the interfacial substrate for binding to the catalytic site of E* to give the Michaelis complex E*S. For example, group IIA secreted phospholipase A_2 does not appreciably bind to vesicles that lack anionic phospholipids (for example, vesicles rich in phosphatidylcholine), yet once bound to anionic phospholipid vesicles, phosphatidylcholine that may be present in the same vesicles are good substrates for the enzyme (thus, $E^* + S \rightleftharpoons E^*S$ is favorable) (Gelb et al., 1995). Often substrate specificity studies of interfacial enzymes are carried out in ways that do not allow the deconvolution of these two processes and are thus almost impossible to interpret and to extrapolate from *in vitro* to *in vivo* settings.

Inhibition of interfacial enzymes is more difficult to study than with noninterfacial enzymes (Gelb et al., 1995). It is possible for inhibitors to act through nonspecific mechanisms, for example, compounds that partition into the membrane interface and change the physical properties of the interface in ways that promote enzyme desorption to the aqueous phase (E* to E). More interesting and useful inhibitors are those that bind specifically to the catalytic site (or other site) on the enzyme (E* + I gives E*I); these are analogous to inhibitors of noninterfacial enzymes. Early studies of phospholipase A_2 inhibitors were plagued by the common occurrence of nonspecific, membrane-perturbing inhibitors. For example, annexins were initially characterized as phospholipase A_2 inhibitors, but later studies showed that annexins form a tightly packed array on the membrane surface and simply occlude the interfacial enzyme from binding to the vesicle to access its substrate. Thus, annexins, at least *in vitro*, are inhibitors of virtually all interfacial enzymes that must undergo an E to E* transition as part of its catalytic cycle.

Volume 583 of *Methods in Enzymology* is focused on interfacial enzymes that act on lipid substrates. Methods for detecting and quantifying the binding of proteins and enzymes to membranes *in vitro* are covered in Chapters 1, 9, 10, and 11. Methods for detection of interfacial binding of proteins and enzymes to membranes in living cells are covered in Chapters 2, 4, 6, 13, and 15. Some interfacial enzymes form protein–protein complexes, and this is described in Chapters 8 and 14. Studies to evaluate the substrate specificity of interfacial enzymes on natural members are covered in Chapter 5. Novel spectroscopic studies for providing structural insight into protein–membrane binding are provided in Chapters 7, 9, 11, and 12. Methods for production of recombinant interfacial enzymes are given in Chapters 3 and 6.

Volume 584 of *Methods in Enzymology* is focused on proteolytic enzymes that access their transmembrane peptide substrates in the membrane phase (interfacial proteases). Biochemical characterization of bacterial and eukaryotic transmembrane proteases is covered in Chapters 1, 2, 4, 5, 6, 7, 10, and 15. Kinetic studies including substrate specificity and inhibition are included in Chapters 2, 4, 5, 6, 9, 11, 12, and 15. Studies that probe the structure of transmembrane proteases are given in Chapters 3, 6, 8, and 13. Production of recombinant transmembrane proteases is the specific focus of Chapters 5 and 10.

In summary, interfacial enzymes are an important subset of enzymes. Early studies were focused on membranes that act at the lipid–water interface. More recently a new class of proteases that act on transmembrane protein segments have been discovered in several organisms. Special methods are required for the characterization of interfacial enzymes including their basic features of substrate specificity and inhibition. Structural studies are challenging because the enzyme acts in an environment that is not amenable to conventional techniques for determining molecular structure.

Michael H. Gelb
University of Washington, Seattle, WA, United States

REFERENCES

Berg, O. G., & Jain, M. K. (2002). *Interfacial enzyme kinetics*. West Sussex, England: Wiley.

Dennis, E. A., Cao, J., Hsu, Y.-H., Magrioti, V., & Kokotos, G. (2011). Phospholipase A2 enzymes: Physical structure, biological function, disease implication, chemical inhibition, and therapeutic intervention. *Chemical Reviews*, *111*, 6130–6185. http://dx.doi.org/10.1021/cr200085w.

Gelb, M. H., Jain, M. K., Hanel, A. M., & Berg, O. (1995). The interfacial enzymology of glycerolipid lipases: Lessons from secreted phospholipases A2. *Annual Review of Biochemistry*, *64*, 653–688.

Jain, M. K., & Berg, O. (1989). The kinetics of interfacial catalysis by phospholipase A2 and regulation of interfacial activation: Hopping versus scooting. *Biochimica et Biophysica Acta*, *1002*, 127–156.

Min, J. H., Jain, M. J., Wilder, C., Paul, L., Apitz-Castro, R., Aspleaf, D. C., et al. (1999). Membrane-bound plasma platelet activating factor acetylhydrolase acts on substrate in the aqueous phase. *Biochemistry*, *38*, 12935–12942.

CHAPTER ONE

A High-Throughput Fluorometric Assay for Lipid–Protein Binding

W. Cho*^,†,1^, Y. Hu*, K. Baek†, H. Kim*

*University of Illinois at Chicago, Chicago, IL, United States
†Kyung Hee University, Yongin, South Korea
[1]Corresponding author: e-mail address: wcho@uic.edu

Contents

Abstract

An increasing number of intracellular and extracellular proteins are shown to interact with membrane lipids under physiological conditions. For rapid and robust quantitative measurement of lipid–protein interaction, we developed a sensitive fluorescence quenching-based assay that is universally applicable to all proteins and lipids. The assay employs fluorescence protein (FP)-tagged proteins whose fluorescence emission intensity is decreased when they bind vesicles containing quenching lipids. This simple assay can be performed with a fluorescence plate reader or a spectrofluorometer and optimized for different proteins with various combinations of FPs and quenching lipids. The assay allows a rapid, sensitive, and accurate determination of lipid specificity and affinity for various lipid-binding proteins, and high-throughput screening of molecules that modulate their membrane binding.

Methods in Enzymology, Volume 583
ISSN 0076-6879
http://dx.doi.org/10.1016/bs.mie.2016.09.004

1. INTRODUCTION

Membrane lipids regulate and coordinate a wide variety of biological processes by interacting with diverse proteins (Cho, 2006; van Meer, Voelker, & Feigenson, 2008). It has been well documented that the lipid–protein interaction is important for the structure and function of transmembrane proteins that constitute about one-third of human proteome (Barrera, Zhou, & Robinson, 2013; Lee, 2011; van Meer et al., 2008). In addition, many intracellular and extracellular proteins that are involved in various processes, including cell signaling, membrane trafficking, blood coagulation, and viral infection, are recruited to cell membranes through lipid–protein interactions (Cho, 2006; Cho & Stahelin, 2005; DiNitto, Cronin, & Lambright, 2003; Lemmon, 2008). Most of these proteins contain lipid-binding domains or motifs that mediate membrane binding either through specific recognition of lipid headgroups or by nonspecific electrostatic and hydrophobic interactions with membrane lipids (Cho & Stahelin, 2005; DiNitto et al., 2003; Lemmon, 2008). Recent bioinformatics calculations have predicted that a large portion of cellular proteins may be able to interact with membrane lipids under physiological conditions because they have structural features necessary for membrane binding (Bhardwaj, Stahelin, Zhao, Cho, & Lu, 2007; Silkov et al., 2011). Consistent with the prediction, recent genome-wide studies of protein domains, including PSD95, Dlg1, ZO-1 (PDZ) and Src-homology 2 (SH2) domains, have demonstrated that a large number of these domains bind membrane lipids with high affinity and specificity (Bhardwaj, Stahelin, Langlois, Cho, & Lu, 2006; Chen et al., 2012; Feng & Zhang, 2009; Park et al., 2016; Sheng et al., 2012, 2014; Silkov et al., 2011; Zimmermann, 2006). These results suggest that the number of membrane-binding intracellular and extracellular proteins is much larger than originally suspected and that membrane binding may be an important regulatory step for these proteins. A rapid, sensitive, and quantitative assay for membrane–protein interaction is thus necessary for further identification and characterization of rapidly growing families of membrane-binding proteins and for identification of small molecules modulating their membrane interaction.

Membrane binding of soluble proteins has been measured by various biochemical and biophysical methods (Cho, Bittova, & Stahelin, 2001; Narayan & Lemmon, 2006). Sedimentation assays using lipid vesicles (Rebecchi, Peterson, & McLaughlin, 1992) or lipid-coated beads (Kim,

Lichtenbergova, Snitko, & Cho, 1997) have been most commonly used to assess membrane binding of proteins. However, difficulties associated with accurately quantifying membrane-bound vs free proteins and variable pelleting efficiency associated with different lipid vesicles have limited their utility. The lipid overlay assay has been commonly used due to its convenience but it suffers from many drawbacks, including low sensitivity, poor reliability, and an inability to yield quantitative information (Dowler, Kular, & Alessi, 2002). Also, lipids are presented in a poorly defined, nonphysiological state in this assay. The surface plasmon resonance (SPR) analysis allows robust quantitative analysis of membrane–protein interactions and has thus been a mainstay in biophysical characterization of membrane-binding proteins (Stahelin, 2013; Stahelin & Cho, 2001). Although this assay offers many advantages, including high sensitivity, no requirement for protein labeling, and an ability to provide kinetic information, it also has drawbacks, including the necessity of expensive instrumentation and rigorous controls to eliminate nonspecific binding, uncertainty about the physical nature of lipids coated on the sensor chip, and binding measurements under nonequilibrium conditions. Due to their high sensitivity, various fluorescence techniques have been extensively employed to monitor membrane–protein interaction. Most commonly, an increase (Kraft, Garrido, Leiva-Vega, & Romero, 2009), quenching (Dua, Wu, & Cho, 1995), or fluorescence resonance energy transfer (Bazzi & Nelsestuen, 1987; Nalefski, Slazas, & Falke, 1997) of Trp fluorescence is monitored during membrane–protein binding. Although rapid and convenient, these methods are not generally applicable to all proteins because many proteins do not have Trp on their membrane-binding surfaces. Although Trp can be genetically introduced to the membrane-binding surface to a protein, this can dramatically change its membrane-binding property (Sumandea, Das, Sumandea, & Cho, 1999). Alternatively, the protein can be labeled with an organic fluorophore, but this approach is limited by experimental inconvenience and the relative low yield of chemical modification (Liu et al., 2014; Yoon, Lee, Kurilova, & Cho, 2011). Fluorescence anisotropy (Miao et al., 2010) and fluorescence correlation spectroscopy analyses (Rusu, Gambhir, McLaughlin, & Radler, 2004) have been also used to measure membrane–protein interaction but with limited applications. Most important, none of these methods are suitable for robust and universal high-throughput analysis.

To overcome these technical limitations and obstacles, we developed a high-throughput membrane-binding assay that is based on fluorescence quenching of fluorescence proteins (FPs), such as enhanced green FP (EGFP), fused to a membrane-binding protein by a dark quencher-containing lipid, such as *N*-dimethylaminoazobenzenesulfony l-phosphatidylethanolamine (dabsyl-PE), incorporated in lipid vesicles (Cho, Kim, & Hu, 2016; Kim, Afsari, & Cho, 2013). Since FPs do not have affinity for membrane lipids (Kim et al., 2013), an FP tag does not affect the membrane binding of a diverse group of proteins tested so far. Furthermore, FPs offer an additional advantage of stabilizing the fused protein or domain (Park et al., 2016). Through the custom selection of an FP and a dark quencher from widely available collections, this simple and rapid assay can be optimized for sensitive, accurate, and reproducible quantitative determination of lipid affinity and specificity of diverse proteins as well as for high-throughput screening of small molecules that can modulate their membrane binding.

2. FLUORESCENCE QUENCHING ASSAY

2.1 Assay Design and Optimization

As described earlier, intrinsic Trp fluorescence cannot be utilized for a sensitive, universal membrane-binding assay. Also, chemical labeling of proteins, although offering higher sensitivity, suffers from low labeling yield and experimental inconvenience. Our strategy for the universal fluorescence assay is to genetically render favorable fluorescence properties to all proteins by expressing them as FP-fusion proteins and then quantify their membrane binding by monitoring the decrease in FP fluorescence as the proteins bind the lipid vesicles incorporating a nonfluorescent quenching dye (dark quencher; see Fig. 1A). FP-fusion proteins have been extensively used to monitor the subcellular localization, dynamics, and interactions of numerous cellular proteins (Miyawaki, 2011; Shaner, Steinbach, & Tsien, 2005) and many FP tags are known to have a minimal effect on the membrane-binding properties of fusion partner proteins when attached either their N- or C-terminus with an appropriate linker (Kim et al., 2013; Park et al., 2016). As is the case with other fusion tag proteins, FP tags also stabilize the partner proteins, especially when they are expressed as isolated protein domains or truncated proteins, thereby allowing higher expression yields of recombinant proteins (Park et al., 2016). EGFP is a preferred FP because it is the most commonly used FP with relatively high quantum yield and with a lesser effect on protein aggregation than other FPs (Miyawaki, 2011;

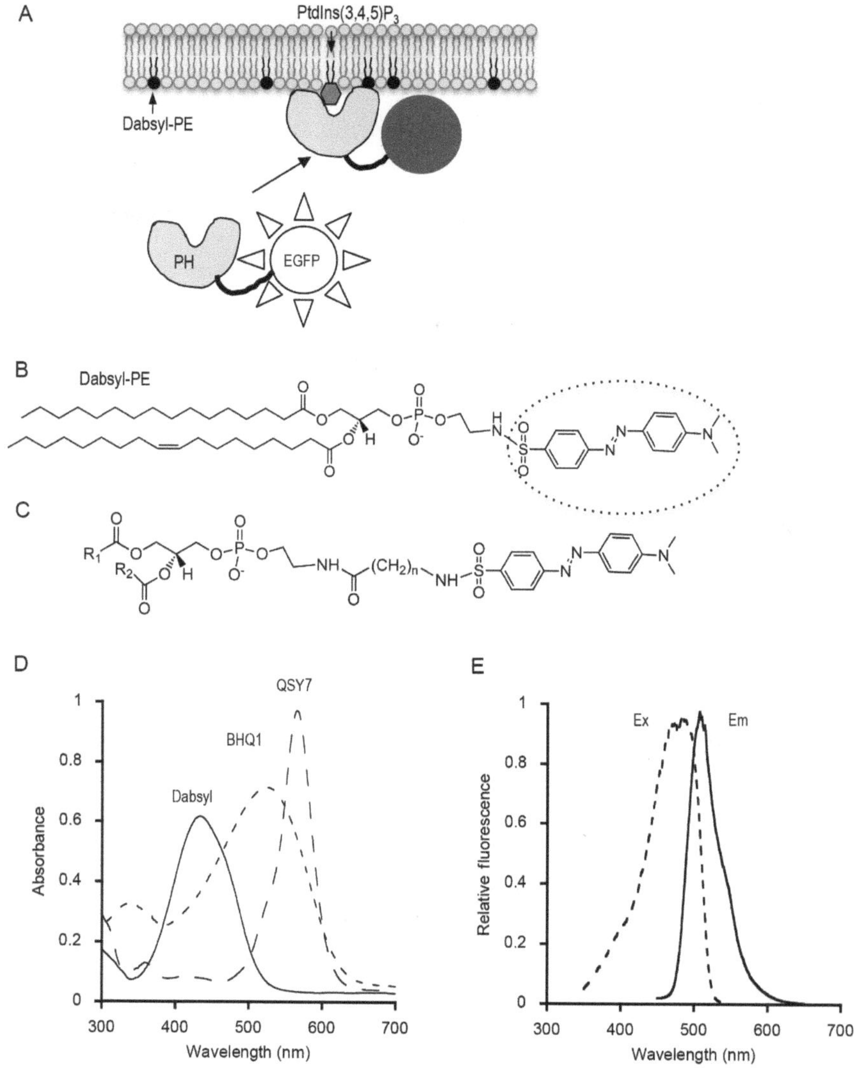

Fig. 1 Principles and components of our fluorescence quenching-based membrane-binding assay. (A) The general strategy of the assay. To the solution of an FP-fusion protein (e.g., Akt1-PH-EGFP) is added vesicles containing their favorite lipids (e.g., PtdIns (3,4,5)P_3) and a dark quencher, such as dabsyl-PE (shown in *black*). As the protein approaches the membrane surface, the fluorescence emission intensity of FP is greatly reduced due to quenching by dabsyl-PE, allowing quantitative analysis of membrane–protein interaction. (B) The chemical structure of dabsyl-PE. The quencher moiety is highlighted by the *dotted oval*. (C) The chemical structures of dabsyl-PE derivatives with extended linkers. Typically, R_1 = palmitoyl, R_2 = oleoyl, n = 2–5. (D) The absorption spectra of dabsyl-PE, BHQ1-PE, and QSY7-PE (all 60 n*M*) in chloroform. (E) Fluorescence excitation (Ex) and emission (Em) spectra of EGFP.

Shaner et al., 2005). In the case higher sensitivity is required for the assay, a brighter FP, such as mNeonGreen (Shaner et al., 2013), can be substituted for EGFP.

Dabsyl-PE (Fig. 1B) has been primarily used as a quenching lipid for routine assays because it is easy and inexpensive to prepare in large quantities. Because its absorption spectrum (Fig. 1D) overlaps more with the excitation spectrum than with the emission spectrum of EGFP (Fig. 1E), however, it is not ideal for the assay employing EGFP-fusion proteins. For high quenching efficiency and assay sensitivity, one can employ black hole quencher® 1 (BHQ1)-PE or QSY®7-PE for dabsyl-PE as a quenching lipid (Fig. 1D). BHQ1-PE typically allows >30% improvement over dabsyl-PE in EGFP quenching efficiency. QSY7-PE (absorption maximum at ≈570 nm; see Fig. 1D) is better suited for YFP (λ_{Ex} = 514 nm; λ_{Em} = 527 nm) and YFP quenching can be measured with λ_{Ex} = 510 nm and λ_{Em} = 540 nm. Despite their improved spectral properties, however, general use of BHQ1-PE and QSY7-PE is limited because they are expensive to prepare in large quantities.

Another factor in assay design is the quenching efficiency that depends on the distance between the lipid quencher and the FP. Ideally, an FP tag should be located close enough to the membrane-binding surface of the host protein to yield >40% of quenching upon membrane binding. One can achieve higher quenching yield by changing the location of the FP tag (i.e., N- or C-terminal) or the length of the linker between the FP and the host protein. Another way to improve the quenching efficiency is to extend the hydrophilic linker between PE and the quencher moiety of the quenching lipid (e.g., see Fig. 1C) so that the quencher can stretch further into the solution.

2.2 Preparation of Quencher Lipid Vesicles

The lipid composition of vesicles and the mole% of a quenching lipid can be adjusted according to the purpose of the assay. All PE-based quenching lipids are anionic lipids and might nonspecifically increase the vesicle binding of proteins with cationic membrane-binding surfaces. For accurate determination of membrane-binding affinity (or specificity) of a protein under physiological conditions, the concentration of a quenching lipid must be thus kept as low as possible (i.e., just high enough to allow robust quenching). Typically, 40% of FP quenching is sufficient for robust and reproducible binding analysis. For initial assessment of membrane-binding affinity of a large number of proteins and high-throughput screening of small molecule, one can increase the concentration of the quenching lipid to maximize the FP quenching efficiency. In all cases, background correction with lipid vesicles containing primarily the quenching lipid is recommended. For example, for phosphatidylinositol-3,4,5-trisphosphate (PtdIns(3,4,5)P_3)-binding proteins that are assayed with 1-palmitoyl-2-oleoyl-*sn*-glycero-3-phosphocholine (POPC)/dabsyl-PE/PtdIns (3,4,5)P_3 (92:5:3) vesicles, background correction with POPC/dabsyl-PE (95:5) is appropriate. Here, POPC is used as a bulk lipid because most proteins do not bind PC.

Large unilamellar vesicles (LUVs) are more uniform and stable than small unilamellar vesicles (SUVs) and they are thus more suitable for accurate binding measurements. The uniformity of vesicles is also important for quenching efficiency because the presence of vesicles lacking quenching lipids would lower the overall quenching efficiency by sequestering FP-tagged proteins. However, it takes much longer to prepare LUVs through extrusion, which becomes a major limiting factor for high-throughput screening that requires a large amount of vesicles. For high-throughput screening, SUVs are thus preferred to LUVs. Due to their low stability, SUVs should be used within a few hours of preparation.

1. POPC (Cat. No. 850457), 1-palmitoyl-2-oleoyl-*sn*-glycero-3-phosphoethanolamine (POPE; Cat. No. 850757), 1-palmitoyl-2-oleoyl-*sn*-glycero-3-phosphoserine (POPS; Cat. No. 840034), or soy phosphatidylinositol (PtdIns, Cat. No. 840044) (all from Avanti Polar Lipids) are dissolved in the highest grade oxygen-free chloroform to yield 10 mg/mL stock solutions and stored in Teflon-sealed vials at −20°C. All organic solvents including chloroform (Cat. No. C607-4), methanol (Cat. No. 53816), triethylamine (Cat. No. O4885-1), and dichloromethane (Cat. No. D37-1) are from ThermoFisher (Waltham, MA, USA).
2. 1,2-Dipalmitoyl derivatives of phosphoinositides (PtdIns*P*s), including PtdIns(3,4,5)P_3 (Cayman Chemical, Cat. No. 64920), are dissolved in the highest grade oxygen-free chloroform/methanol/water (3:3:1 v/v/v) to yield 0.5 mg/mL solutions and also stored in Teflon-sealed vials at −20°C.
3. Dabsyl-PE is synthesized from POPE as follows: POPE (50 mg) is dissolved in chloroform (2 mL) and the solution is added to a solution of dabsyl chloride (Tokyo Chemical Industry, Cat. No. D1382; 22.6 mg) and triethylamine (0.2 mL) in chloroform (5 mL). The mixture is stirred for 6 h at room temperature in the dark and the solvent

removed in vacuo. The residue is dissolved in dichloromethane/methanol (9:1) and purified by silica column chromatography using the same solvent mixture as eluent. The solvent is evaporated in vacuo to afford dabsyl-PE as an orange solid. A stock solution of dabsyl-PE in chloroform (10 mg/mL) is prepared and stored in a Teflon-sealed vial at −20°C.

4. BHQ1-PE and QSY7-PE are synthesized as follows: 8 mg of POPE and 5 mg of QSY7 carboxylic acid succinimidyl ester (ThermoFisher, Waltham, MA, USA, Cat. No. Q10193) or BHQ1 carboxylic acid succinimidyl ester (LGC Biosearch Technologies, Cat. No. BHQ-1000S-5) are mixed in 800 μL of chloroform and 200 μL of triethylamine. The reaction and purification are performed as described for dabsyl-PE. 10 mg/mL stock solutions of BHQ1-PE and QSY7-PE in chloroform are prepared and stored in Teflon-sealed vials at −20°C.
5. LUVs (100-nm diameter) are prepared by mixing the lipid solutions in chloroform according to the final lipid composition (e.g., POPC/dabsyl-PE/PtdIns(3,4,5)P_3 = 92:5:3) and the solvent is evaporated under the gentle stream of nitrogen gas. 20 m*M* Tris (ThermoFisher, Waltham, MA, USA, Cat. No. 142521) buffer, pH 7.4, containing 0.16 *M* NaCl (ThermoFisher, Waltham, MA, USA, Cat. No. 156697) is added to the lipid film to adjust the final lipid concentration, and the mixture is vortexed for 1 min, and sonicated in a sonicating bath for 1 min to break multilamellar vesicles. LUVs are prepared by multiple extrusion through a 100-nm polycarbonate filter (Avanti, Cat. No. 610005) using a Mini-Extruder (Avanti, Cat. No. 610000).
6. SUVs (<15-nm diameter) are prepared by mixing the lipid solutions in chloroform according to the final lipid composition and the solvent is evaporated under the gentle stream of nitrogen gas. 20 m*M* Tris buffer, pH 7.4, containing 0.16 *M* NaCl is added to the lipid film to adjust the final lipid concentration, and the mixture is vortexed for 1 min, and sonicated using a Branson Sonifier until the solution becomes clear.

2.3 Preparation of FP-Tagged Proteins

For most proteins, N-terminal and C-terminal FP tags have essentially the same effect: i.e., they improve the protein expression yield without affecting membrane-binding properties of the fusion partner proteins (Park et al., 2016). Depending on the structure of the protein, however, N-terminal and C-terminal FP tags may differentially affect the structure and function of the proteins and also the quenching efficiency of FP in the assay. The location of the FP tag and the sequence and length of the linker between FP and the host protein should thus be adjusted to optimize the assay condition. FP-tagged proteins are typically expressed with a His_6 tag and purified by affinity chromatography using a Ni-chelate column. For some proteins, most notably electrostatically neutral proteins, the His_6 tag may significantly enhance the binding of a neutral to anionic protein to anionic vesicles. In such a case, the His_6 tag should be enzymatically removed after affinity purification or the assay should be performed at slightly elevated pH (e.g., pH 7.8) to ensure that all His side chains are deprotonated. Also, the protein can be expressed as a glutathione-*S*-transferase (GST)-tagged protein, purified by glutathione-based affinity chromatography, and the GST tag removed before the assay. A typical procedure for bacterial expression of EGFP- and His_6-tagged protein is as follows:

1. 500 mL of Luria (BD, Cat. No. 244610) broth containing 50 μg/mL kanamycin (ThermoFisher, Waltham, MA, USA, Cat. No. 155479) or 100 μg/mL ampicillin (ThermoFisher, Waltham, MA, USA, Cat. No. 160561) is inoculated with BL21 RIL colonies expressing each FP-tagged protein.
2. Cells are allowed to grow in the medium at 37°C until an absorbance at 600 nm reaches 0.6.
3. Protein expression is induced with the addition of 100 μ*M* isopropyl 1-thio-β-D-galactopyranoside (Gold Biotechnology, St. Louis, MO, USA, Cat. No. 12481C5), and cells are moved to a 25°C shaker for 14 h incubation.
4. Cells are harvested by centrifugation (2500 × *g* for 10 min at 4°C), and the pellet is resuspended in 20 mL of the lysis buffer (50 m*M* Tris–HCl, 300 m*M* NaCl, 10 m*M* imidazole (ACROS Organics, Cat. No. B0141064), and 10% (*v*/*v*) glycerol (ThermoFisher, Waltham, MA, USA, Cat. No. 155081), pH 7.9).
5. The solution is sonicated for 5 min with 30-s intervals and the mixture is centrifuged for 30 min (39,000 × *g* at 4°C). The supernatant is transferred to a 50-mL Falcon tube and 1 mL of Ni-NTA agarose (Qiagen, Cat. No. 151032765) is added to it.
6. The supernatant is allowed to equilibrate with the resin for 30 min at 4°C with gentle mixing. The supernatant is poured onto a column, and the resin is washed with 50 mL of 50 m*M* Tris buffer, pH 7.4, containing 20 m*M* imidazole and another 50 mL of 50 m*M* Tris buffer, pH 7.4, containing 40 m*M* imidazole. The protein is eluted using 50 m*M* Tris buffer, pH 7.4, with 300 m*M* imidazole.
7. The purity of the eluted protein is checked by sodium-dodecylsulfate (ThermoFisher, Waltham, MA, USA, Cat. No. BP166-500) gel

electrophoresis using an 18% polyacrylamide gel, and the protein concentration is determined during the Bradford reagents (BioRad, Hercules, CA, USA, Cat. No. 5000006). The protein is frozen in liquid nitrogen and stored at −20°C.

2.4 Plate Reader-Based Quantitative Lipid Specificity Assay

Many membrane-binding proteins coincidently interact with both bulk lipids and a specific lipid under physiological conditions (Cho & Stahelin, 2005). For example, most of pleckstrin homology (PH) domains interact with multiple anionic lipid molecules, most notably phosphatidylserine (PS), via their membrane-binding surface and a PtdIns*P* molecule in their lipid-binding pocket (Lucas & Cho, 2011; Manna, Albanese, Park, & Cho, 2007; Manna et al., 2008; Stahelin et al., 2003). For accurate, physiologically relevant determination of lipid specificity, a two-step assay is thus recommended. First, any requirement for a bulk lipid for a protein is determined using POPC/dabsyl-PE/L_1 ($95-x$:5:x) (L_1 = PS, cholesterol, etc.; x = 0–30 mole%) vesicles. Then, using POPC/dabsyl-PE/L_1/L_2 ($75-x$:5:20:x) (L_2 = PtdIns*P*, phosphatidic acid, diacylglycerol, etc.; x = 0–10 mole%; 20 mole% is an arbitrary concentration for a bulk lipid) vesicles, specificity (headgroup or acyl chain) of its lipid-binding pocket can be determined. This two-step assay is preferred to a single-step assay employing, e.g., POPC/dabsyl-PE/L (70:5:25) (L = all lipids) because it is not physiologically meaningful to directly compare the affinity for bulk lipids, such as PS, and that for signaling lipids, such as PtdIns(3,4,5)P_3, which exist in much lower concentrations in cell membranes.

1. A fixed concentration of protein (e.g., 100 n*M*) is added to each well of a Corning nontreated black polystyrene 96-well plate (ThermoFisher, Waltham, MA, USA, Cat. No. 07-200-590).
2. To each row vesicles with varying lipid composition (e.g., POPC/dabsyl-PE/L_1 ($95-x$:5:x) (L_1 = any signaling lipid such as PtdIns*P*, phosphatidic acid, and etc.; x = 0–10 mole%)) are added. The total lipid concentration (e.g., 10 μ*M*) is kept constant and the final volume of the mixture in the assay buffer is adjusted to 200 μL.
3. Other rows are filled with different lipid mixtures (e.g., POPC/dabsyl-PE/L_2 ($95-x$:5:x), POPC/dabsyl-PE/L_3 ($95-x$:5:x), etc.).
4. One row is reserved for background correction for nonspecific quenching using the same protein and POPC/dabsyl-PE (95:5) vesicles.
5. The plate is incubated at 25°C with extremely gentle shaking for 5 min.

6. The decrease in EGFP fluorescence emission intensity at 509 nm is monitored with excitation set at 488 nm using a fluorescence plate reader. Excitation and emission wavelengths are adjusted for other FPs.

2.5 Plate Reader-Based Membrane-Binding Affinity Assay

Once lipid specificity of a protein is known, its affinity for the favorite lipid vesicles, such as POPC/POPS/PtdIns(3,4,5)P_3/dabsyl-PE (72:20:3:5), is determined by varying the total concentration of the same lipid vesicles with a fixed concentration of the protein. An alternative way is to employ the fixed total concentration of lipid vesicles and varying concentrations of the protein. This approach is useful when the nonspecific quenching becomes significant because of low quenching efficiency of specific binding.

1. To each well of a given row of a 96-well plate a fixed concentration of protein (e.g., 100 n*M*) and an increasing concentration of lipid vesicles with fixed composition (e.g., 0–150 μ*M* of POPC/PtdIns(3,4,5)P_3/dabsyl-PE (92:3:5) vesicles) are added. The final volume of the assay mixture is adjusted to 200 μL.
2. Multiple (i.e., ≥triplicate) rows are filled with the same protein and lipid mixtures for multiplex determination.
3. One row is selected for background correction for nonspecific quenching using the same protein and dabsyl-PE vesicles (e.g., POPC/dabsyl-PE (95:5) vesicles).
4. The plate is incubated at 25°C with extremely gentle shaking for 5 min.
5. The decrease in EGFP fluorescence emission intensity at 509 nm is monitored with excitation set at 488 nm using a fluorescence plate reader. Excitation and emission wavelengths are adjusted for other FPs.

2.6 Spectrofluorometric Membrane-Binding Affinity Assay

The cuvette-based spectrofluorometric assay is recommended when accurate and robust quantification of membrane affinity of proteins is required for biophysical analysis (e.g., binding energy calculation).

1. 10–20 μL of the protein stock solution is added to 2 mL of the assay buffer equilibrated at 25°C to the final concentration (typically 100 n*M*) depending on the K_d value for the protein and the particular vesicles (e.g., POPC/PtdIns(3,4,5)P_3/dabsyl-PE (92:3:5) vesicles). The protein solution is transferred to a 3-mL (or smaller) quartz cuvette in a thermostated (25°C) spectrofluorometer.

2. The equilibrated value of EGFP fluorescence emission intensity at 509 nm is monitored with excitation set at 488 nm after 5 μL incremental addition of the lipid vesicle solution.
3. Lipid addition is continued until binding reaches saturation.
4. The same experiment is repeated with dabsyl-PE vesicles (e.g., POPC/dabsyl-PE (95:5) vesicles) for background correction for nonspecific quenching.

2.7 High-Throughput Screening of Membrane-Binding Inhibitors

This assay has been also used to screen small molecules that specifically modulate the membrane binding of various proteins through competition analysis (Cho et al., 2016; Kim et al., 2013). In most cases, molecules are screened for inhibitory activity by monitoring dequenching of FP fluorescence as a consequence of membrane unbinding of proteins. Since the inhibition assay monitors the increase of FP fluorescence, it allows high sensitivity. It should be noted that some active small molecules are fluorescent and may have spectral overlap with FP and dabsyl-PE. Therefore, rigorous controls and background correction are necessary to minimize false-positive hits. The assay is typically run in two steps: an initial high-throughput assay using a single-point lipid determination (e.g., a fixed vesicle concentration giving a half-maximal quenching of EGFP in the absence of an inhibitor), followed by a more accurate determination with multiple lipid concentrations. The assay can also be used to screen molecules that allosterically enhance the membrane binding of a protein. In this case, increased quenching of FP signals (i.e., a further decease in fluorescence intensity) is monitored.

2.7.1 High-Throughput Inhibitor Screening Protocol

1. Fixed concentrations of a protein and a lipid vesicle in 200 μL of the assay buffer (20 m*M* Tris–HCl buffer, pH 7.4, containing 0.16 *M* NaCl) are added to each well of the plate. Typically, the protein concentration is 100 n*M* and the lipid vesicle concentration is adjusted to give ≈80% of maximal FP quenching.
2. A fixed concentration of different inhibitors is added to each well. The volume of dimethyl sulfoxide (DMSO; ThermoFisher, Waltham, MA, USA, Cat. No. D128-500) (i.e., ≤2%, v/v) is minimized to avoid protein denaturation. The inhibitor concentration is typically set at 20 μ*M*

for initial screening and is gradually lowered for the following rounds of screening to increase the detection threshold.

3. For background correction for each row, the assay is run with the mixture minus lipid vesicles in the next row.
4. The plate is incubated at 25°C with extremely gentle shaking for 10 min.
5. The increase in EGFP fluorescence intensity at 509 nm is monitored with excitation set at 488 nm.

2.7.2 Determination of Inhibition Constant

1. Fixed concentrations of a protein and a lipid vesicle in 200 μL of the assay buffer are added to each well of the plate. Typically, the protein concentration is 100 *nM* and the lipid vesicle concentration is adjusted to give ≈ 80% of maximal FP quenching.
2. An increasing concentration of an inhibitor is added to each well of a given row. The total volume of DMSO (i.e., ≤2%, v/v) is kept constant for all wells.
3. The plate is incubated at 25°C with extremely gentle shaking for 10 min.
4. The increase in EGFP fluorescence intensity at 509 nm is monitored with excitation set at 488 nm.
5. For background correction for each row, the assay is run with the mixture minus lipid vesicles in the next row.

3. DATA ANALYSIS

Since all lipid molecules are associated with membranes, protein–lipid interaction is a complex process that involves membrane adsorption of protein molecules, lateral diffusion of membrane-bound proteins and lipids, binding of a lipid(s) into a lipid pocket, reorientation and/or aggregation of membrane-bound proteins, membrane penetration of proteins, etc. (Cho & Stahelin, 2005). Thus, unlike protein–ligand binding that can be described in a straightforward manner, protein–lipid binding must be analyzed using simplified models. The simplest model for membrane binding is to assume that each protein (P) binds independently to a site on the membrane composed of n lipids (L) (i.e., $[P]+[L]/n \leftrightarrow$ complex) with the dissociation constant as $K_d=[P]_{free}$ [free sites on membrane]/[occupied site on membrane]. Then, binding can be analyzed by a Langmuir adsorption isotherm using a quadratic equation (Eq. 1; Cho et al., 2001):

$$[P_b]/[P]_0 = \frac{[P]_0 + K_d + [L]_0/n - \sqrt{([P]_0 + K_d + [L]_0/n)^2 - 4[P]_0[L]_0/n}}{2[P]_0} \quad (1)$$

where $[L]_0$, $[P]_0$, and $[P]_b$ are total lipid, total protein, and bound protein concentrations, respectively. In our fluorometric assay, $\Delta F/\Delta F_{max} = [P]_b/[P]_0$, where ΔF and ΔF_{max} are the fluorescence quenching and the maximal quenching, respectively. It should be noted that this simplified model works successfully when the surface coverage of proteins is low enough to avoid a significant degree of protein–protein interaction. Nonlinear least-squares analysis of either the $\Delta F/\Delta F_{max}$ vs $[L]_0$ plot with a fixed value of $[P]_0$ or the $\Delta F/\Delta F_{max}$ vs $[P]_0$ plot with a fixed value of $[L]_0$ using Eq. (1) will yield values of K_d and n. The two plots are obtained by titrating lipid vesicles into the protein solution and by titrating a protein into lipids, respectively. The selection of the two approaches depends on several experimental factors, including the extent of the background signal. Finally, it should be noted that since K_d is expressed in terms of molarity of protein-binding sites on the membrane, nK_d is the dissociation constant in terms of molarity of lipid monomers. If two related proteins (e.g., wild type and a mutant) bind the vesicles in essentially the same modes, their n values should be comparable and their relative membrane affinity can be assessed from the ratio of K_d values. Otherwise, the ratio of nK_d values should indicate the relative affinity, as the nK_d to $[L]_0$ ratio, but not the K_d to $[L]_0$ ratio, determines the relative population of membrane-bound protein ($=[P]_b/[P]_0$) at a given $[L]_0$ (e.g., Eq. 1 becomes $[P]_b/[P]_0 = 1/(1 + nK_d/[P]_0)$ when $[L]_0 > n[P]_0$).

It should be stressed that the above quantitative treatment is only an approximation, as the protein-binding site on the membrane has no permanent reality, and therefore binding parameters should be interpreted with caution. Furthermore, binding of a protein to heterogeneous lipid mixtures, including biological membranes, could be much more complex. More advanced isotherms for membrane binding of proteins have been reported (Heimburg, 2007), but they are not practical for the large-scale and high-throughput analysis of membrane-binding data. For those proteins whose membrane-binding isotherms cannot be explained by a Langmuir isotherm, apparent dissociation constant ($(K_d)_{app}$) is defined as the lipid concentration giving rise to half-maximal membrane binding of protein and is estimated graphically from the binding curve. When a protein selectively binds a specific lipid (e.g., PtdInsP) in the membrane, interfacial dissociation constant

(K_s) for the lipid (s) can be determined by measuring K_d (or $(K_d)_{app}$) values at different mol% of s (X_s) (11). Curve fitting of the $1/(K_d)$ vs X_s plot to an equation, $1/(K_d) = (1 + X_s/K_s)/(K_d)_0$, where $(K_d)_0$ is K_d in the absence of s, yields the value of K_s.

3.1 Specificity Data Analysis

1. The background fluorescence values are subtracted from binding data.
2. Resulting membrane-binding data of proteins are analyzed using the equation: $\Delta F/\Delta F_{max} = 1/(1 + [L]_{\%}/K)$. $[L]_{\%}$ and K are mole% of a particular lipid and the $[L]_{\%}$ value causing half-maximal quenching, respectively.
3. Lipid specificity of a protein can be expressed in terms of relative K values. For example, if the protein is specific for PtdIns(3,4,5)P_3, K for PtdIns(3,4,5)P_3 should be much (e.g., >5-fold) lower than those for other PtdIns*P*s.

3.2 Binding Data Analysis

1. The background fluorescence values are subtracted from binding data.
2. Values of n and K_d are determined by nonlinear least-squares analysis of the $\Delta F/\Delta F_{max}$ vs $[P]_0$ or $\Delta F/\Delta F_{max}$ vs $[L]_0$ plot using Eq. (1).
3. K_s for a specific lipid is determined by nonlinear least-squares analysis of the $1/(K_d)$ vs X_s plot using the equation, $1/(K_d) = (1 + X_s/K_s)/(K_d)_0$ (see earlier).

3.3 Inhibition Data Analysis

1. The background fluorescence values are subtracted from inhibition data.
2. Inhibition of membrane binding of a protein by an inhibitor is analyzed using an equation: $\Delta F = \Delta F_0/(1 + [I]/K_i)$ (Dua & Cho, 1994). ΔF and ΔF_0 indicate the fluorescence intensity decrease of EGFP by dabsyl-PE-containing vesicles in the presence and the absence of a given concentration of inhibitor, respectively. $[I]$ and K_i are the free inhibitor concentration and the inhibition constant.

ACKNOWLEDGMENTS

The work is supported by National Institutes of Health Grants GM68849 and GM110128.

REFERENCES

Barrera, N. P., Zhou, M., & Robinson, C. V. (2013). The role of lipids in defining membrane protein interactions: Insights from mass spectrometry. *Trends in Cell Biology*, *23*, 1–8.

Bazzi, M. D., & Nelsestuen, G. L. (1987). Association of protein kinase C with phospholipid vesicles. *Biochemistry*, *26*, 115–122.

Bhardwaj, N., Stahelin, R. V., Langlois, R. E., Cho, W., & Lu, H. (2006). Structural bioinformatics prediction of membrane-binding proteins. *Journal of Molecular Biology*, *359*, 486–495.

Bhardwaj, N., Stahelin, R. V., Zhao, G., Cho, W., & Lu, H. (2007). MeTaDoR: A comprehensive resource for membrane targeting domains and their host proteins. *Bioinformatics*, *23*, 3110–3112.

Chen, Y., Sheng, R., Kallberg, M., Silkov, A., Tun, M. P., Bhardwaj, N., et al. (2012). Genome-wide functional annotation of dual-specificity protein- and lipid-binding modules that regulate protein interactions. *Molecular Cell*, *46*, 226–237.

Cho, W. (2006). Building signaling complexes at the membrane. *Science's STKE*, *2006*, pe7.

Cho, W., Bittova, L., & Stahelin, R. V. (2001). Membrane binding assays for peripheral proteins. *Analytical Biochemistry*, *296*, 153–161.

Cho, W., Kim, H., & Hu, Y. (2016). High-throughput fluorometric assay for membrane-protein interaction. *Methods in Molecular Biology*, *1376*, 163–174.

Cho, W., & Stahelin, R. V. (2005). Membrane-protein interactions in cell signaling and membrane trafficking. *Annual Review of Biophysics and Biomolecular Structure*, *34*, 119–151.

DiNitto, J. P., Cronin, T. C., & Lambright, D. G. (2003). Membrane recognition and targeting by lipid-binding domains. *Science's STKE*, *2003*, re16.

Dowler, S., Kular, G., & Alessi, D. R. (2002). Protein lipid overlay assay. *Science's STKE*, *2002*, pl6.

Dua, R., & Cho, W. (1994). Inhibition of human secretory class II phospholipase A2 by heparin. *European Journal of Biochemistry*, *221*, 481–490.

Dua, R., Wu, S. K., & Cho, W. (1995). A structure-function study of bovine pancreatic phospholipase A2 using polymerized mixed liposomes. *The Journal of Biological Chemistry*, *270*, 263–268.

Feng, W., & Zhang, M. (2009). Organization and dynamics of PDZ-domain-related supramodules in the postsynaptic density. *Nature Reviews. Neuroscience*, *10*, 87–99.

Heimburg, T. (2007). *Thermal biophysics of membranes*. Weinheim, Germany: Wiley-VCH.

Kim, H., Afsari, H. S., & Cho, W. (2013). High-throughput fluorescence assay for membrane-protein interaction. *Journal of Lipid Research*, *54*, 3531–3538.

Kim, Y., Lichtenbergova, L., Snitko, Y., & Cho, W. (1997). A phospholipase A2 kinetic and binding assay using phospholipid-coated hydrophobic beads. *Analytical Biochemistry*, *250*, 109–116.

Kraft, C. A., Garrido, J. L., Leiva-Vega, L., & Romero, G. (2009). Quantitative analysis of protein-lipid interactions using tryptophan fluorescence. *Science Signaling*, *2*, p14.

Lee, A. G. (2011). Biological membranes: The importance of molecular detail. *Trends in Biochemical Sciences*, *36*, 493–500.

Lemmon, M. A. (2008). Membrane recognition by phospholipid-binding domains. *Nature Reviews. Molecular Cell Biology*, *9*, 99–111.

Liu, S. L., Sheng, R., O'Connor, M. J., Cui, Y., Yoon, Y., Kurilova, S., et al. (2014). Simultaneous in situ quantification of two cellular lipid pools using orthogonal fluorescent sensors. *Angewandte Chemie (International Ed. in English)*, *53*, 14387–14391.

Lucas, N., & Cho, W. (2011). Phosphatidylserine binding is essential for plasma membrane recruitment and signaling function of 3-phosphoinositide-dependent kinase-1. *The Journal of Biological Chemistry*, *286*, 41265–41272.

Manna, D., Albanese, A., Park, W. S., & Cho, W. (2007). Mechanistic basis of differential cellular responses of phosphatidylinositol 3,4-bisphosphate- and phosphatidylinositol 3,4,5-trisphosphate-binding pleckstrin homology domains. *The Journal of Biological Chemistry*, *282*, 32093–32105.

Manna, D., Bhardwaj, N., Vora, M. S., Stahelin, R. V., Lu, H., & Cho, W. (2008). Differential roles of phosphatidylserine, PtdIns(4,5)P2, and PtdIns(3,4,5)P3 in plasma membrane targeting of C2 domains. Molecular dynamics simulation, membrane binding, and cell translocation studies of the PKCalpha C2 domain. *The Journal of Biological Chemistry*, *283*, 26047–26058.

Miao, B., Skidan, I., Yang, J., Lugovskoy, A., Reibarkh, M., Long, K., et al. (2010). Small molecule inhibition of phosphatidylinositol-3,4,5-triphosphate (PIP3) binding to pleckstrin homology domains. *Proceedings of the National Academy of Sciences of the United States of America*, *107*, 20126–20131.

Miyawaki, A. (2011). Proteins on the move: Insights gained from fluorescent protein technologies. *Nature Reviews Molecular Cell Biology*, *12*, 656–668.

Nalefski, E. A., Slazas, M. M., & Falke, J. J. (1997). Ca^{2+}-signaling cycle of a membrane-docking C2 domain. *Biochemistry*, *36*, 12011–12018.

Narayan, K., & Lemmon, M. A. (2006). Determining selectivity of phosphoinositide-binding domains. *Methods*, *39*, 122–133.

Park, M. J., Sheng, R., Silkov, A., Jung, D. J., Wang, Z. G., Xin, Y., et al. (2016). SH2 domains serve as lipid-binding modules for pTyr-signaling proteins. *Molecular Cell*, *62*, 7–20.

Rebecchi, M., Peterson, A., & McLaughlin, S. (1992). Phosphoinositide-specific phospholipase C-delta 1 binds with high affinity to phospholipid vesicles containing phosphatidylinositol 4,5-bisphosphate. *Biochemistry*, *31*, 12742–12747.

Rusu, L., Gambhir, A., McLaughlin, S., & Radler, J. (2004). Fluorescence correlation spectroscopy studies of peptide and protein binding to phospholipid vesicles. *Biophysical Journal*, *87*, 1044–1053.

Shaner, N. C., Lambert, G. G., Chammas, A., Ni, Y., Cranfill, P. J., Baird, M. A., et al. (2013). A bright monomeric green fluorescent protein derived from Branchiostoma lanceolatum. *Nature Methods*, *10*, 407–409.

Shaner, N. C., Steinbach, P. A., & Tsien, R. Y. (2005). A guide to choosing fluorescent proteins. *Nature Methods*, *2*, 905–909.

Sheng, R., Chen, Y., Yung Gee, H., Stec, E., Melowic, H. R., Blatner, N. R., et al. (2012). Cholesterol modulates cell signaling and protein networking by specifically interacting with PDZ domain-containing scaffold proteins. *Nature Communications*, *3*, 1249.

Sheng, R., Kim, H., Lee, H., Xin, Y., Chen, Y., Tian, W., et al. (2014). Cholesterol selectively activates canonical Wnt signalling over non-canonical Wnt signalling. *Nature Communications*, *5*, 4393.

Silkov, A., Yoon, Y., Lee, H., Gokhale, N., Adu-Gyamfi, E., Stahelin, R. V., et al. (2011). Genome-wide structural analysis reveals novel membrane binding properties of AP180 N-terminal homology (ANTH) domains. *The Journal of Biological Chemistry*, *286*, 34155–34163.

Stahelin, R. V. (2013). Surface plasmon resonance: A useful technique for cell biologists to characterize biomolecular interactions. *Molecular Biology of the Cell*, *24*, 883–886.

Stahelin, R. V., & Cho, W. (2001). Differential roles of ionic, aliphatic, and aromatic residues in membrane-protein interactions: A surface plasmon resonance study on phospholipases A2. *Biochemistry*, *40*, 4672–4678.

Stahelin, R. V., Long, F., Peter, B. J., Murray, D., De Camilli, P., McMahon, H. T., et al. (2003). Contrasting membrane interaction mechanisms of AP180 N-terminal homology (ANTH) and epsin N-terminal homology (ENTH) domains. *The Journal of Biological Chemistry*, *278*, 28993–28999.

Sumandea, M., Das, S., Sumandea, C., & Cho, W. (1999). Roles of aromatic residues in high interfacial activity of Naja naja atra phospholipase A2. *Biochemistry*, *38*, 16290–16297.

van Meer, G., Voelker, D. R., & Feigenson, G. W. (2008). Membrane lipids: Where they are and how they behave. *Nature Reviews. Molecular Cell Biology*, *9*, 112–124.

Yoon, Y., Lee, P. J., Kurilova, S., & Cho, W. (2011). In situ quantitative imaging of cellular lipids using molecular sensors. *Nature Chemistry*, *3*, 868–874.

Zimmermann, P. (2006). The prevalence and significance of PDZ domain-phosphoinositide interactions. *Biochimica et Biophysica Acta*, *1761*, 947–956.

CHAPTER TWO

Fluorescence-Based *In Situ* Quantitative Imaging for Cellular Lipids

W. Cho*,†,1, Y. Yoon‡, S.-L. Liu*, K. Baek†, R. Sheng*

*University of Illinois at Chicago, Chicago, IL, United States
†Kyung Hee University, Yongin, South Korea
‡Konkuk University, Seoul, South Korea
[1]Corresponding author: e-mail address: wcho@uic.edu

Contents

Abstract

Membrane lipids are dynamic molecules and their local concentrations serve as regulatory signals for diverse biological processes. To achieve quantitative *in situ* imaging of various lipids, we developed a ratiometric analysis using fluorescence biosensors, each of which is composed of an engineered lipid-binding protein and a covalently attached solvatochromic fluorophore. To cover a wide range of lipid concentration, lipid-binding proteins are engineered to have variable dynamic ranges. These tunable sensors allow robust and sensitive *in situ* quantitative lipid imaging in mammalian cells, providing new insight into the spatiotemporal dynamics and fluctuation of key signaling lipids. The sensor strategy is also applicable to *in situ* quantification of multiple cellular lipids or a single lipid in the opposing leaflets of cell membranes.

Methods in Enzymology, Volume 583
ISSN 0076-6879
http://dx.doi.org/10.1016/bs.mie.2016.09.005

1. INTRODUCTION

Many membrane lipids, such as phosphoinositides (PtdInsPs) and sphingolipids, control diverse cellular processes by regulating the localization, activities, and mutual interactions of their effector proteins (Chen et al., 2012; Di Paolo & De Camilli, 2006; Hannun & Obeid, 2008; Park et al., 2016). These lipids and their downstream targets constitute complex signaling networks and perturbations in these networks contribute to the pathogenesis of diverse human diseases (Cho, 2006; Wymann & Schneiter, 2008). Since lipids are dynamic molecules that are continuously produced, degraded, and transported in a tightly controlled manner (van Meer, Voelker, & Feigenson, 2008) and their local concentrations may serve as thresholds for triggering diverse cellular processes (Yoon, Lee, Kurilova, & Cho, 2011), determination of their concentrations in a spatiotemporally resolved manner is a key step toward the understanding of myriad of lipid-mediated biological processes and the development of new strategies to diagnose, treat, and prevent human diseases caused by dysfunctional membrane-associated processes. However, real-time *in situ* lipid quantification in living cells poses technical challenges because lipids exist not as monodispersed entity but as part of large aggregates, such as cell membranes.

Most regulatory lipids are specifically recognized by cellular proteins containing modular lipid-binding domains (Cho & Stahelin, 2005; Lemmon, 2008). Thus, lipid-binding domains tagged with a fluorescence protein (FP; or two FPs for fluorescence resonance energy transfer analysis) have been widely used to visualize the spatiotemporal dynamics of various cellular lipids (Downes, Gray, & Lucocq, 2005; Varnai & Balla, 2006). Although these genetically incorporated probes offer experimental convenience, they can provide only semiquantitative information due to inherent problems associated with FP-based imaging (Irvine, 2004; Varnai & Balla, 2006). Also, overexpression of FP-tagged proteins may interfere with cellular processes. Furthermore, lipid sensors made of naturally occurring lipid-binding domains present many practical problems because of their relatively low lipid specificity and affinity as well as potential interaction with other cellular proteins (Irvine, 2004).

To overcome the limitations of currently available lipid imaging analyses, we developed a new strategy in which a lipid-binding domain (or protein) is first engineered for optimal lipid-binding properties and minimal affinity for cellular proteins (Yoon et al., 2011). The engineered protein is then converted into a "ratiometric" fluorescence sensor by single-site chemical labeling with a solvatochromic fluorophore that exhibits large changes in fluorescence emission wavelength and quantum yield upon lipid binding (see Fig. 1A and B). After *in vitro* calibration of lipid sensors using giant unilamellar vesicles (GUVs) with proper lipid composition, they are either added to the cells or delivered into cells for *in situ* lipid quantification. The same strategy can be applied to the *in situ* quantification of all cellular lipids. It can be used for quantifying a single or multiple species, and for quantifying a single lipid species in two opposing layers of cellular membranes to

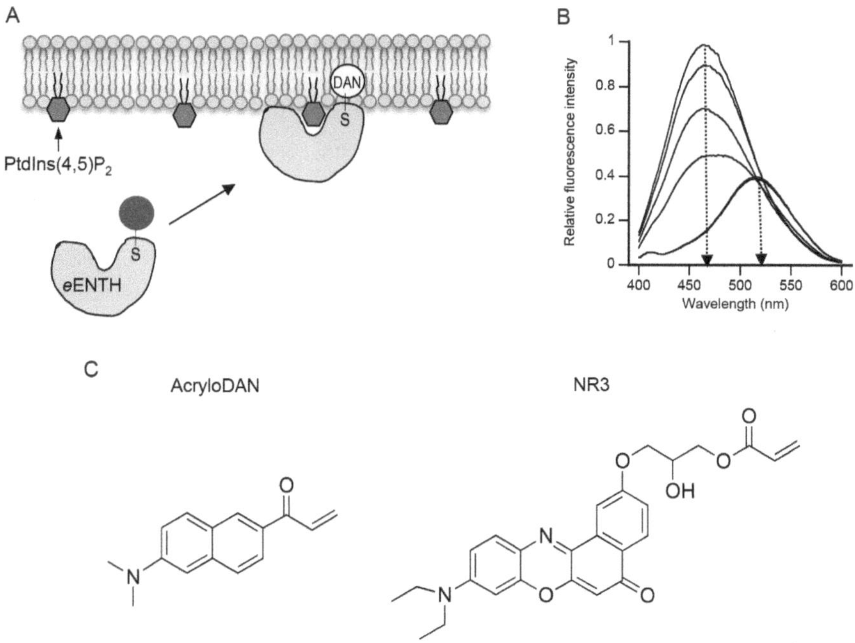

Fig. 1 Principles and components of quantitative lipid imaging. (A) The general imaging strategy. A lipid-binding domain specific for a particular lipid is engineered to improve lipid affinity and specificity and to introduce a single cysteine on the membrane-binding surface. Chemical labeling of the cysteine with a solvatochromic fluorophore produces a lipid sensor. As an example, an engineered ENTH domain (*e*ENTH), which is highly specific for PtdIns(4,5)P_2, labeled with a DAN fluorophore is shown. As the DAN-*e*ENTH binds the PtdIns(4,5)P_2-containing membrane, DAN exhibits major changes in fluorescence emission wavelength and quantum yield. (B) Fluorescence emission spectra of DAN-*e*ENTH in the presence of POPC/POPS/PtdIns(4,5)P_2 ($80-x$:20:x in mole%) vesicles. The excitation wavelength is set at 390 nm. DAN-*e*ENTH shows dramatic solvatochromism as a function of PtdIns(4,5)P_2 concentration in the vesicles. Fluorescence intensity values at two wavelengths (*dotted arrows*), 520 and 460 nm, are used for ratiometric data analysis. (C) The chemical structures of acrylodan and NR3.

determine the dynamic transbilayer asymmetry and movement (Liu et al., 2014). For simultaneous *in situ* quantification of multiple lipids, orthogonal sensors are prepared by labeling individual lipid-binding domains with spectrally orthogonal solvatochromic dyes (Liu et al., 2014).

2. PREPARATION OF TUNABLE ORTHOGONAL LIPID SENSORS

2.1 General Strategy

Our lipid sensors have two components: a lipid-binding protein module and a covalently attached solvatochromic fluorophore (see Fig. 1A). The protein module is prepared by genetically engineering a selected lipid-binding domain to improve its lipid affinity and specificity and to create a single cysteine on its membrane-binding surface for chemical labeling. The fluorophore is chemically attached to the protein using a cysteine-selective reagent derived from a wide range of solvatochromic fluorophores (Loving, Sainlos, & Imperiali, 2010). The protein module determines the specificity, the dynamic range, and the stability of a sensor, whereas the fluorophore controls its spectral properties (i.e., spectral range, signal intensity, and signal-to-noise ratio) and also significantly contributes to its membrane-binding properties (Liu et al., 2014; Yoon et al., 2011). *In situ* quantification of a lipid whose concentration varies broadly would require a panel of tunable lipid sensors with different dynamic ranges that are produced by systematic protein engineering of a lipid-binding domain. Simultaneous *in situ* quantification of multiple lipids or a single lipid in different locations, most notably two layers of cell membranes, employs bio-orthogonal lipid sensors that are prepared using spectrally orthogonal fluorophores (Liu et al., 2014).

2.2 Engineering Lipid-Binding Domains

2.2.1 Lipid Headgroup Specificity

High lipid headgroup specificity is the most important requirement for lipid sensors. Some lipid-binding domains (or proteins), such as FYVE domain, have high lipid headgroup stereospecificity (Blatner et al., 2004; Kutateladze, 2006), but a large majority of them show significant affinity for multiple lipid species. For example, the Akt1 pleckstrin homology (PH) domain binds both phosphatidylinositol-3,4,5-trisphosphate (PtdIns(3,4,5)P_3) and phosphatidylinositol-3,4-bisphosphate (PtdIns(3,4)P_2) tightly and has considerable affinity for other PtdInsPs (Manna, Albanese, Park, & Cho, 2007). Also, phospholipase C (PLC) δ1 PH domain, which has been commonly used as a phosphatidylinositol-4,5-bisphosphate (PtdIns(4,5)P_2) probe, can also bind other PtdInsPs, including PtdIns(3,4,5)P_3 (Yoon et al., 2011). This promiscuous specificity becomes particularly problematic if the natural abundance of other lipids is higher than the lipid of primary interest. Because lipid headgroup specificity, which is achieved through a complex array of short-range interactions (DiNitto, Cronin, & Lambright, 2003; DiNitto & Lambright, 2006), is hard to improve through directed evolution or rational design, identifying a candidate protein with high specificity for each lipid is a critical first step in lipid sensor design. Minor improvement of lipid headgroup specificity has been achieved through mutations when secondary lipids have a larger headgroup than the primary lipid (e.g., eliminating residual PtdIns(3,4,5)P_3 affinity from a PtdIns (4,5)P_2-binding protein; Yoon et al., 2011). An increasing number of cellular proteins have been reported to be recruited to cell membranes by coincident recognition of a lipid and a protein (Chen et al., 2012; Di Paolo & De Camilli, 2006; Park et al., 2016; Sheng et al., 2012, 2014). Thus, the possibility of protein-mediated membrane recruitment must be precluded for any lipid sensor by demonstrating that mutation of its lipid-binding site completely abrogates its cellular membrane localization (Yoon et al., 2011).

2.2.2 Lipid Affinity

In addition to high lipid headgroup specificity, an ideal lipid sensor would have fast membrane binding and unbinding kinetics in response to changes in lipid concentration under physiological conditions (Manna et al., 2007; Yoon et al., 2011). Most lipid-binding domains lack such ideal membrane-binding properties and they often do not have high enough affinity for their specific lipids to compete with endogenous intact proteins. The most common approach to enhancing membrane affinity of domains is to construct a tandem repeat of domains that achieve higher affinity through avidity effect (Lemmon, 2008; Oancea, Teruel, Quest, & Meyer, 1998). However, we prefer point mutations to domain duplication because membrane binding of a protein with two membrane-binding sites is not always straightforward to analyze. In general, membrane binding of soluble proteins involves initial membrane adsorption, mostly driven by nonspecific electrostatic interactions between cationic residues of proteins and anionic lipids, and subsequent specific short-range interactions, including hydrogen bonds and hydrophobic interactions (Cho & Stahelin, 2005). In particular, aromatic residues, most notably tryptophan, play a crucial role in membrane binding of proteins through complex interactions including π-cationic interaction (Han et al., 1999; He et al., 2015; Sumandea, Das, Sumandea, & Cho, 1999). We have shown that addition of tryptophan and/or cationic residues to the membrane-binding surface of proteins greatly improves their membrane affinity and membrane-binding kinetics without influencing their lipid specificity (Cho & Stahelin, 2005). We thus use this approach as a general strategy for preparing tunable lipid sensors with incrementally higher membrane affinity. However, the number and the molecular location of genetically introduced residues must be carefully controlled through systematic structure–function analysis because surface-exposed tryptophan (and cationic residues to a lesser extent) residues may induce nonspecific membrane binding and protein aggregation (Han, Yoon, & Cho, 1998).

2.2.3 Introduction of a Single Cysteine

Cysteine has enjoyed widespread use in selective protein modification because of its high reactivity at neutral pH and relatively low abundance in intracellular proteins (Basle, Joubert, & Pucheault, 2010; Chalker, Bernardes, Lin, & Davis, 2009). For those lipid-binding proteins without endogenous cysteine, a single cysteine is genetically introduced to a membrane-binding surface. For those with one or more endogenous cysteine, a systematic approach should be taken to determine if any of them is located on the membrane-binding surface and can thus be directly used for chemical labeling and if extra cysteine residues can be mutated to alanine or other residues without causing deleterious structural and functional effects. It should be noted that those cysteine residues that are directly involved in metal coordination and are thus essential for protein stability (e.g., Btk1 PH domain and many C1 domains) should remain intact.

To achieve a strong solvatochromic shift of the covalently attached fluorophore, which is critical for high signal intensity and high sensitivity of quantitative lipid imaging, it must experience a maximal degree of polarity change during membrane binding of the host protein. It is thus important to place the cysteine in such a position that allows the covalently attached fluorophore to stay in maximal contact with water or polar protein residues before membrane binding and then to insert itself into the hydrophobic core of the membrane during lipid binding. Lipid-binding proteins with high headgroup specificity have a well-defined lipid-binding pocket surrounded by a rim of cationic and aromatic/hydrophobic residues (i.e., membrane-binding surface; Cho & Stahelin, 2005). The cysteine for labeling is usually introduced to this membrane-binding surface, replacing a residue that is preferentially located in a flexible loop region and is not essential for membrane binding.

2.2.4 Protein Stability

Stability of the protein component of the sensor is essential for robust and reproducible quantitative imaging of cellular lipids as it affects the protein expression yield, chemical coupling efficiency, and signal reliability. Many isolated lipid-binding domains have surface-exposed aromatic and hydrophobic residues, which are normally buried in the domain–domain interfaces in the intact proteins, and they may cause protein aggregation and instability. If so, these residues should be mutated to a smaller alanine or a polar residue to improve protein stability as long as they are not directly involved in membrane binding of the protein. Also, flexible loops can be shortened or removed to improve protein stability. Furthermore, the length of N- and C-terminal extensions of the domains can be systematically varied for maximal protein stability.

2.3 Cysteine-Selective Solvatochromic Fluorophores

Solvatochromic fluorophores are fluorescent molecules whose emission properties, including the emission wavelength, the quantum yields, and the fluorescence lifetime, are highly sensitive to their microenvironment (Loving et al., 2010). Our lipid sensors employ polarity-sensitive fluorophores that exhibit solvatochromism as the sensors migrate from the aqueous phase to the hydrophobic membrane environment. Among a large number of solvatochromic fluorophores reported to date (Loving et al., 2010), only a small number of them meet the requirements for quantitative lipid imaging (Liu et al., 2014; Yoon et al., 2011). These requirements include amphiphilicity, relatively small size, and synthetic accessibility, in addition to strong solvatochromism (i.e., a large change in emission wavelength and quantum yield) and photostability. Strong solvatochromism is essential for ratiometric data analysis (Yoon et al., 2011; see Fig. 1B), whereas photostability is important for long-term imaging. The fluorophore in the lipid sensor shuttles between the polar and the nonpolar environment, and thus its amphiphilicity is essential for both solubility of the sensor in the aqueous solution and its efficient partitioning into the membrane. Also, because the fluorophore would partially penetrate into the hydrophobic core of the membrane, its size should not be too bulky. Finally, synthesis of cysteine-selective reagents containing these fluorophores should be relatively straightforward. We found that 2-dimethylaminonaphthalene (DAN; Weber & Farris, 1979) and Nile red (Jose & Burgess, 2006) fluorophores most satisfactorily meet these requirements and have little spectral overlap. Thus, their cysteine-selective acrylic derivatives, acrylodan (6-acryloyl-2-dimethylaminonaphthalene) and NR3, have been preferentially used for preparing bio-orthogonal lipid sensors (see Fig. 1C; Liu et al., 2014; Yoon et al., 2011). DAN is excited at wavelengths below 400 nm and has an appreciably smaller extinction coefficient (<20,000 M^{-1} cm^{-1} in methanol; Yoon et al., 2011), hence the enhanced possibility of cell damage and autofluorescence. However, its small size, availability, excellent amphiphilicity, and spectral orthogonality with red fluorophores make it an excellent choice for multiplex lipid imaging, especially when its photostability is enhanced under two-photon excitation conditions (Yoon et al., 2011). NR3 has a high extinction coefficient (>40,000 M^{-1} cm^{-1} in methanol) and good amphiphilicity (Liu et al., 2014) although Nile red derivatives are in general not good two-photon fluorophores. These compounds have high enough solubility in water to allow efficient coupling to proteins. Also, DAN- and NR3-labeled proteins generally have significantly higher affinity and faster membrane-binding kinetics than their parent proteins because of excellent amphiphilicity of DAN and NR3 (Liu et al., 2014; Yoon et al., 2011).

2.4 Preparation of Lipid Sensors

2.4.1 Preparation of Acrylodan and NR3

Acrylodan (ThermoFisher, Waltham, MA, Cat. No. A433) is commercially available and NR3 is prepared by a four-step synthesis from 5-diethylamino-2-nitrosophenol hydrochloride (Sigma-Aldrich, St. Louis, MO, Cat. No. PH012666) and 1,6-dihydroxynaphthalene (Sigma-Aldrich, Cat. No. 274127) as described previously (Liu et al., 2014). Stock solutions of acrylodan or NR3 are prepared in dimethyl sulfoxide (DMSO; ThermoFisher, Cat. No. D128-500).

2.4.2 Protein Expression, Purification, and Labeling

In general, a lipid sensor protein is expressed as a glutathione-*S*-transferase (GST)-tagged protein, purified by glutathione-based affinity chromatography, labeled on the column with acrylodan or NR3, and recovered from the column after removing the GST tag. A typical procedure for the preparation of the engineered ENTH domain-based (*e*ENTH) PtdIns(4,5)P_2 sensor (Yoon et al., 2011) is as follows:

1. 500 mL of Luria (BD, Cat. No. 244610) broth containing 50 μg/mL (ThermoFisher, Cat. No. 155479) or 100 μg/mL ampicillin (ThermoFisher, Cat. No. 160561) is inoculated with BL21 RIL codon plus (Agilent, Santa Clara, CA, Cat. No. 230240) colonies expressing the sensor protein (*e*ENTH).
2. 0.5 m*M* of isopropyl β-D-1-thiogalactopyranoside (Gold Biotechnology, St. Louis, MO, Cat. No. 12481C5) is added to cells to induce overexpression of the recombinant protein when the absorbance at 600 nm of the media reaches 0.5–0.8 and cells are grown for additional 6–10 h at 25°C.
3. Cells are harvested by centrifugation and cell pellets are resuspended in 50 m*M* Tris–HCl buffer (pH 7.9) (ThermoFisher, Cat. No. 142521) containing 0.3 *M* KCl (ThermoFisher, Cat. No. P217), 1 m*M* phenylmethanesulfonylfluoride (ThermoFisher, Cat. No. ICN19538105), and 1 m*M* of 2-mercaptoethanol (ThermoFisher, Cat. No. ICN19024280).
4. Cells are lysed by sonication and the lysate is collected by centrifugation at 4°C.
5. GST affinity resin (GenScript, Piscataway, NJ, Cat. No. C40141509) is added into the cell lysate and the mixture is gently shaken for 2 h at 4°C. The mixture is applied to a column and the column is washed several times with 20 m*M* Tris–HCl buffer (pH 7.4) containing 0.16 *M* NaCl (ThermoFisher, Cat. No. 156697).
6. For labeling of *e*ENTH at the single cysteine site, an excess amount of acrylodan or NR3 (>500 μg) in DMSO (ThermoFisher, Cat. No. D128-500) is incrementally added to the resin and the mixture is incubated at room temperature for 2 h with gentle shaking. For less stable proteins, the reaction is carried out at 4°C for 8–10 h. The excess reagent is removed by washing the resin with the buffer solution extensively.
7. The resin is incubated with thrombin (GE Healthcare, Pittsburgh, PA, Cat. No. 27-0846-01) at 4°C overnight to remove the GST tag from the labeled protein.
8. The labeled protein is eluted from the resin. Purity and the concentration of the recombinant proteins are determined by sodium dodecylsulfate-polyacrylamide gel electrophoresis and Bradford (BioRad, Hercules, CA, Cat. No. 5000006) assay, respectively. The labeling yield of each sensor is then calculated spectrophotometrically using the extinction coefficients of DAN and NR3 (Liu et al., 2014; Yoon et al., 2011). The labeling step is repeated until >70% labeling yield is achieved.

3. *IN SITU* QUANTIFICATION OF CELLULAR LIPIDS

3.1 Sensor Calibration

Ideally, a sensor should be calibrated in cellulo to allow accurate *in situ* quantification of cellular lipids. However, it is impractical to systematically vary the concentration of a single lipid in cell membranes. It is therefore necessary to carry out *in vitro* calibration of lipid sensors using cell membrane-mimicking GUVs with systematic variation of the concentration of a specific lipid component. Since lipids are dynamic molecules that are rapidly metabolized and transferred, exact lipid compositions of various cell membranes are difficult to determine (Sprong, van der Sluijs, & van Meer, 2001; Stahelin, Rafter, Das, & Cho, 2003; van Meer et al., 2008). One thus has to use approximate compositions for cell membranes in GUV calibration. For instance, a PtdIns(4,5)P_2 sensor is calibrated using an inner plasma membrane-mimetic vesicles (1-palmitoyl-2-oleoyl-*sn*-glycero-3-phosphocholine (POPC)/1-palmitoyl-2-oleoyl-*sn*-glycero-3-phosphoethanolamine (POPE)/1-palmitoyl-2-oleoyl-*sn*-glycero-3-phosphoserine (POPS)/phosphatidylinositol (PI)/cholesterol/PtdIns(4,5)P_2 (12/36−x/22/8/22/x: x=0–3 mole%)), whereas a sphingomyelin (SM) sensor is calibrated using an outer plasma membrane-mimicking vesicles (POPC/SM/cholesterol/ganglioside (57−x/x/40/3: x=0–30 mole%)) (Sprong et al., 2001; Stahelin et al., 2003; van Meer et al., 2008). Notice that these are approximate lipid compositions and membrane binding of a lipid sensor may be sensitive not only to the concentration of a specific lipid in GUV but also to the composition of other lipids. It is thus recommended that sensor calibration be performed with multiple GUVs with different lipid compositions, including variation of acyl chains of lipids, to confirm that sensor calibration is not biased by the selected lipid composition of GUVs. Finally, *in vitro* sensor calibration and cellular lipid quantification should be performed under identical imaging conditions (e.g., light source, filter, and detector settings). Fluorescence data from both measurements are analyzed by the ratiometric analysis at two different wavelengths, which greatly simplifies the overall data analysis because ratiometric determination is independent of the sensor concentration.

3.1.1 Preparation of GUVs

GUVs with 5–30 μm diameter are routinely prepared by electroformation in a nonelectrolyte solution (Bagatolli & Gratton, 1999; Yoon et al., 2010;

Yoon, Zhang, & Cho, 2012). To allow osmotic balance of GUV in a physiological buffer solution, GUVs are first grown in 350 m*M* sucrose solution and transferred to the physiological buffer solution. Sucrose-loaded GUVs also provide experimental convenience in imaging due to their low mobility (Yoon et al., 2010).

1. POPC (Cat. No. 850457), POPE (Cat. No. 850757), POPS (Cat. No. 840034), soy PI (Cat. No. 840044), and cholesterol (Cat. No. 700000) (all from Avanti Polar Lipids) are dissolved in the highest grade oxygen-free chloroform to yield 10 mg/mL stock solutions and stored in Teflon-sealed vials at −20°C. All organic solvents including chloroform (Cat. No. C607-4) and methanol (Cat. No. 53816) are from ThermoFisher.
2. 1,2-Dipalmitoyl derivatives of PtdInsPs, including PtdIns(4,5)P_2 (Cayman Chemical, Cat. No. 64924) and PtdIns(3,4,5)P_3 (Cayman Chemical, Cat. No. 64920), are dissolved in the highest grade oxygen-free chloroform/methanol/water (3:3:1 v/v/v) to yield 0.5 mg/mL solutions and also stored in Teflon-sealed vials at −20°C.
3. Lipid mixtures are prepared in chloroform/methanol (3:1) from the lipid stock solutions to a total concentration of 0.4 mg/mL.
4. The lipid solution is spread onto the homemade indium tin oxide electrode surface and dried under vacuum to form a uniform lipid film.
5. Vesicles are grown in a sucrose solution (350 m*M*), while an electric field (3 V, 20 Hz frequency) is applied for 5 h at room temperature.
6. 1–2 μL of sucrose-loaded GUV solution is added into a well glued onto a coverslip that is placed on the microscope stage. The well contains 300 μL of 20 m*M* Tris–HCl buffer, pH 7.4, with 0.16 *M* KCl.

3.1.2 Lipid Sensor Calibration Using GUV

Lipid sensors can be calibrated in either one-photon excitation mode or two-photon excitation mode. We prefer the latter because it allows easier separation of excitation and emission signals and milder illumination conditions. However, the same level of resolution and sensitivity can be achieved with proper one-photon excitation setup. A typical procedure for calibration of the PtdIns(4,5)P_2 sensor (i.e., DAN or NR3-*e*ENTH) with a two-photon microscope is described here.

1. POPC/POPE/POPS/PI/cholesterol/PtdIns(4,5)P_2 (12/36 − x/22/8/22/x: x = 0–3 mole%) GUVs are mixed with DAN-*e*ENTH or NR3-*e*ENTH in the concentration range of 0–500 n*M*.

2. DAN-*e*ENTH and NR3-*e*ENTH are two-photon excited at 780 and 900 nm, respectively. 480 ± 40 and 560 ± 50 nm band pass filters are employed for the blue channel and the green channel, respectively, whereas 620 ± 60 and 670 ± 50 nm band pass filters are used for the orange channel and the red channel, respectively. For DAN-*e*ENTH, blue channel fluorescence signals derive from membrane-bound sensors only, whereas green channel signals are from both membrane bound and sensors. Likewise, orange channel fluorescence signals derive from membrane-bound sensors only, whereas red channel signals are from both membrane bound and sensors for NR3-*e*ENTH.
3. At each time point, an image of 512×512 pixels is collected with the pixel dwell time of 20 ms using Hamamatsu photomultiplier tubes.
4. For each PtdIns(4,5)P_2 concentration, 10 GUVs are selected for image analysis by Image-Pro Plus (Media Cybernetics, Inc.).
5. For data analysis, the region of interest (the membrane in our case) is selected by setting a threshold intensity (or brightness) value on the basis of the intensity distribution profile of the image. Since the orange channel (or blue channel for DAN-*e*ENTH) always gives stronger membrane signals than the red channel (or green channel for DAN-*e*ENTH) for NR3-*e*ENTH, the mask from the orange (or blue) channel is first selected and superimposed onto the same image in the red (or green) channel. The estimated membrane region of the vesicle is validated by comparing it with the membrane region in the differential interference contrast image of the vesicle.
6. The total intensity of GUV ($F_{B(total)}$ and $F_{G(total)}$ for DAN-*e*ENTH and $F_{O(total)}$ and $F_{R(total)}$ for NR3-*e*ENTH) is divided by the total area of the pixels that constitute each GUV to yield the average intensities, F_B and F_G for DAN-*e*ENTH and F_O and F_R for NR3-*e*ENTH (counts/m^2), which are then used to prepare the calibration curves for ratiometric analyses. For DAN-*e*ENTH, nonlinear least-squares analysis of the (F_B/F_G) vs the PtdIns(4,5)P_2 concentration ([PtdIns(4,5)P_2]) plot using the equation: $F_B/F_G = (F_B/F_G)_{min} + ((F_B/F_G)_{max} - (F_B/F_G)_{min})/(1 + [\text{PtdIns}(4,5)P_2]/K_{1/2})$ yielded $K_{1/2}$, $(F_B/F_G)_{max}$, and $(F_B/F_G)_{min}$ values. $K_{1/2}$, $(F_B/F_G)_{max}$, and $(F_B/F_G)_{min}$ are [PtdIns(4,5)P_2] yielding half maximal binding (in mole%), the maximal F_B/F_G value, and the minimal F_B/F_G value, respectively. The theoretical calibration curve is then constructed using these values and [PtdIns(4,5)P_2] from an unknown sample is calculated using the calibration curve. The same calibration is performed for NR3-*e*ENTH.

3.2 *In Situ* Lipid Imaging and Analysis

Once the lipid sensor is calibrated using GUVs, it can be delivered into the cells for intracellular lipid quantification or added to the media for outer plasma membrane lipid quantification. Since these sensors cannot normally cross the plasma membrane, they have to be microinjected into individual cells or delivered to the cells with lipid-based transfection reagents. Although laborious, we prefer the former method because it allows higher quality imaging and lower background than the latter. Successful cell-penetrating peptide-based cell delivery of lipid sensors has not been achieved to date. A typical procedure for *in situ* quantification of PtdIns(4,5)P_2 by DAN-*e*ENTH in NIH 3T3 cells is described here.

1. DAN-*e*ENTH is delivered into the NIH 3T3 cells (ATCC® Cat. No. CRL-1658™) using the Eppendorf InjectMan NI 2 Micromanipulator system. P-97 Micropipette Puller (Sutter Instrument) is used for micro-injection micropipette preparation. To minimize cell damages by micro-injection, calcium-free media are used. Also, enough time is given after injection for cells to recover and only healthy cells are selected for further imaging.
2. The DAN-*e*ENTH concentration is kept as low as possible but high enough to give strong membrane signals for robust data analysis. Typically, 20–30 fL of 0.5–1 μ*M* sensor solution is microinjected into the cell.
3. All microscopy measurements for *in vitro* GUV calibration and cell imaging are performed under the same conditions.
4. The local PtdIns(4,5)P_2 concentration on the plasma membrane is determined from the observed local (F_B/F_G) values using the *in vitro* calibration curves determined using GUVs.
5. The three-dimensional display of local PtdIns(4,5)P_2 concentration is obtained using the surf function in MATLAB.
6. To display spatial distribution of PtdIns(4,5)P_2 in the plasma membrane, the angular profile of photon counts along the plasma membrane is calculated by Image-Pro Plus software.

4. SUMMARY

Quantitative lipid imaging is a powerful, universally applicable technique that provides the detailed spatiotemporal information about how local lipid concentrations change in response to various cell stimuli. Quantitative correlation of spatiotemporal changes in lipid concentrations with the membrane localization and activities of their effector proteins also provide new insight into how different lipids regulate and coordinate complex biological processes with high specificity and fidelity. All this information is critically important for systems-level understanding of lipid-mediated cell regulation and function.

ACKNOWLEDGMENTS

The work is supported by National Institutes of Health Grants GM68849 and GM110128.

REFERENCES

Bagatolli, L. A., & Gratton, E. (1999). Two-photon fluorescence microscopy observation of shape changes at the phase transition in phospholipid giant unilamellar vesicles. *Biophysical Journal, 77*, 2090–2101.

Basle, E., Joubert, N., & Pucheault, M. (2010). Protein chemical modification on endogenous amino acids. *Chemistry & Biology, 17*, 213–227.

Blatner, N. R., Stahelin, R. V., Diraviyam, K., Hawkins, P. T., Hong, W., Murray, D., et al. (2004). The molecular basis of the differential subcellular localization of FYVE domains. *The Journal of Biological Chemistry, 279*, 53818–53827.

Chalker, J. M., Bernardes, G. J., Lin, Y. A., & Davis, B. G. (2009). Chemical modification of proteins at cysteine: Opportunities in chemistry and biology. *Chemistry, an Asian Journal, 4*, 630–640.

Chen, Y., Sheng, R., Kallberg, M., Silkov, A., Tun, M. P., Bhardwaj, N., et al. (2012). Genome-wide functional annotation of dual-specificity protein- and lipid-binding modules that regulate protein interactions. *Molecular Cell, 46*, 226–237.

Cho, W. (2006). Building signaling complexes at the membrane. *Science's STKE, 2006*, pe7.

Cho, W., & Stahelin, R. V. (2005). Membrane-protein interactions in cell signaling and membrane trafficking. *Annual Review of Biophysics and Biomolecular Structure, 34*, 119–151.

DiNitto, J. P., Cronin, T. C., & Lambright, D. G. (2003). Membrane recognition and targeting by lipid-binding domains. *Science's STKE, 2003*, re16.

DiNitto, J. P., & Lambright, D. G. (2006). Membrane and juxtamembrane targeting by PH and PTB domains. *Biochimica et Biophysica Acta, 1761*, 850–867.

Di Paolo, G., & De Camilli, P. (2006). Phosphoinositides in cell regulation and membrane dynamics. *Nature, 443*, 651–657.

Downes, C. P., Gray, A., & Lucocq, J. M. (2005). Probing phosphoinositide functions in signaling and membrane trafficking. *Trends in Cell Biology, 15*, 259–268.

Han, S. K., Kim, K. P., Koduri, R., Bittova, L., Munoz, N. M., Leff, A. R., et al. (1999). Roles of Trp31 in high membrane binding and proinflammatory activity of human group V phospholipase A2. *The Journal of Biological Chemistry, 274*, 11881–11888.

Han, S. K., Yoon, E. T., & Cho, W. (1998). Bacterial expression and characterization of human secretory class V phospholipase A2. *The Biochemical Journal, 331*(Pt. 2), 353–357.

Hannun, Y. A., & Obeid, L. M. (2008). Principles of bioactive lipid signalling: Lessons from sphingolipids. *Nature Reviews. Molecular Cell Biology, 9*, 139–150.

He, T., Gershenson, A., Eyles, S. J., Lee, Y. J., Liu, W. R., Wang, J., et al. (2015). Fluorinated aromatic amino acids distinguish cation-pi interactions from membrane insertion. *The Journal of Biological Chemistry, 290*, 19334–19342.

Irvine, R. (2004). Inositol lipids: To PHix or not to PHix? *Current Biology, 14*, R308–R310.

Jose, J., & Burgess, K. (2006). Benzophenoxazine-based fluorescent dyes for labeling biomolecules. *Tetrahedron, 62*, 11021–11037.
Kutateladze, T. G. (2006). Phosphatidylinositol 3-phosphate recognition and membrane docking by the FYVE domain. *Biochimica et Biophysica Acta, 1761*, 868–877.
Lemmon, M. A. (2008). Membrane recognition by phospholipid-binding domains. *Nature Reviews. Molecular Cell Biology, 9*, 99–111.
Liu, S. L., Sheng, R., O'Connor, M. J., Cui, Y., Yoon, Y., Kurilova, S., et al. (2014). Simultaneous in situ quantification of two cellular lipid pools using orthogonal fluorescent sensors. *Angewandte Chemie (International Ed. in English), 53*, 14387–14391.
Loving, G. S., Sainlos, M., & Imperiali, B. (2010). Monitoring protein interactions and dynamics with solvatochromic fluorophores. *Trends in Biotechnology, 28*, 73–83.
Manna, D., Albanese, A., Park, W. S., & Cho, W. (2007). Mechanistic basis of differential cellular responses of phosphatidylinositol 3,4-bisphosphate- and phosphatidylinositol 3,4,5-trisphosphate-binding pleckstrin homology domains. *The Journal of Biological Chemistry, 282*, 32093–32105.
Oancea, E., Teruel, M. N., Quest, A. F., & Meyer, T. (1998). Green fluorescent protein (GFP)-tagged cysteine-rich domains from protein kinase C as fluorescent indicators for diacylglycerol signaling in living cells. *The Journal of Cell Biology, 140*, 485–498.
Park, M. J., Sheng, R., Silkov, A., Jung, D. J., Wang, Z. G., Xin, Y., et al. (2016). SH2 domains serve as lipid-binding modules for pTyr-signaling proteins. *Molecular Cell, 62*, 7–20.
Sheng, R., Chen, Y., Yung Gee, H., Stec, E., Melowic, H. R., Blatner, N. R., et al. (2012). Cholesterol modulates cell signaling and protein networking by specifically interacting with PDZ domain-containing scaffold proteins. *Nature Communications, 3*, 1249.
Sheng, R., Kim, H., Lee, H., Xin, Y., Chen, Y., Tian, W., et al. (2014). Cholesterol selectively activates canonical Wnt signalling over non-canonical Wnt signalling. *Nature Communications, 5*, 4393.
Sprong, H., van der Sluijs, P., & van Meer, G. (2001). How proteins move lipids and lipids move proteins. *Nature Reviews. Molecular Cell Biology, 2*, 504–513.
Stahelin, R. V., Rafter, J. D., Das, S., & Cho, W. (2003). The molecular basis of differential subcellular localization of C2 domains of protein kinase C-alpha and group IVa cytosolic phospholipase A2. *The Journal of Biological Chemistry, 278*, 12452–12460.
Sumandea, M., Das, S., Sumandea, C., & Cho, W. (1999). Roles of aromatic residues in high interfacial activity of Naja naja atra phospholipase A2. *Biochemistry, 38*, 16290–16297.
van Meer, G., Voelker, D. R., & Feigenson, G. W. (2008). Membrane lipids: Where they are and how they behave. *Nature Reviews. Molecular Cell Biology, 9*, 112–124.
Varnai, P., & Balla, T. (2006). Live cell imaging of phosphoinositide dynamics with fluorescent protein domains. *Biochimica et Biophysica Acta, 1761*, 957–967.
Weber, G., & Farris, F. J. (1979). Synthesis and spectral properties of a hydrophobic fluorescent probe: 6-Propionyl-2-(dimethylamino)naphthalene. *Biochemistry, 18*, 3075–3078.
Wymann, M. P., & Schneiter, R. (2008). Lipid signalling in disease. *Nature Reviews Molecular Cell Biology, 9*, 162–176.
Yoon, Y., Lee, P. J., Kurilova, S., & Cho, W. (2011). In situ quantitative imaging of cellular lipids using molecular sensors. *Nature Chemistry, 3*, 868–874.
Yoon, Y., Tong, J., Lee, P. J., Albanese, A., Bhardwaj, N., Kallberg, M., et al. (2010). Molecular basis of the potent membrane-remodeling activity of the epsin 1N-terminal homology domain. *The Journal of Biological Chemistry, 285*, 531–540.
Yoon, Y., Zhang, X., & Cho, W. (2012). Phosphatidylinositol 4,5-bisphosphate (PtdIns (4,5)P2) specifically induces membrane penetration and deformation by Bin/amphiphysin/Rvs (BAR) domains. *The Journal of Biological Chemistry, 287*, 34078–34090.

CHAPTER THREE

Preparation of the Full Set of Recombinant Mouse- and Human-Secreted Phospholipases A_2

F. Ghomashchi*, V. Brglez†, C. Payré†, L. Jeammet†, S. Bezzine‡, M.H. Gelb*[,1], G. Lambeau[†,1]

*University of Washington, Seattle, WA, United States
†Institut de Pharmacologie Moléculaire et Cellulaire, Université de Nice Sophia Antipolis and Centre National de la Recherche Scientifique (CNRS), Valbonne, France
‡University of Sfax, Laboratoire de Biochimie et de Génie Enzymatique des Lipases, Sfax, Tunisia
[1]Corresponding authors: e-mail address: gelb@uw.edu; lambeau@ipmc.cnrs.fr

Contents

Methods in Enzymology, Volume 583
ISSN 0076-6879
http://dx.doi.org/10.1016/bs.mie.2016.10.034

Abstract

A family of 14–20 kDa, disulfide-rich, calcium-dependent secreted phospholipases A_2 ($sPLA_2s$) that release fatty acids from the *sn*-2 position of glycerophospholipids can be found in mammals. They have a diverse array of tissue distribution and biological functions. In this chapter we provide detailed protocols for production of nearly all of the mouse and human $sPLA_2s$ mainly by expression in bacteria and *in vitro* refolding or by expression in insect cells. High-resolution mass spectrometry and enzymatic assays were, respectively, used to show that all disulfides are formed and that the enzymes are active, strongly suggesting that each $sPLA_2$ was prepared in the structurally native form. The availability of these proteins has allowed kinetic studies to be carried out, to prepare highly selective antisera, to screen for selective inhibitors, to study receptor binding, and to study the action of each enzyme on mammalian cell membranes and their *in vivo* biological roles.

1. INTRODUCTION

Secreted phospholipases A_2, $sPLA_2s$, like all PLA_2s hydrolyze the *sn*-2 ester bond of glycerophospholipids to produce a free fatty acid and a lysophospholipid. $sPLA_2s$ contain an N-terminal signal peptide for secretion from cells via the classical secretory pathway. In addition, some $sPLA_2s$ have a propeptide that needs to be cleaved intracellularly or extracellularly to get the fully active enzyme. All $sPLA_2s$ use a single Ca^{2+} ion as a catalytic cofactor along with a conserved histidine-aspartate catalytic diad (Lambeau & Gelb, 2008).

The primary structures of mammalian $sPLA_2s$ are summarized in Fig. 1 (Valentin & Lambeau, 2000). The naming of $sPLA_2s$ is based on the order of discovery as well as the pattern of disulfide bond number and location (group numbering system of PLA_2s; Dennis, Cao, Hsu, Magrioti, & Kokotos, 2011). We abbreviate each $sPLA_2$ starting with m or h for mouse and human, respectively, followed by the group name (i.e., mGIIC for mouse group IIC, hGIIE for human group IIE, etc.). The first mammalian $sPLA_2$ to be discovered was the pancreatic $sPLA_2$ group IB (group IA is reserved for a structurally similar $sPLA_2$ found in some snake venoms). The next mammalian $sPLA_2$ discovered is sometimes called nonpancreatic, platelet, or synovial fluid $sPLA_2$ (group IIA) because it was originally found in intestine, platelets, macrophages, and synovial fluid from the joints of patients with arthritis (Kramer et al., 1989). Other mammalian group II $sPLA_2s$ are the groups IIC, IID, IIE, and IIF enzymes. The group IIC $sPLA_2$ occurs as a

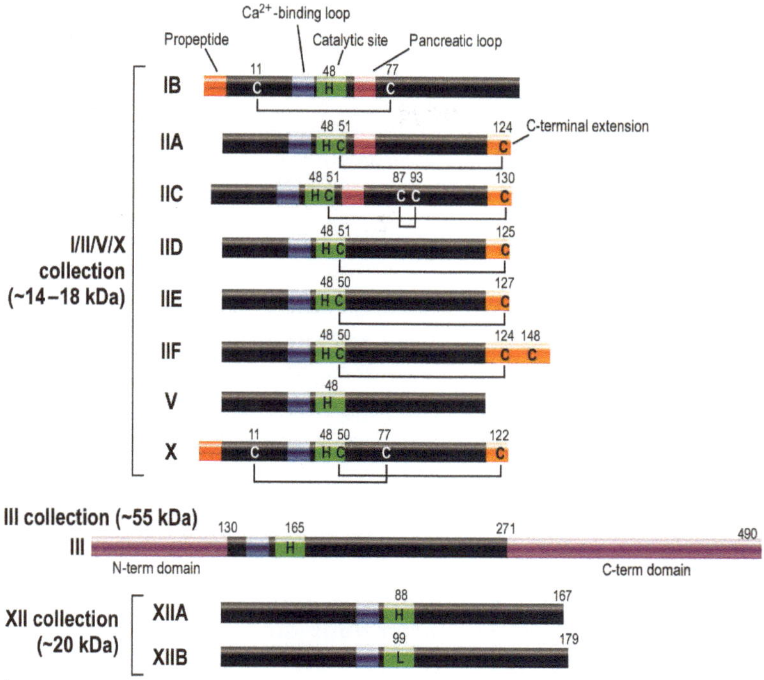

Fig. 1 Schematic of the primary structures of the mammalian $sPLA_2$ family. The calcium-binding loop and active site with the catalytic dyad HD are conserved in all active $sPLA_2s$ belonging to the group I/II/V/X collection as well other group III and XII collections. The H48 catalytic residue is replaced by a leucine in $sPLA_2$ group XIIB, which is the only enzymatically inactive $sPLA_2$. Group IB and X $sPLA_2s$ have a N-terminal short propeptide, which has to be cleaved for enzyme activation. Group IB $sPLA_2$ has a unique insertion called the pancreatic loop. A short C-terminal extension is present in all group II $sPLA_2s$ (IIA, IIC, IID, IIE, and IIF) and group X $sPLA_2$. Only group IIF $sPLA_2$ has a longer C-terminal extension with an odd cysteine likely involved in an interchain disulfide bond with an unknown subunit or in a posttranslational modification. Group V $sPLA_2$ is the smallest enzyme with no N-terminal propeptide, no insertion, no C-terminal extension, and only six disulfides. The unique disulfides of each $sPLA_2$ in the group I/II/V/X collection are indicated. Group III $sPLA_2$, the only enzyme in the group III collection, has a central catalytic domain with only five disulfides, and unique N- and C-terminal domains, which may act as prodomains or chaperones. The group XII collection comprises XIIA and XIIB whose the primary sequences markedly differ from other $sPLA_2s$ except for the conserved calcium-binding loop and active site.

nonfunctional pseudogene in humans but as a functional gene in mice (Chen, Engle, Seilhamer, & Tischfield, 1994). Group IID, IIE, and IIF enzymes are expected to be expressed and functional in both mouse and humans (Valentin, Koduri, et al., 1999), yet the tissue expression of hGIIE in human tissues is unclear, and likely very unique. Group IIF $sPLA_2$

contains a ~20 amino acid C-terminal extension of unknown function (Valentin, Singer, et al., 2000). The extension contains a single cysteine that is not involved in an intramolecular disulfide (all other cysteines in all mammalian $sPLA_2s$ are involved in disulfides). It is thus possible that group IIF forms a disulfide with itself (homodimer) or with a different protein, or is posttranslationally modified on the cysteine, but evidence for this is lacking. Group V $sPLA_2$ lacks the structural features of both group I and II $sPLA_2s$, i.e., it lacks the group II-specific C-terminal extension and the group I and II-specific disulfide. Opposedly, group X $sPLA_2$ has most of the structural features of both group I and II $sPLA_2s$, with a propeptide, the group II-specific C-terminal extension and the group I and II-specific disulfides (Cupillard, Koumanov, Mattéi, Lazdunski, & Lambeau, 1997). Group IB and X $sPLA_2s$ have short N-terminal propeptides and are thus produced as enzymatically inactive zymogens that require proteolysis for generation of the active $sPLA_2$ enzyme (Murakami, Sato, Miki, Yamamoto, & Taketomi, 2015). In the case of group IB pancreatic $sPLA_2$, trypsin is the physiologically relevant activator for enzyme generation in the digestive tract and plasmin may be relevant in some other tissues like lung (Abita, Lazdunski, Bonsen, Pieterson, & de Haas, 1972); the protease(s) responsible for group X activation is not known with certainty, but proprotein convertases like furin are candidates (Jemel et al., 2011). Group III $sPLA_2$ contains long N- and C-terminal extensions of unknown function (Valentin, Ghomashchi, Gelb, Lazdunski, & Lambeau, 2000). These extensions may correspond to prodomains removed by proteases from the proprotein convertase family, but direct evidence is lacking. The central $sPLA_2$ domain is similar to the group III enzyme from honeybee venom. Groups XIIA and XIIB $sPLA_2$ are structurally similar to each other but have very limited sequence similarity to other mammalian $sPLA_2s$ (Gelb, Valentin, Ghomashchi, Lazdunski, & Lambeau, 2000). Interestingly, group XIIB is naturally enzymatically inactive owing to the replacement of the active site histidine of the His-Asp catalytic diad with a leucine. Other phospholipase-inactive proteins with similar primary structure to $sPLA_2s$ include the otoconins which are exclusively expressed in inner ears, are devoid of enzymatic activity, and are major protein components of otoliths (Wang et al., 1998).

The functions of each of the mammalian $sPLA_2s$ are under active investigation. It is clear that they are differentially expressed in mammalian cells and thus do not have redundant functions. Reviews covering the functions of the $sPLA_2s$ have been recently published (Eerola et al., 2006; Murakami et al., 2015; Murakami, Taketomi, Sato, & Yamamoto, 2011; Nevalainen, Haapamaki, & Gronroos, 2000).

The protocols described in the current chapter allow for the production of recombinant, fully folded mouse and human $sPLA_2s$, which have been useful in the preparation of polyclonal antisera for each species (for example, Degousee et al., 2001), as well as the study of their enzymatic properties (for example, Singer et al., 2002), their receptor-binding properties (Rouault et al., 2007), the search for selective inhibitors (Oslund & Gelb, 2008; Smart et al., 2004), and the discovery of their biological functions (see chapter "Secreted Phospholipase A_2 Specificity on Natural Membrane Phospholipids" by Murakami and coworkers). Of importance, these protocols describe the preparation of fully folded and mature $sPLA_2s$, with their wild-type sequence (except for hGIIA with the N1A mutation which has been proved to be without effect on enzymatic activity), and devoid of any additional tags at the N-terminus or C-terminus, which are known to modify the enzymatic properties (especially at the N-terminus of $sPLA_2s$ I/II/V/X) and other biological activities. Of note, some $sPLA_2s$ exist as natural SNP variants. For example, hGIID exists as two natural SNP variants S60 or G60 (Takabatake et al., 2005).

2. STRATEGIES FOR PRODUCTION OF RECOMBINANT $sPLA_2s$

Most of the $sPLA_2s$ were produced by expression of unfolded protein (inclusion bodies) in *Escherichia coli* using two types of bacterial expression plasmids. In many cases, the $sPLA_2$ was produced as a fusion protein with a truncated glutathione-*S*-transferase protein (8.7 kDa, pAB3 plasmid) (Singer et al., 2002; Valentin, Ghomashchi, Gelb, Lazdunski, & Lambeau, 1999) and with insertion of a factor Xa cleavage site (IEGR) right before the mature $sPLA_2$. This cleavage site allows the use of either factor Xa or trypsin to cleave the fusion protein. In most cases, the $sPLA_2$ is quite resistant to trypsin digestion (i.e., under mild conditions like cleavage at room temperature and at a ratio of 1/300 to 1/1000 (trypsin/$sPLA_2$)) and this latter is used to cleave the fusion protein and release the native-length $sPLA_2$. Trypsin also has two additional advantages over factor Xa. It allows the removal of misfolded $sPLA_2$ conformers which are more sensitive to trypsin than the fully folded $sPLA_2$, and trypsin can be used in the presence of urea but not guanidine as denaturant to help cleavage when the IEGR cleavage site is buried. On the other hand, factor Xa can be used when the $sPLA_2$ is very sensitive to trypsin (like for hGXIIA $sPLA_2$). The use of this pAB3 plasmid virtually guarantees high-level protein expression in bacteria. In some cases, when the natural N-terminal amino acid of the $sPLA_2$ is a glycine (hGIID, mGIID, hGV, mGV, hGX, mGX) (Jemel et al., 2011; Rouault et al., 2007; Singer et al., 2002) or when this N-terminal amino acid can be mutated into alanine without a significant loss of enzymatic activity (one example is the hGIIA mutant N1A) (Snitko, Koduri, et al., 1997), the $sPLA_2$ can be produced using the pET21a plasmid where the Met initiator is directly followed by the mature $sPLA_2$. In this case, the strategy relies on the effective removal of the methionine initiator by the bacterial Met-aminopeptidase right after translation of the $sPLA_2$ and before the protein starts to precipitate as an inclusion body. This removal is tightly controlled by the nature of the amino acid following the Met initiator, and glycine is the preferred amino acid (Hirel, Schmitter, Dessen, Fayat, & Blanquet, 1989). Here, mass spectrometry analysis of the dissolved inclusion body is critical to verify the effective cleavage of the Met initiator before attempting to refold the $sPLA_2$, as it will not be possible to remove the Met initiator later on. For some $sPLA_2$s, we found the two forms of the $sPLA_2$, with and without the Met initiator, depending on the $sPLA_2$, the bacterial strain used, and the level of overexpression. In this case, refolding can still be performed starting from the two variants, but they will need to be separated by high-resolution HPLC to get the final pure mature $sPLA_2$ (see example for hGX; Jemel et al., 2011). In a few cases, bacterial expression was problematic and we resorted to expression in insect cells (S2 cells) or mammalian cells (HEK293 cells). Table 1 gives the expression plasmid and expression cell type for each $sPLA_2$ produced using the detailed protocols given later.

In the cases where bacterial inclusion bodies were produced, the latter were solubilized in denaturant and all cysteines were reduced and sulfonated by treatment with the Thannhauser reagent (see protocols below). The protein was allowed to refold *in vitro* at lower concentrations of denaturant and in the presence of sulfhydryl reagents and oxidants (dissolved O_2) allowing removal of the sulfonate groups and intramolecular disulfide bond formation. Of importance, the refolding buffer and all subsequently used buffers and HPLC solvents also contained 5 m*M* L-methionine to prevent oxidation of $sPLA_2$ methionines. A substantial amount of protein remained unfolded and typically remained insoluble (removed by centrifugation) or was misfolded, either insoluble or soluble; however, milligram quantities of pure $sPLA_2$s are obtained in the end owing to a large amount of inclusion body

Table 1 $sPLA_2$ Expression Plasmids, Expression Hosts, Apparent MW on Denaturing SDS-PAGE (Reducing Conditions), Mass Spectrometry Characterization, and Overall Yield of $sPLA_2$ Production Per Liter of Culture Medium

$sPLA_2$	Expression Plasmid and Host Cells	Approx. MW (SDS-PAGE) (kDa)	Measured MW (Mass Spectrometry) (Da)	Overall Yield of Production (mg/L of Culture)
hGIB	pAB3/BL21 (DE3) pET21a/BL21 (DE3)	14	14,124.99	1.7 15
mGIB	pVL1392/ baculovirus/ Tn5 insect cells	14	14,074.79	1.5
hGIIA (N1A) hGIIA WT	pET21a/BL21 (DE3) pVL1392/ baculovirus/ Tn5 pS2neo/S2 insect cells	14 14	13,859.7 (N1A) 13,904 (WT)	4 0.3–1
mGIIA	pAB3/BL21 (DE3) pVL1392/ baculovirus/ Tn5	16 16	13,958.19 13,957.71	2 0.1
mGIIC	pS2neo/S2 insect cells	17/16[a]	15,701.84 (N-glycosylated)/ 14,661.68[a]	0.15
hGIID[b]	pAB3/BL21 (DE3) (G60) pS2neo/S2 cells (S60)	14	14,492.6 14,522.9[a]	1 0.05
mGIID	pAB3/BL21 (DE3) pS2neo/S2 insect cells	14	14,252.41 14,254.0	2 0.1
hGIIE	pAB3/BL21 (DE3)	14	13,952.07	3
mGIIE	pAB3/BL21 (DE3)	14	14,066.09	1

Continued

Table 1 $sPLA_2$ Expression Plasmids, Expression Hosts, Apparent MW on Denaturing SDS-PAGE (Reducing Conditions), Mass Spectrometry Characterization, and Overall Yield of $sPLA_2$ Production Per Liter of Culture Medium—cont'd

$sPLA_2$	Expression Plasmid and Host Cells	Approx. MW (SDS-PAGE) (kDa)	Measured MW (Mass Spectrometry) (Da)	Overall Yield of Production (mg/L of Culture)
hGIIF WT and ΔCter	pAB3/BL21 (DE3) pET21a/BL21 (DE3)	23	16,583.4 (hGIIF WT no extra Cys) 14,171.2 (hGIIFΔCter)	3/2
mGIIF	pAB3/BL21 (DE3) pET21a/BL21 (DE3)	23	16,805.96	2
hGIII $sPLA_2$ domain	pET21a/BL21 (DE3) pS2neo/S2 insect cells	22	16,034.2 16,040[a]	0.5 0.05
mGIII $sPLA_2$ domain	pS2neo/S2 insect cells	~22	15,908.38[a] 17,989.91 (major N-glycosylated form)	0.07
hGV	pET21a/BL21 (DE3)	13	13,578.6	3
mGV	pET21a/BL21 (DE3)	14	13,790.8	3
hGX	pAB3/BL21 (DE3) pET21a/BL21 (DE3)	14	13,615.52	2.5 5
mGX	pAB3/BL21 (DE3) pET21a/BL21 (DE3)	14	13,854.87	3 3
hGXIIA (EAC variant)	pAB3/BL21 (DE3)	20	18,702.15	3

Table 1 sPLA$_2$ Expression Plasmids, Expression Hosts, Apparent MW on Denaturing SDS-PAGE (Reducing Conditions), Mass Spectrometry Characterization, and Overall Yield of sPLA$_2$ Production Per Liter of Culture Medium—cont'd

sPLA$_2$	Expression Plasmid and Host Cells	Approx. MW (SDS-PAGE) (kDa)	Measured MW (Mass Spectrometry) (Da)	Overall Yield of Production (mg/L of Culture)
mGXIIA	pAB3/BL21 (DE3)	18	18,724.29	1
hGXIIB	pAB3/BL21 (DE3)	23	19,658.69	2
mGXIIB	pAB3/BL21 (DE3)	22	19,402.93	2

[a]After deglycosylation with *N*-glycosidase F (see Rouault et al., 2007; Singer et al., 2002 for human and mouse enzymes, respectively).
[b]hGIID has a polymorphism at position 60 (S60G) (Takabatake et al., 2005). Hence, natural variants S60 or G60 were produced depending on the expression vector and host.

protein available from bacterial expression, optimized refolding procedures for each sPLA$_2$ and HPLC purification with shallow gradients to separate fully folded and misfolded enzymes.

3. GROWTH OF BACTERIA

Bacteria (BL21(DE3)pLysS, Novagen, MilliporeSigma, Billerica, MA) were transformed with the appropriate sPLA$_2$ expression plasmid (pAB3 or pET21a), and the ampicillin-resistant bacteria were grown in Luria Broth (1 L culture in 2 L Erlenmeyer flask) with orbital shaking at ~250 rpm at 37°C. Note that we routinely observed that pAB3 vectors produced more sPLA$_2$ protein (as inclusion body) than pET21a vectors. Accordingly, we usually scaled up the culture volume for pET21a. When the OD_{600} of an aliquot reached ~0.7, IPTG was added to 1 m*M*, and culturing was continued for 3–4 h. Cells were pelleted by centrifugation at 6500 rpm in a Beckman GSA rotor for 15–20 min at 4°C. Culture medium was decanted, and bacteria cell pellets were stored at −80°C until processed. To make sure that the sPLA$_2$ protein was induced and is present in the cell pellet, a small spatula tip of the cell pellet can be dissolved in Laemmli loading buffer and analyzed by SDS-PAGE, along with a control lane using bacteria not induced with IPTG.

4. ISOLATION OF INCLUSION BODY

The frozen cell pellet was thawed and resuspended in cold resuspension buffer (50 m*M* Tris–HCl, pH 8.0, 50 m*M* NaCl, 1 m*M* EDTA, 0.5 m*M* freshly added PMSF (from a stock in ethanol)) containing 0.4% (v/v) Triton X-100 and 0.4% (w/v) sodium deoxycholate. For the pellet from 1 L of bacterial culture, 50 mL of the resuspension buffer was used. Depending on the type of bacteria used, cells can be resuspended in resuspension buffer containing 10 m*M* $MgCl_2$, lysozyme (0.1 mg/mL), and DNase I (0.1 mg/mL). The suspension was stirred on ice for 20 min in a beaker. With cooling on ice, the suspension was sonicated for 15 cycles (15 s on/15 s off) at power setting 8 using a Heat Systems Ultrasonics Inc., Model W-225 sonicator with a 1.25-cm diameter probe. Alternatively, lysis can be made with a French press. The mixture was then centrifuged at 4°C for 10 min at 10,000 rpm in a Beckman GSA rotor. Tared centrifuge bottles were used so that the final weight of inclusion body pellet could be determined. The supernatant was decanted, and 50 mL of cold resuspension buffer with 0.8% (v/v) Triton X-100 and 0.8% (w/v) sodium deoxycholate was added to the pellet. The suspension was stirred on ice for 20 min and then sonicated and centrifuged as above. The pellet was stirred at room temperature for 20 min in 50 mL of resuspension buffer containing 1% (v/v) Triton X-100. After centrifugation as above, the pellet was stirred for 20 min at room temperature in 50 mL of resuspension buffer without detergent and then centrifuged as above. After decantation, the bottles were weighed to determine the weight of wet inclusion body pellet. The pellets were stored at −80°C until processed. Typically 200 mg of $sPLA_2$ protein as inclusion body was obtained with the pAB3 plasmid per liter of bacterial culture and about 50 mg with the pET21a plasmid, but yields depend on the $sPLA_2$ species produced.

5. PREPARATION OF SULFONATED PROTEIN

The inclusion body pellet (typically 50–100 mg of $sPLA_2$ protein) was dissolved in 35 mL of 6 *M* guanidine-HCl, 0.3 *M* Na_2SO_3, 50 m*M* Tris–HCl, pH 8.0 and was stirred in a beaker using a magnetic stir bar at room temperature for 30–45 min. In the case of difficult to solubilize pellets, the $sPLA_2$ protein can be further dissolved by sonication and dounce. The solution should appear slightly brown. Thannhauser reagent

(Thannhauser, Konishi, & Scheraga, 1984) was added to a final concentration of 15 m*M* (Singer et al., 2002), and the solution was stirred at room temperature for at least 1 h to overnight, or until the solution turned from dark orange to faint yellow.

Sulfonated protein was clarified from the remaining insoluble impurities by centrifugation at 10,000 rpm in a Beckman GSA rotor at 4°C for 20 min. The supernatant containing the soluble sulfonated protein was extensively dialyzed against 8 L of 0.1% acetic acid at 4°C (6–8 kDa dialysis tubing) for 2–3 h, then overnight with fresh acetic acid. The solution turns from yellow to white and the protein precipitates as a dense flocculated white powder. The precipitate was collected by centrifugation at 10,000 rpm in a Beckman GSA rotor at 4°C for 20 min. The pellet of sulfonated protein can be placed in a desiccator hooked up to a vacuum pump (lyophilizer can be used instead as a vacuum source), dried overnight at room temperature, then stored at −80°C until further processed.

6. PREPARATION OF hGIB

Inclusion body protein was prepared and sulfonated as described earlier from a hGIB/pAB3 vector. The sulfonated protein (80–100 mg) was dissolved in 200–250 mL of 6 *M* guanidine-HCl, 50 m*M* Tris–HCl, pH 8.0 at room temperature by stirring. Any remaining insoluble protein was removed by centrifugation at 14,000 × *g* for 15 min at 4°C, and the supernatant was used for refolding. Dialysis tubing (6–8 kDa MW cutoff, SpectraPor) was filled with the protein solution and dialyzed against 7 L of prechilled 0.9 *M* guanidine-HCl, 50 m*M* Tris–HCl, pH 8.0, 5 m*M* freshly added L-cysteine, 5 m*M* EDTA, 5 m*M* L-methionine (to avoid oxidation of the $sPLA_2$ methionines), and 5 m*M* SB12 detergent (*N*-dodecyl-*N*,*N*-dimethyl-3-ammonio-1-propanesulfonate, Sigma 40232) at 4°C with magnetic stirring. The next day, the enzymatic activity of a small aliquot was checked with the pyrene-PG assay, and it was checked again in the afternoon. If the activity was increasing, the dialysis was continued until the pyrene-PG activity stopped increasing (~1–2 days). The dialysis bag was transferred into 6.5 L of trypsin buffer (50 m*M* Tris–HCl, pH 8.0, 100 m*M* NaCl, 1 m*M* $CaCl_2$ with 5 m*M* SB12 detergent), and dialysis was continued at 4°C overnight. The bag contents were transferred to a centrifuge bottle, which was spun at 15,000 × *g* for 20 min at 4°C to pellet any insoluble material.

The supernatant was tested for trypsinization by placing 0.1–0.2 mL of solution into each of five tubes. To each of the five tubes, trypsin (Sigma Cat. T-1426 or T1005) was added in 1–2 μL of water to give 0.0, 0.25, 0.5, 1, and 1.5 μg/mL. The tubes were incubated at room temperature for 1.5 h, and the enzymatic activity was checked every 20 min with the pyrene-PG assay. The trypsin concentration that gave the highest activity was chosen for large-scale trypsinization. Trypsin solution (1–2 μg/μL in water) was added dropwise to a large volume of refolded protein solution over 1–2 min with magnetic stirring. The sample was incubated for 60–90 min, and activity was checked with pyrene-PG every 20–30 min until it stopped increasing. PMSF (Sigma Cat. p7626) in ethanol was added to give 0.5 m*M*, and the pH was lowered to 2–3 with dilute acetic acid. The solution was kept at 4°C prior to HPLC. The protein solution was centrifuged at 15,000 × *g* for 20 min at 4°C to pellet any insoluble material. The supernatant was filtered through a 0.45-μm syringe cartridge filter unit, and the filtrate corresponding to 20–25 mg sulfonated protein was loaded (using the pump directly) onto the HPLC column (Vydac 218TP1010, C18, 1 × 25 cm, 10 μm, 300 Å) preequilibrated in solvent A (water, 0.08% trifluoroacetic acid). The column was washed with ~1 column volume of solvent A before starting a linear gradient from 0% to 25% solvent B (acetonitrile, 0.08% trifluoroacetic acid) over 40 min and from 25% B to 60% B over the next 110 min at a flow rate of 1.5 mL/min. The $sPLA_2$ elutes as a sharp peak at about 49 min. Combined HPLC fractions were diluted with water to reduce the acetonitrile to 15–20%, and the protein was lyophilized to give a white powder. In more recent refolding assays, hGIB was also produced from the pET21a vector with similar yield and no need for trypsin treatment.

7. PREPARATION OF mGIB

mGIB $sPLA_2$ can likely be produced in *E. coli* as above for hGIB, but this enzyme was originally produced in insect cells as follows (Cupillard et al., 1999). The full-length cDNA including the prepropeptide sequence was subcloned into the baculovirus transfer vector pVL 1392 and transfected into Spodoptera frugiperda cells (Sf9; ATCC CRL 1711) using the BaculoGoldTM transfection kit (Pharmingen). After two rounds of virus amplification into Sf9 cells, Trichoplusia ni High Five insect cells (Tn5) were used for the production of a recombinant $sPLA_2$, since preliminary experiments have shown a threefold higher yield of $sPLA_2$ production compared with Sf9 cells. Furthermore, preliminary experiments indicated that infection of cells with mGIB baculovirus in growth medium containing fetal bovine serum resulted in the secretion of a mixture of proenzyme and mature forms of mGIB, since the $sPLA_2$ activity of cell supernatant can be increased by the addition of trypsin, while fully activated mGIB was recovered from supernatant of cells that were cultivated in protein-free Insect-Xpress medium (BioWhittaker). Large-scale $sPLA_2$ productions were thus performed with Tn5-infected cells (2.106 cells/mL) grown in spinner culture bottles in protein-free Insect-Xpress medium for 5 days. Cell-free supernatants of infected cells (1 L) were diluted twice in 1% (v/v) acetic acid and incubated batchwise for 2 h, at 4°C, and under continuous agitation with 150 mL of SP Sephadex C-25 gel (Amersham Pharmacia Biotech), which had been preequilibrated with 1% acetic acid. The gel was washed with 1% acetic acid and 1% acetic acid containing 100 m*M* ammonium acetate. Bound proteins were then eluted stepwise with 1% acetic acid containing 350 m*M* ammonium acetate. $sPLA_2$-containing fractions were lyophilized and applied to a C18 reverse-phase HPLC column (Beckman, 10 × 250 mm, 19.6 mL, 5 μm, 100 Å). Elution was performed using an acetonitrile linear gradient in 0.1% trifluoroacetic acid, 10–60% acetonitrile for 40 min at a flow rate of 4.5 mL/min. The peak-containing $sPLA_2$ activity was lyophilized and applied to the Spherogel TSK-SP-5PW HPLC column (TosoHaas, 10 μm, 75 × 7.5 mm, 3.3 mL) equilibrated in acetic acid 1% (v/v), and eluted using a linear gradient of ammonium acetate (0–2 *M*, pH 6.8 in 70 min). The $sPLA_2$ peak was finally applied on a C18 Nucleosil reverse-phase HPLC column (Macherey-Nagel, 4.6 × 250 mm, 4.2 mL, 5 μm, 300 Å) that was eluted using an acetonitrile linear gradient in 0.1% trifluoroacetic acid, 15–25% acetonitrile for 10 min followed by 25–45% acetonitrile for 100 min at 1 mL/min. Combined fractions were diluted with water to reduce the acetonitrile concentration to about 15–20%, and the solution was lyophilized to give a white powder.

8. PREPARATION OF hGIIA

hGIIA was prepared in the pET21a vector as a N-terminal mutant, N1A, allowing the removal of the N-terminal methionine (see earlier) (Snitko, Yoon, & Cho, 1997). In one protocol, inclusion body protein was prepared and sulfonated as above. However, the protein can also be refolded as a nonsulfonated protein with similar yield as follows. Inclusion body protein from 1 L of bacterial culture was stirred for 3 h at room temperature in 25 mL of 50 m*M* Tris, pH 8.0, 50 m*M* NaCl,

1 m*M* EDTA, 0.5 m*M* PMSF (freshly added), 6 *M* guanidine-HCl, 5% β-mercaptoethanol. The mixture was centrifuged at 14,000 × *g* for 15 min at 4°C. The clarified supernatant was dialyzed against 2 L of 25 m*M* Tris, pH 8.0, 5 m*M* $CaCl_2$, 5 m*M* L-cysteine, 0.9 *M* guanidine-HCl, 5 m*M* L-methionine over 2–3 days at 4°C with three changes of buffer. $sPLA_2$ activity was monitored with the pyrene-PG assay. The solution was centrifuged as above, and the supernatant was dialyzed against 2.5 L of 20% acetonitrile, 0.08% trifluoroacetic acid for 3–4 h at 4°C, and then against 2.5 L of the same solvent overnight at 4°C. The dialyzed protein was centrifuged as above, and the clear supernatant was diluted with an equal volume of solvent A (water, 0.08% trifluoroacetic acid). The solution was filtered through a 0.45-μm syringe cartridge filter, and the filtrate was loaded onto the C18 Vydac HPLC column (see earlier for hGIB). The solvent program was 0% solvent B (acetonitrile, 0.08% trifluoroacetic acid) to 60% B over 60 min with a flow rate of 1.5 mL/min. The $sPLA_2$ elutes as a sharp peak at 40 min. Combined fractions were lyophilized to give a white powder.

hGIIA was also produced as a wild-type protein (N-terminal is asparagine) in insect cells using either the Sf9/baculovirus system (Cupillard et al., 1999) or *Drosophila* S2 cells (Singer et al., 2002).

9. PREPARATION OF mGIIA

The sulfonated protein (80–100 mg, prepared from a pAB3 vector) was dissolved in 200–250 mL of 6 *M* guanidine-HCl, 50 m*M* Tris–HCl, pH 8.0 at room temperature by stirring for at least 1 h. The protein was refolded and trypsinized as described for hGIB (see earlier).

The protein solution was filtered through a 0.45-μm syringe filter unit, and the filtrate loaded onto the C18 Vydac HPLC column (see earlier for hGIB) preequilibrated in solvent A (water, 0.08% trifluoroacetic acid). The column was washed with ~1 column volume of solvent A, before starting a linear gradient from 0% to 30% solvent B (acetonitrile, 0.08% trifluoroacetic acid) over 40 min at a flow rate of 1.5 mL/min. Over the next 90 min, the solvent increased from 30% to 60% B. The $sPLA_2$ elutes as a sharp peak at about 45 min. Combined HPLC fractions were lyophilized to give a white powder.

mGIIA was also produced in Sf9/baculovirus and *Drosophila* S2 cells as an alternative to the *E. coli* expression system described earlier with similar yields (Cupillard et al., 1999; Rouault et al., 2007).

10. PREPARATION OF mGIIC

hGIIC is a nonfunctional pseudogene, but the mouse genome contains a functional group IIC $sPLA_2$ gene. Recombinant mGIIC was produced in *Drosophila* S2 cells. The cells were grown at 24–27°C in HyClone SFX-insect cell culture medium supplemented with 1% fetal bovine serum and antibiotic–antimycotic cocktail from ThermoFisher Scientific (Gibco #15240062). Cells were transfected using the calcium phosphate method. The calcium phosphate–DNA precipitate was incubated with cells for 4 h, after which the cells were washed by centrifugation and incubated in fresh medium. Two days after transfection, 2 mg/mL G418 sulfate was added to the cell culture medium to select for stably transfected cells. Fresh selection medium was added every 4–5 days and selection lasted for at least 3 weeks. mGIIC expression was examined by measuring membrane-bound and soluble $sPLA_2$ activity present in cell medium, as described previously for hGIIA produced in Sf9 insect cells (Cupillard et al., 1999). $sPLA_2$ activity was measured using labeled *E. coli* membranes as substrate. For these assays, cells were plated at 3×10^6 cells/mL in 24-well plates and induced the day after with 500 μ*M* $CuSO_4$ for 3–7 days. For large-scale production, stably transfected cells were seeded into 500 mL of complete medium in 2 L spinner culture flasks. Cells were induced at a density of 3–4 $\times 10^6$ cells/mL with 500 μ*M* $CuSO_4$ for up to 7 days, after which cells were pelleted and the medium containing $sPLA_2$ activity was collected. Unlike mGIIA and mGIID, no cell-associated $sPLA_2$ activity was found for mGIIC. To purify mGIIC, 2 L batches of cell medium were diluted to 4 L with 1% acetic acid and loaded onto 500 mL of Sephadex SPC-25 gel (Amersham Biosciences) which had been equilibrated with 1% acetic acid and poured into a 500-mL glass-sintered funnel. The gel was washed with 1 L of 1% acetic acid and 1 L of 1% acetic acid containing 400 m*M* ammonium acetate. Bound proteins were eluted with 4 L of 1% acetic acid containing 1.5 *M* ammonium acetate. Fractions of 500 mL were collected and assayed for $sPLA_2$ activity using the *E. coli* assay. $sPLA_2$-containing fractions were concentrated to 30 mL by ultrafiltration using an Amicon stirred cell concentrator with a YM-10 membrane, and the buffer was exchanged with 1% acetic acid containing 10% acetonitrile. The resulting fraction was filtered on 0.45 μm hydrophilic PVDF membrane filters (Millipore #SLHVBZ5NZ) and loaded onto a Spherogel TSK-SP-5PW column (see earlier for mGIB) equilibrated in 1% acetic acid with 200 m*M* ammonium acetate and 10% acetonitrile. The column was eluted at 8 mL/min using a linear gradient of ammonium acetate (from 0.2 to 2 *M*, at pH 6.8, over 90 min) in 10% acetonitrile. Fractions containing $sPLA_2$ activity were pooled, lyophilized, resuspended in 10% acetonitrile and 0.1% trifluoroacetic acid, and applied to C18 Beckman HPLC column (see earlier for mGIB). Elution was performed at 4 mL/min using a water/acetonitrile gradient in 0.1% trifluoroacetic acid (from 10% to 20% over 10 min and from 20% to 40% over 60 min). mGIIC eluted at 39% acetonitrile and was recovered as a pure protein after lyophilization. As previously found for hGIID (Singer et al., 2002), mGIIC $sPLA_2$ was found to be glycosylated and was produced as a homogeneously glycosylated protein with a molecular mass of 15,700 Da corresponding to a sugar moiety of 1038 Da. N-terminal sequencing and analysis of $sPLA_2$ trypsin digests by MALDI-TOF mass spectrometry showed that the glycosylation occurred only on Asn-72 in mGIIC. To determine the molecular mass of the protein moiety (Table 1), deglycosylation experiments were performed by overnight treatment of 3 μg of purified $sPLA_2$ at 20°C with 0.5 U of *N*-glycosidase F (Sigma-Aldrich, original Roche #1365169) in 40 μL of 20 m*M* NaH_2PO_4, pH 6.5, and complete mini protease inhibitor cocktail (Sigma-Aldrich, original Roche #1836153). The purified $sPLA_2$ was lyophilized to give a white powder.

11. PREPARATION OF hGIID

The sulfonated protein (80–100 mg, prepared from a pAB3 vector) was dissolved in 80–100 mL of 8 *M* urea, 50 m*M* Tris–HCl, pH 8.0, 0.1 *M* NH_4Cl, 5 m*M* calcium chloride, 0.5 *M* L-arginine, 2.5 *M* glycerol, 5 m*M* L-methionine at room temperature by stirring. Dialysis tubing (6–8 kDa MWCO, SpectraPor) was filled with protein solution and dialyzed against 2–2.5 L of prechilled 1.6 *M* urea, 50 m*M* Tris–HCl pH 8.0, 0.1 *M* NH_4Cl, 5 m*M* $CaCl_2$, 8 m*M* L-cysteine, 0.5 *M* L-arginine, and 2.5 *M* glycerol at 4°C without changing the buffer. The next day, the enzymatic activity of a small aliquot was checked with the pyrene-PG assay, and it was checked again in the afternoon. If the activity was increasing, the dialysis was continued until the pyrene-PG activity stopped increasing (~2 days). The bag contents were transferred to a centrifuge bottle, which was spun at 15,000 × *g* for 20 min at 4°C to pellet any insoluble material. The supernatant was dialyzed against modified trypsin buffer, 50 m*M* Tris–HCl, pH 8.0, 1 m*M* $CaCl_2$, 0.1 *M* NaCl, 0.25 *M* L-arginine, 1.25 *M* glycerol, and 0.1 *M* NDSB-201 (Fluka 82804), at 4°C for overnight. The protein solution was cleared at 15,000 × *g* for 20 min at 4°C to remove any insoluble material. The $sPLA_2$ was trypsinized as described for hGIB with the exception of allowing the protein to be digested for 4–5 h with up to 3 μg trypsin/mL refolded protein. The enzymatic activity of small aliquot was checked with pyrene-PG assay at various time points. The amount of trypsin giving the highest enzymatic activity was chosen for large-scale digestion.

The protein solution was spun at 15,000 × *g* for 20 min at 4°C to pellet any insoluble material. The supernatant was filtered through a 0.45-μm syringe filter unit, and the filtrate corresponding to about 20–30 mg of originally sulfonated protein was loaded onto the C18 Vydac HPLC column (see earlier for hGIB) preequilibrated in solvent A (water, 0.08% trifluoroacetic acid). The column was washed with ~1 column volume of solvent A, before starting the linear gradient from 0% to 20% B (acetonitrile, 0.08% trifluoroacetic acid) over 10 min and from 20% B to 80% B over the next 80 min at 1.5 mL/min. Protein elutes at ~40 min. The peak of $sPLA_2$ was repurified with the same gradient program to obtain fully pure $sPLA_2$. The combined purified protein was lyophilized as white powder.

hGIID exists as two natural SNP variants at the amino acid position 60 (S60 or G60). The G60 variant was produced in *E. coli* (above), while the S60 variant was produced in S2 cells (Singer et al., 2002).

12. PREPARATION OF mGIID

The sulfonated protein (80–100 mg, prepared from a pAB3 vector) was dissolved in 200–250 mL of 6 *M* guanidine-HCl, 50 m*M* Tris–HCl, pH 8.0 at room temperature by stirring. The protein was refolded and trypsinized as described for hGIB.

After trypsinization, the solution was centrifuged at 15,000 × *g* for 20 min at 4°C to remove insoluble material. The supernatant was filtered through a 0.45-μm syringe filter unit, and the filtrate corresponding to 20–30 mg sulfonated protein was directly pumped onto the C18 Vydac HPLC column (see earlier for hGIB) preequilibrated in solvent A (water, 0.08% trifluoroacetic acid). The column was washed with ~1 column volume of solvent A, before starting a linear gradient from 0% to 20% solvent B (acetonitrile, 0.08% trifluoroacetic acid) over 20 min at a flow rate of 1.5 mL/min. Elution continued with a linear gradient from 30% to 60%

B over 110 min. The $sPLA_2$ elutes as a sharp peak at ~28 min. Combined HPLC fractions were pooled and lyophilized to give a white powder.

Because of a putative N-glycosylation site previously found on Asn-80 of mGIID (Valentinis et al., 1999), we also expressed mGIID in *Drosophila* S2 cells and found that this $sPLA_2$ is not glycosylated in this expression system (Rouault et al., 2007).

13. PREPARATION OF hGIIE

The sulfonated protein (80–100 mg, prepared from a pAB3 vector) was dissolved in 200–250 mL of 6 *M* guanidine-HCl, 50 m*M* Tris–HCl, pH 8.0 at room temperature by stirring. The protein was refolded and trypsinized as described for hGIB. The protein solution was spun at 15,000 × *g* for 20 min at 4°C to pellet any insoluble material. The supernatant was filtered through a 0.45-μm syringe cartridge filter unit, and the filtrate corresponding to 20–30 mg sulfonated protein was loaded (using the pump directly) onto the C18 Vydac HPLC column (see earlier for hGIB) preequilibrated in 10% B (acetonitrile, 0.08% trifluoroacetic acid). The column was washed with ~1 column volume of solvent A (water, 0.08% trifluoroacetic acid), before starting a linear gradient from 10% B to 30% B in 20 min and 30% B to 45% B over the next 110 min. The $sPLA_2$ elutes at ~45 min. The combined $sPLA_2$ peak was lyophilized as white powder.

14. PREPARATION OF mGIIE

The sulfonated protein (80–100 mg, prepared from a pAB3 vector) was dissolved in 200–250 mL of 6 *M* guanidine-HCl, 50 m*M* Tris–HCl, pH 8.0 at room temperature by stirring. The protein was refolded and trypsinized as described for hGIB. The protein solution was centrifuged at 15,000 × *g* for 20 min at 4°C to pellet any insoluble material. The supernatant was filtered through a 0.45-μm syringe filter unit, and the filtrate corresponding to 20–30 mg sulfonated protein was loaded (using the pump directly) onto the C18 Vydac HPLC column (see earlier for hGIB) preequilibrated in solvent A (water, 0.08% trifluoroacetic acid). The column was washed with ~1 column volume of solvent A before starting a linear gradient from 10% to 30% solvent B (acetonitrile, 0.08% trifluoroacetic acid) over 20 min at a flow rate of 1.5 mL/min, then 30% B to 45% B over the next 110 min. The $sPLA_2$ elutes as a sharp peak at about 49 min. Combined HPLC fractions were pooled and lyophilized to give a white powder.

15. PREPARATION OF hGIIF

A particular feature of hGIIF and mGIIF $sPLA_2$ is the presence of a 23 amino acid C-terminal extension containing an extra cysteine, in addition to the typical group II-specific C-terminal extension of seven residues (Fig. 1). The presence of this odd cysteine raises the possibility that endogenous hGIIF and mGIIF $sPLA_2$ occurs as a covalent homodimer or a heterodimer with an unknown protein partner or has a special posttranslational modification on this cysteine. The C-terminal extension also contains several proline and acidic residues which may participate in its enzymatic properties (Wijewickrama et al., 2006). We have thus produced recombinant hGIIF $sPLA_2$ with (WT) and without the C-terminal extension (hGIIFΔCter).

Sulfonated hGIIF fusion protein (80–100 mg, prepared from a pAB3 vector) was dissolved to 10 mg/mL in 6 *M* guanidine-HCl, 50 m*M* Tris–HCl, pH 8.0 (this and all subsequently used buffers and HPLC solvents also contained 5 m*M* L-methionine to prevent oxidation of protein methionines), by stirring for 2 h at room temperature or overnight at 4°C. The sample was centrifuged at 4°C at 12,000 rpm for 20 min to remove undissolved protein. Four milliliter of protein solution was added dropwise (~1 drop/s) to 2 L of refolding buffer (50 m*M* Tris–HCl, pH 8.0, 0.9 *M* guanidine-HCl, 10 m*M* $CaCl_2$, 8 m*M* freshly added cysteine, 5 m*M* L-methionine, 30% acetonitrile (by volume), acetonitrile added last to buffer preadjusted to pH 8.0) with stirring in an Erlenmeyer flask at room temperature. Stirring was continued for a few min, and the solution was kept at room temperature without stirring until the $sPLA_2$ activity reached a maximum (~2–3 days). The volume of the solution was reduced to 1.4 L by rotary evaporation at 30°C. Lauryl sulfobetaine was added to a concentration of 5 m*M* followed by the addition of 14 mL of 100 m*M* L-methionine (previously filtered through a 0.45-μm syringe filter). The solution was concentrated to 40–50 mL in an ultrafiltration cell (Amicon, YM-10 membrane) with stirring under 30–40 psi of N_2 pressure. The solution was dialyzed against trypsin buffer (50 m*M* Tris–HCl, pH 8.0, 100 m*M* NaCl, 1 m*M* $CaCl_2$, 5 m*M* L-methionine) at 4°C, three cycles, each with 40 volumes of buffer. Trypsin was added to a final concentration of 0.2 μg/mL, and the sample was left at room temperature overnight. PMSF was added to a final concentration of 0.5 m*M*, and the solution was filtered through a 0.45-μm nylon HPLC solvent filter. An aliquot (12.5 mL) was pumped onto the Vydac reverse-phase HPLC column (see earlier for hGIB, preequilibrated with 20% acetonitrile, 0.1% trifluoroacetic acid) at 3 mL/min. The column was washed with 15 mL of solvent A (water, 0.1% trifluoroacetic acid) followed by a gradient of 0–6.3% B (acetonitrile, 0.1% trifluoroacetic acid) in 2 min, then to 27.5% B in 42 min, then to 100% B in 4 min. hGIIF (i.e., recombinant hGIIF with its C-terminal cysteine disulfide linked to the cysteine amino acid from the refolding buffer) elutes at ~36 min. The purified protein solution was neutralized by adding Tris to a final concentration of 20 m*M* (10 μL of 2 *M* Tris/mL of HPLC fraction), and lauryl sulfobetaine was added to a concentration of 5 m*M*. The sample was concentrated in a Centriprep-10 (Amicon) and then dialyzed against 40 volumes of 10 m*M* Tris, 0.1 m*M* DTT, pH 8.0, at 4°C for one cycle, and then 40 volumes of 10 m*M* Tris–HCl, pH 8.0, for two cycles. The protein solution was stored at −20°C. To ensure complete removal of buffer L-cysteine that may be disulfide linked to the hGIIF cysteine residue in its C-terminal extension, 0.64 mg of hGIIF in 1 mL of 20 m*M* Tris, pH 8.0, was treated with 1 m*M* DTT for 1 h at room temperature. The sample was diluted to 1.5 mL with 20% B and loaded onto an analytical HPLC column (Supelco Discovery BIO wide pore C5, 5 μm, 4.6 × 250 mm). The column was developed for 5 min in 20% B followed by a gradient to 50% B over 120 min. hGIIF eluted at 69.4 min, and the fraction was diluted with water and lyophilized. The protein was resuspended in 31.5% B, and the solution was loaded onto the same HPLC column. After 5 min at 31.5% B, a gradient to 36.5% acetonitrile and 0.1% trifluoroacetic acid was started. hGIIF eluted at 46.3 min and was lyophilized.

hGIIFΔCter was refolded as follows. Sulfonated hGIIFΔCter protein (80–100 mg, prepared from a pAB3 vector) was dissolved at 0.2 mg of protein/mL in 500 mL of 6 *M* guanidine-HCl, 50 m*M* Tris–HCl, pH 8.0, and refolded by dialysis against 8 L of 0.7 *M* guanidine-HCl, 50 m*M* Tris–HCl, pH 8.0, 5 m*M* EDTA, 5 m*M* L-methionine, and 5 m*M* L-cysteine for 48 h at 4°C. The refolded protein was dialyzed twice against 8 L of 2 *M* urea, 0.1 *M* NH_4Cl, 5 m*M* $CaCl_2$, 50 m*M* Tris–HCl, pH 8.0. The protein solution was filtered through Sephadex G-50 and subjected to digestion with trypsin (0.025 μg of trypsin/μg of protein) at room temperature. The cleavage of the fusion protein was followed by measuring the increase of $sPLA_2$ activity using *E. coli* membranes as a substrate. After about 2 h of incubation with trypsin, the reaction was stopped by acidification with 0.1% trifluoroacetic acid, and the mixture was concentrated to 40 mL using an Amicon stirred cell concentrator with a YM-10 membrane.

The solution was filtered and loaded directly onto a Beckman C18 HPLC column (see earlier for mGIB). Elution was performed at 4 mL/min using a linear gradient of solvent B (20–35% over 150 min). Fractions containing $sPLA_2$ activity were lyophilized, dissolved in 23% B, and loaded on a Waters Symmetry Shield C8 column (4.6 × 250, 4.6 mL, 100 Å, 5 μm) equilibrated with the same buffer. hGIIFΔCter was eluted with a shallow gradient of solvent B (23–35% over 120 min). The hGIIFΔCter protein was eluted at 92.7 min. Combined fractions were pooled and lyophilized to give a white powder.

16. PREPARATION OF mGIIF

Sulfonated protein (80–100 mg, prepared from a pAB3 vector) was resuspended in 500 mL of 8 *M* urea, 50 m*M* Tris–HCl pH 8.0, 5 m*M* $CaCl_2$, 0.1 *M* NH_4Cl, and 5 m*M* L-methionine and refolded by dialysis at 4°C for 3 days using a 8-kDa membrane tubing (SpectraPor, diameter of 76 mm) against 8 L of 1.6 *M* urea, 50 m*M* Tris–HCl pH 8.0, 0.1 *M* NH_4Cl, 5 m*M* $CaCl_2$, 5 m*M* L-methionine, and 8 m*M* L-cysteine. The dialyzed solution (560 mL) was centrifuged at 10,000 × *g* for 30 min to remove aggregates and subjected to digestion by trypsin (1/300 ratio by weight, Sigma T-1005) for 90 min at room temperature. The cleaved mGIIF protein was concentrated by ultrafiltration using an Amicon YM-10 cell membrane, buffer exchanged against 10% solvent B (acetonitrile, 0.1% trifluoroacetic acid), filtered, and then directly loaded onto a C5 semi-preparative HPLC column (see earlier for hGIIF). The elution was performed at 3 mL/min using a gradient of solvent B (10% for 5 min, from 10% to 30% over 20 min, and then from 30% to 35% over 100 min). The mGIIF Cys monomer (i.e., recombinant mGIIF with its C-terminal cysteine disulfide linked to the cysteine amino acid from the refolding buffer) eluted at 32% B and was recovered as a pure protein after lyophilization. If desired, the mGIIF monomer without the additional cysteine from the refolding buffer can be prepared as above for hGIIF.

17. PREPARATION OF hGIII $sPLA_2$ DOMAIN

Both hGIII and mGIII full-length proteins are much larger than the other $sPLA_2$s (Fig. 1, about 55 kDa, Valentin, Ghomashchi, et al., 2000), and their mRNA predicts a protein consisting of three individual domains, each containing several disulfides (Valentin, Ghomashchi, et al., 2000). The central acidic $sPLA_2$ domain is flanked by two N- and C-terminal basic domains which contain putative cleavage sites by proprotein convertases such as furin (Valentin, Ghomashchi, et al., 2000). It is unclear whether the GIII $sPLA_2$ is produced as a proenzyme with the N- and C-terminal domains acting as prodomains that need to be cleaved to control and release the catalytically active domain or whether these additional domains play a role in its biological role, for instance, by directing the enzyme to target cells and specific substrates after secretion (Murakami et al., 2005).

Our initial attempts to prepare full-length (~55 kDa) hGIII or mGIII in *E. coli*, and other expression systems including baculovirus/Sf9 and S2 cells were unsuccessful. We then tried to express the isolated hGIII and mGIII $sPLA_2$ domains in *E. coli* as a ΔGST fusion protein using the pAB3 vector and as a nonfusion Met-$sPLA_2$ protein using the pET21a vector. In all cases, we could not refold in significant yield the sulfonated or the nonsulfonated hGIII and mGIII proteins prepared from inclusion body and using the various above refolding buffers which were successful for the other $sPLA_2$s. This led us to produce both hGIII and mGIII $sPLA_2$ domains in *Drosophila* S2 cells as secreted proteins (the protocol indicated below for mGIII was also used for hGIII), after fusion of the catalytic domain with the hGIIA signal peptide (Rouault et al., 2007). Both hGIII and mGIII proteins purified from S2 cells were found to be glycosylated at two positions, and as active as many other recombinant $sPLA_2$s. However, the overall yield of production in S2 cells for both hGIII and mGIII was consistently low (about 0.05 mg/L), prompting us to reinvestigate several refolding conditions from inclusion bodies obtained in the *E. coli* expression system. After several trials, we ended up with the following protocol starting from the sulfonated hGIII $sPLA_2$ domain.

Sulfonated hGIII $sPLA_2$ domain (10 mg, from the pET21a vector) was dissolved in 6 mL of 6 *M* guanidine-HCl, 5 m*M* L-methionine and refolded by rapid dilution in a refolding buffer (1 L) consisting of 0.9 *M* guanidine-HCl, 50 m*M* Tris pH 7.0, 0.8 *M* L-arginine, 8 m*M* L-cysteine, 5 m*M* L-methionine, 1 m*M* EDTA, and 0.1 *M* sulfobetaine SB-12 (Sigma-Aldrich, St. Louis, MO). After 1–3 days of refolding at room temperature, the mixture was concentrated by ultracentrifugation using a YM-10 membrane (EMD Millipore, Billerica, MA). The concentrate was dialyzed against 20% acetonitrile/0.1% trifluoroacetic acid/5 m*M* L-methionine to remove SB-12. The resulting fraction was filtered on 0.45 μm hydrophilic PVDF membrane filters (Millipore #SLHVBZ5NZ) and submitted to HPLC purification on a C18 Symmetry Shield column (Waters Corp.,

Milford, MA) using a gradient of 20–30% solvent B (acetonitrile, 0.1% trifluoroacetic acid) in 60 min. The $sPLA_2$ eluted at ~40 min. Combined fractions were lyophilized to give a white powder.

18. PREPARATION OF mGIII $sPLA_2$ DOMAIN

Recombinant mGIII $sPLA_2$ domain was produced in *Drosophila* S2 cells as above for mGIIC but with the following modifications. No cell-associated $sPLA_2$ activity was found for the mGIII $sPLA_2$ domain, and most of the $sPLA_2$ activity was present in the clarified cell-free culture medium. To purify the mGIII $sPLA_2$ domain, 2 L batches of cell medium were diluted with 2 L of water, adjusted to pH 7.0, and loaded at room temperature onto 300 mL of Q Sepharose fast flow gel (GE Healthcare Life Sciences) poured into a 500-mL glass-sintered funnel and preequilibrated with 20 m*M* Tris–HCl pH 8.0. The gel was washed with 1 L of 20 m*M* Tris–HCl pH 8.0 and then with 1 L of 20 m*M* Tris–HCl pH 8.0 and 0.5 *M* NaCl. Interestingly, the mGIII protein did not elute under these conditions. Bound proteins were eluted stepwise with 1 L of 1% acetic acid, 1 L of a 1% acetic acid/20% acetonitrile/0.5 *M* NaCl mixture, and finally with 1 L of 1% acetic acid containing 80% acetonitrile. $sPLA_2$-containing fractions were pooled, evaporated with a rotary evaporator to remove acetonitrile, and then concentrated to 30 mL by ultrafiltration using an Amicon stirred cell concentrator with a YM-10 membrane, and the buffer was exchanged with 20 m*M* Tris–HCl pH 8.0 containing 30% acetonitrile. The $sPLA_2$ was then loaded onto a Spherogel TSK-DEAE-5PW column (see earlier for mGIB) equilibrated in 20 m*M* Tris–HCl pH 8.0 and 30% acetonitrile. The protein was eluted from the column at 1 mL/min using a linear gradient of NaCl (from 0 to 1 *M* over 70 min) in 20 m*M* Tris–HCl pH 8.0 and 30% acetonitrile. Fractions containing $sPLA_2$ activity were pooled, diluted to 20% acetonitrile and 0.1% trifluoroacetic acid, and applied to a C8 Symmetry Shield HPLC column (see earlier for hGIIF). Elution was performed at 1 mL/min using a water/acetonitrile gradient in 0.1% trifluoroacetic acid (from 20% to 30% over 20 min and from 30% to 40% over 100 min). The mGIII $sPLA_2$ domain eluted as a broad peak at 37% acetonitrile and was recovered as a pure protein after lyophilization. As previously found for hGIID (Singer et al., 2002), the recombinant mGIII $sPLA_2$ domain was heterogeneously glycosylated with measured molecular masses of 17,989.91 (major peak), 18,192.01 (minor peak), and 18,315.74 Da (minor peak). The Δmass between the major peak and the calculated sequence of the mGIII sPLA$_2$ domain suggests that the protein is glycosylated at two sites, each with a sugar moiety of 1038 Da. N-terminal sequencing and analysis of sPLA$_2$ trypsin digests by MALDI-TOF mass spectrometry showed that the glycosylation occurred on both Asn-18 and Asn-131 in the mGIII sPLA$_2$ domain. To determine the molecular mass of the protein moiety (Table 1), deglycosylation experiments were performed by overnight treatment of 3 μg of purified sPLA$_2$ at 20°C with 0.5 U of *N*-glycosidase F (Sigma-Aldrich, original Roche #1365169) in 40 μL of 20 m*M* NaH_2PO_4, pH 6.5, and complete mini protease inhibitor cocktail (Sigma-Aldrich, original Roche #1836153). The purified sPLA$_2$ was lyophilized to give a white powder.

19. PREPARATION OF hGV

The sulfonated protein (40 mg, prepared from a pET21a vector) was dissolved at 10 mg/mL in 6 *M* guanidine-HCl, 50 m*M* Tris–HCl, pH 8.0 by stirring for 2–3 h at room temperature or overnight at 4°C. The protein solution was spun at 15,000 × *g* for 20 min at 4°C to pellet any insoluble material. The cleared solution was added into 2 L of refolding buffer (50 m*M* Tris–HCl, pH 8.0, 0.9 *M* guanidine-HCl, 10 m*M* $CaCl_2$, 5 m*M* freshly added L-cysteine, 5 m*M* L-methionine, 30% acetonitrile (v/v)), dropwise at ~1 drop per 1–2 s while stirring at room temperature for a few min, then at 4°C overnight without stirring. The pH of the buffer was adjusted before acetonitrile was added. The enzymatic activity was checked with pyrene-PG assay for 2 days until the activity stopped increasing. Detergent (*N*-dodecyl-*N*,*N*-dimethyl-3-ammonio-1-propanesulfonate, Sigma 40232) at 5 m*M* final concentration was added to the refolded protein solution, and the volume was reduced by rotary evaporation at 30°C to ~75%, and further concentrated with Amicon stirred cell with a YM-10 membrane (Millipore 13651) to a final volume of 40–50 mL at room temperature. The refolded protein was dialyzed against 2 L of solvent (20% acetonitrile, 0.1% trifluoroacetic acid) at 4°C overnight with one buffer change after 4–6 h. The protein solution is cleared as before, and half of the prep is loaded onto C18 Vydac HPLC column (see earlier for hGIB) preequilibrated in 20% solvent B (acetonitrile, 0.1% trifluoroacetic acid). The linear gradient program immediately started from 100% A (water, 0.1% trifluoroacetic acid) to 12.5% B in 2 min and to 22.5% B in 32 min and finally to 100% B in 6 min at 3 mL/min. The hGV elutes at ~25 min. Protein solution can be stored as is at −20°C or lyophilized and stored as white powder. The hGV should be resolubilized in 30% solvent B as needed.

20. PREPARATION OF mGV

Sulfonated protein (40 mg, prepared from a pET21a vector) was dissolved at 4 mg/mL in 6 *M* guanidine-HCl, 50 m*M* Tris–HCl pH 8.0 by stirring for 30 min at room temperature. The protein solution was centrifuged at 12,000 rpm (Beckmann SS-34 rotor) for 20 min at 4°C to remove insoluble material. The supernatant was added dropwise (1 drop per 2–3 s) to the prechilled refolding buffer (1 volume protein solution per 9 volumes refolding buffer; 0.6 *M* guanidine-HCl, 50 m*M* Tris–HCl pH 8.0, 100 m*M* NaCl, 10 m*M* $CaCl_2$, 5 m*M* freshly added L-cysteine, 5 m*M* L-methionine, 30% acetonitrile (by volume)) at 4°C with magnetic stirring. The pH of the buffer was adjusted before acetonitrile was added. Stirring was continued for 30 min at room temperature, then at 4°C overnight without stirring. The next day, the enzymatic activity was checked with the pyrene-PG assay until activity stopped increasing (~30 h). The refolded protein solution was centrifuged as before, and the supernatant was dialyzed overnight at 4°C against solvent (20 m*M* Hepes pH 7.5, 100 m*M*, NaCl, 30% acetonitrile (by volume), pH was adjusted before acetonitrile was added). The beaker was covered with foil to prevent acetonitrile loss during dialysis. The protein solution was spun again as before to remove the insoluble material. The supernatant was filtered through a 0.45-μm PVDF syringe filter unit, and the filtrate corresponding to 10 mg of initial sulfonated protein was loaded on to a cation exchange Spherogel TSK-SP-5PW column (see earlier for mGIB) preequilibrated in solvent A (20 m*M* Hepes pH 7.0, 30% acetonitrile). The solvent program was from 0% solvent B (20 m*M* Hepes pH 7.0, 1 *M* NaCl, 30% acetonitrile) to 33% B over 20 min and to 100% B over 10 min at 1 mL/min. The mGV elutes at ~17 min. The $sPLA_2$ peak was collected and diluted with the C18 reverse-phase solvent A (water, 0.1% trifluoroacetic acid) to ~10% acetonitrile before being loaded on C18 Vydac column (see earlier for hGIB). Solvent program was 0–30% B (acetonitrile, 0.08% trifluoroacetic acid) over 20 min and 30–40% B in the next 40 min and finally 40–100% B over 10 min at 1.5 mL/min. The $sPLA_2$ elutes at ~40% B as two close peaks. Electrospray ionization mass spectrometry showed that the peak at 36% B lacks the N-terminal methionine, whereas the nearby peak contains protein containing the N-terminal methionine. The protein without methionine was lyophilized to give a white powder that was stored at −20°C.

21. PREPARATION OF hGX

Sulfonated fusion protein (80–100 mg, prepared from a pAB3 vector) was dissolved in 200 mL (~0.4–0.5 mg/mL) of 8 *M* urea, 50 m*M* Tris, pH 8.5, 0.1 *M* NH_4Cl, 5 m*M* $CaCl_2$, 5 m*M* L-methionine. After 20–30 min stirring at room temperature, the solution was placed in dialysis tubing (6–8 kDa MW cutoff) and dialyzed against 7 L of 1.6 *M* urea, 50 m*M* Tris, pH 8.5, 0.1 *M* NH_4Cl, 5 m*M* $CaCl_2$, 5 m*M* L-methionine, 8 m*M* L-cysteine, and 5 m*M* detergent (*N*-Dodecyl-*N*,*N*-dimethyl-3-ammonio-1-propanesulfonate, Sigma 40232) at 4°C for 2 days without changing the buffer. The content in the dialysis bag was centrifuged at 10,000 rpm in a Beckman GSA rotor at 4°C. The supernatant-containing refolded protein was trypsinized as described for hGIB. The digested protein solution was cleared by centrifugation as above and concentrated by ultrafiltration (Amicon concentrator with a YM-10 membrane, Millipore 13651) to about 30–40 mL. The protein solution was diluted with 20% acetonitrile, 1% acetic acid, 5 m*M* L-methionine to give 200 mL, and then concentrated back to 30–40 mL. This step was repeated once to obtain 30–40 mL concentrated protein solution, then the protein solution was centrifuged again and filtered through a 0.45-μm PVDF syringe filter unit before being loaded (one-third to one-half of the solution) onto a Spherogel TSK-SP-5PW column (see earlier for mGIB), which was preequilibrated in solvent A (1% acetic acid, 50% acetonitrile, 5 m*M* L-methionine) at 1 mL/min. The solvent program started from 0% B (25 m*M* ammonium acetate, pH 6.8, 0.5 *M* NaCl, 50% acetonitrile, 5 m*M* L-methionine) to 70% B over 140 min at 1 mL/min. The hGX protein eluted as a sharp peak at ~50 min. The hGX peak was collected and diluted with solvent C18 reverse-phase solvent A (water, 0.1% trifluoroacetic acid) to 20% acetonitrile, and the solution was loaded onto the C18 column (C18 Vydac, as above for hGIB) preequilibrated with 20% B (acetonitrile, 0.1% trifluoroacetic acid) at a flow rate of 1.5 mL/min. The solvent program was 20–50% B over 120 min at 1.5 mL/min. Protein hGX elutes at ~48 min as a sharp peak. The hGX fraction was lyophilized and the resultant white powder was stored at −20°C.

hGX $sPLA_2$ and prohGX were also prepared from the pET21a expression vector with a similar protocol, avoiding the trypsin digestion step (Jemel et al., 2011). hGX $sPLA_2$ was also demonstrated to be glycosylated when produced in HEK293 cells (Hanasaki et al., 1999) and S2 cells (our unpublished data).

22. PREPARATION OF mGX

mGX was prepared as for hGX from pAB3 vector. The $sPLA_2$ elutes from the Spherogel TSK-SP-5PW column (see earlier for hGX) at 75 min and from the C18 Vydac column (see earlier for hGX) at 38% solvent B. mGX and promGX $sPLA_2s$ were also prepared from the pET21a and pAB3 expression vectors, respectively (Jemel et al., 2011).

23. PREPARATION OF hGXIIA

The sulfonated protein (120 mg, prepared from the pAB3 vector) was dissolved in 240 mL of 6 *M* guanidine-HCl, 50 m*M* Tris–HCl, pH 8.0 at room temperature by stirring (1–2 h). In some cases, solid remains, but this was not removed. Dialysis tubing (6–8 kDa MWCO, SpectraPor) was filled with protein solution and dialyzed against 6.5 L of 0.9 *M* guanidine-HCl, 50 m*M* Tris–HCl, pH 8.0, 5 m*M* freshly added L-cysteine, 5 m*M* L-methionine, 5 m*M* EDTA at 4°C. After 5–6 h, the buffer was replaced with 6.5 L of fresh refolding buffer and dialysis continued for about two more days at 4°C. The protein solution was spun at 15,000 × *g* for 20 min at 4°C to pellet any insoluble material. The supernatant was dialyzed against trypsin buffer (50 m*M* Tris–HCl, pH 8.0, 100 m*M* NaCl, 1 m*M* $CaCl_2$ containing 5 m*M* detergent (*N*-dodecyl-*N*,*N*-dimethyl-3-ammonio-1-propanesulfonate, Sigma 40232)) for 5–6 h at 4°C, then the buffer was replaced and dialysis continued overnight. The next day, the protein solution was spun at 15,000 × *g* for 20 min at 4°C to pellet any insoluble material. Supernatant was stored at 4°C, while ~50–100 mL of trypsin buffer was added to the pellet and stirred for another 2 h at 4°C and pelleted once again as before. The supernatants were combined and concentrated with Amicon stirred cell with a YM-10 membrane to half the volume.

The concentrated protein solution was incubated with 100 U of factor Xa (GE Healthcare, 27-0849-01), then stored at 4°C until HPLC. Alternatively, cleavage can be made with trypsin (using the trypsin treatment protocols given above for other $sPLA_2s$), but the cleavage has to be carefully monitored as hGXIIA was found to be very sensitive to trypsin. The protein solution was filtered through 0.45-μm syringe cartridge filter unit, and the filtrate was pumped directly onto the C18 Vydac HPLC column (see earlier for hGIB) preequilibrated in 10% B (acetonitrile, 0.1% trifluoroacetic acid). The column was washed with ~10 mL of 10% B, before starting the linear gradient from 10% B to 30% B over 20 min, to 35% B over 25 min, to 60% B over the next 30 min, and finally to 100% B over 10 min at 3 mL/min. The sPLA$_2$ elutes at ~57 min. The sPLA$_2$ peaks were combined, acetonitrile adjusted to ~30% and pH to ~5, then half of the protein solution was loaded on a second column (Spherogel TSK-SP-5PW, see earlier for mGIB) preequilibrated with solvent A (20 m*M* Na-acetate, pH 5.0, 30% acetonitrile) at 1 mL/min. The linear gradient started with 0% solvent B (solvent B is solvent A + 0.5 *M* NaCl) for 15 min, then to 100% B over 30 min. hGXIIA elutes at ~37 min. The combined hGXIIA fractions were diluted with water to ~10% acetonitrile and were desalted by loading half of the prep on C18 Vydac column as above. The column was washed with ~1 column volume of solvent A (water, 0.1% trifluoroacetic acid) before starting the linear acetonitrile gradient as above to obtain pure hGXIIA. The protein peaks were combined and lyophilized as a white powder.

24. PREPARATION OF mGXIIA, hGXIIB, AND mGXIIB

Sulfonated mGXIIA, hGXIIB (variants EAC and EAAC) (Rouault, Bollinger, Lazdunski, Gelb, & Lambeau, 2003), and mGXIIB (100 mg, prepared from the pAB3 vector) were resuspended at 0.2 mg/mL in 500 mL of 50 m*M* Tris–HCl, pH 8.0, 8 *M* urea, and 0.1 *M* NH_4Cl. The denatured sPLA$_2$ was refolded by dialysis against 8 L of 50 m*M* Tris–HCl, pH 8.0, 2 *M* urea, 0.1 *M* NH_4Cl, 5 m*M* $CaCl_2$, 5 m*M* L-cysteine, and 5 m*M* L-methionine for 2–4 days at 4°C. The refolded mixture was filtered through a layer of Sephadex G-50 to remove large aggregates and subjected to extensive digestion at room temperature with 400 U of factor Xa (GE Healthcare, 27-0849-01). The time course of cleavage (up to 3 days) was followed by sPLA$_2$ activity on *E. coli* membranes (mGXIIA) or MALDI-TOF mass spectrometry (hGXIIB and mGXIIB which are devoid of enzymatic activity). The protein solution was concentrated to 20 mL by ultrafiltration using an Amicon stirred cell concentrator with a YM-10 membrane, and the buffer was then exchanged to 1% acetic acid (10% acetonitrile was added for GXIIB sPLA$_2$s). The solution was filtered and loaded onto a Spherogel TSK-SP-5PW column (see earlier for mGIB) equilibrated in 1% acetic acid (and 10% acetonitrile for GXIIB sPLA$_2$s). The protein was eluted from the column at 1 mL/min using a linear gradient of ammonium acetate (0–2 *M*, pH 6.8, over 100 min) in 10% acetonitrile. The fractions containing cleaved $sPLA_2$ were pooled, lyophilized, resuspended in 10% solvent B (acetonitrile, 0.1% trifluoroacetic acid), and applied to a C18 Nucleosil reverse-phase HPLC column (see earlier for mGIB). Elution was performed at 1 mL/min using a gradient of solvent B (10–30% over 20 min, then 30–35% over 25 min, and 35–60% over 30 min). The protein peaks were combined and lyophilized as a white powder.

25. QUALITY CONTROL AND CHARACTERIZATION OF REFOLDED $sPLA_2s$ BY MASS SPECTROMETRY

Quantity of the purified recombinant $sPLA_2s$ was assessed by measuring the OD_{280} value and converting the value into mg amount using their calculated specific absorption coefficients. Purity of each $sPLA_2$ was assessed by SDS-PAGE (14% gel, under both reducing and nonreducing conditions) and mass spectrometry analysis. Mass spectrometry was used also to confirm the correct folding of the $sPLA_2$, the absence of any unwanted modification of the recombinant proteins, such as oxidation or carbamylation (due to urea degradation), presence of the Met initiator, uncomplete oxidation of cysteines into disulfide, unwanted proteolysis by trypsin, or factor Xa, as well as the presence of glycosylation (GIIC, GIID, and GIII enzymes).

All $sPLA_2s$ were characterized by accurate mass spectrometry with a mass resolution sufficient to confirm that all disulfides are fully formed. Some measurements were carried out on a Bruker EsquireLC ion trap mass spectrometer. The $sPLA_2$ at a concentration of about 0.1 mg/mL in 50/50 MeOH/water with 1% acetic acid was infused into the mass spectrometer using a syringe pump at a flow rate of 1.5 μL/min. Data were collected in positive ion mode. The observed charge state distribution of m/z peaks was deconvoluted using the Bruker Data Analysis software package to give the observed MW. The calculated MW was obtained using the amino acid sequence and subtracting 2 Da for each pair of cysteines to account for disulfide formation.

For some $sPLA_2s$, the molecular mass determination was carried out by MALDI-TOF analysis on a Voyager DE-Pro mass spectrometer (Applied Biosystems) or a MALDI-TOF/TOF mass spectrometer (4800 plus, Applied Biosystems). The molecular mass was measured in linear mode using sinapinic acid as a matrix and with internal calibration. When needed, $sPLA_2s$ were purified on a C18 ZipTip (Millipore) to remove traces of salts prior to MALDI mass spectrometry. The tip was rinsed twice with 50%

acetonitrile and 0.1% trifluoroacetic acid, then three times with water and 0.1% trifluoroacetic acid. The $sPLA_2$ solution (10 μL) was loaded onto the tip, which was washed twice with water and 0.1% trifluoroacetic acid and then with 80% acetonitrile to elute the protein.

26. ENZYMATIC ASSAYS

Some measurements were carried out with a real-time fluorimetric $sPLA_2$ assay. The pyrene-PG substrate, 1-hexadecanoyl-2-(1-pyrenedecanoyl)-*sn*-glycero-3-phosphoglycerol ammonium salt (Life Technologies, H-3809), was prepared as follows: the lipid (1 mg) in the commercial vial was solubilized in ~40 μL of toluene by mixing on a vortex mixer. Ten microliter were transferred into a new 1.5-mL glass vial containing 1.25 mL of absolute ethanol to give a 200-μ*M* stock solution. The solution was mixed briefly on a vortex mixer and then placed in a 37°C water bath for ~5 min, then in a sonicator water bath for ~10–15 s to ensure complete dissolution of phospholipid. The above phospholipid solution was stored at −10°C; the ethanol solution needed to be resonicated as above after thawing for use.

To 1 mL of assay buffer (50 m*M* Tris pH 8.0, 50 m*M* KCl, 2 m*M* $CaCl_2$) in a 1-cm fluorescence cuvette was added 10 μL of 30 mg/mL bovine serum albumin (Sigma A6003) in water (filtered through a 0.45-μm syringe filter unit) and 10 μL of 200 μ*M* pyrene-PG in ethanol. The cuvette was gently shaken by hand a few times to ensure complete mixing of components. The cuvette was placed in the sample chamber of the fluorimeter (excitation 342 nm, emission 392 nm) at room temperature. The no-enzyme baseline fluorescence was recorded for a few minutes to ensure that a stable-flat baseline was achieved. If several assays were to be done, an appropriate volume of assay cocktail could be mixed at once in a plastic tube. The cocktail was kept at room temperature during use (up to 4 h). Before charging the cuvette with 1 mL of the cocktail, the tube of cocktail was inverted a few times (to avoid foaming). After baseline recording, the $sPLA_2$ aliquot (typically 1–2 μL containing a few ng of $sPLA_2$; dilutions of $sPLA_2$ solutions should be made in bovine serum albumin-containing buffer to avoid the loss of $sPLA_2$ by its absorption to the walls of tubes) was added, the cuvette was shaken briefly, and the fluorescence was recorded for a few minutes to obtain an adequate slope. If the calcium dependence of the reaction is sought, $CaCl_2$ in the buffer can be replaced with 2 m*M* EGTA. The assay response can be converted to moles of $sPLA_2$ product by adding known amounts of pyrene-decanoic acid product (Sigma Cat. 82660) to the assay cocktail in the absence of $sPLA_2$ to generate a standard curve. This assay can also be carried out in 96-well plate format using a black 96-well plate appropriate for a plate reader fluorimeter.

Some $sPLA_2$ enzymatic assays were carried out using [^{3}H]-oleate-radiolabeled *E. coli* membranes. This assay has been found to be as sensitive as the above fluorimetric assay and can be even more sensitive when small amounts of endogenous $sPLA_2$s present in biological samples such as body fluids and tissues have to be detected. All $sPLA_2$s tend to bind tightly to the vesicles made of anionic phospholipids, including pyrene-PG described earlier and the *E. coli* membranes which are rich in anionic phospholipids including phosphatidylglycerol and cardiolipin.

The following protocol was adapted from Franson, Patriarca, and Elsbach (1974). One milliliter of an overnight culture of *E. coli* XL1 (Stratagene Inc.) or DH10B (Invitrogen Life Technologies) strain was diluted in 50 mL of LB medium containing 500 μCi of [9,10-^{3}H]-oleic acid (NEN Radiochemicals, NET289, Perkin Elmer) and incubated for 4 h at 37°C. Cells were pelleted for 10 min at 3000 × *g*, resuspended in 50 mL of LB medium, and allowed to chase for 30 min at 37°C to ensure complete incorporation of labeled oleic acid into phospholipids. After centrifugation (10 min at 3000 × *g*), the cell pellet was washed with 5 mL of washing buffer (0.1 *M* Tris–HCl, pH 8.0, 1 m*M* EDTA, 1% fatty acid-free bovine serum albumin (Sigma-Aldrich A7030)), centrifuged again (10 min at 3000 × *g*), resuspended in 5 mL of washing buffer without BSA, and autoclaved in a Pyrex glass tube. After autoclaving, the lysed *E. coli* cells were dounce homogenized. The resulting radiolabeled *E. coli* membranes were washed five times by successive centrifugation at 13,000 × *g* for 1 min and resuspension in washing buffer. The final pellet was resuspended in 5 mL of washing buffer and counted to evaluate the yield of [^{3}H]-oleic acid incorporation into *E. coli* membrane phospholipids. Typically, the incorporated radioactivity was more than 30–40% of the input radioactivity. The membrane suspension was diluted to 100,000 dpm/μL, aliquoted, and stored at −20°C.

To prepare the *E. coli* substrate pool, the required total amount of radioactivity (100,000 dpm of radiolabeled membranes/assay × the number of assays) was pipetted into an eppendorf tube and brought up to 1.5 mL with PLA_2 activity buffer (0.1 *M* Tris–HCl, pH 8.0, 10 m*M* $CaCl_2$, 0.1% BSA), then centrifuged down for 1.5 min at 14,000 rpm at room temperature. The supernatant was discarded and the pellet carefully resuspended in 150 μL of PLA_2 activity buffer. The volume was then brought up to the volume required for the total number of assays (100 μL × the number of assays). Hundred microliter of this substrate pool was counted to determine the total amount of radioactivity added per assay. A typical reaction was made into an eppendorf tube and consisted of a total volume of 200 μL made with 50 μL PLA_2 activity buffer, 50 μL enzyme solution and/or inhibitor solution as needed, and 100 μL of the above substrate pool (addition of a quite large volume of substrate with a multipipette results in enough mixing so that it is not necessary to vortex after substrate addition, allowing multiple very short term assays if needed). Reaction mixtures were incubated for up to 1 h at either room temperature or at 37°C and with different amounts of enzyme. Incubation time, incubation temperature, and sample volume were adjusted to ensure hydrolysis rates within the linear range of the enzymatic assay (typically 10–20% of total substrate hydrolysis over background). Control incubations in the absence of added $sPLA_2$ (background) were carried out in parallel (usually less than 3–5% of total added radioactivity) and used to calculate specific hydrolysis. The reaction was stopped by adding 300 μL of the stop buffer (0.1 *M* EDTA pH 8.0 containing 0.2% fatty acid-free BSA). The tubes were centrifuged for 3 min at 14,000 rpm at room temperature. Four-hundred microliter of the supernatant containing released labeled oleic acid was carefully removed without disturbing the pellet, and counted. The total released radioactivity (dpm) equals 500/400 × counts (cpm converted into dpm) given by 400 μL of the supernatant. Note that we routinely considered that the counts in the above supernatants correspond to free [^{3}H]-oleate released from membrane phospholipids. To verify that these counts are free oleic acid, a thin-layer chromatography on silica gel 60 can be performed in conditions where free oleic acid and phospholipids can be separated. Note also that this protocol does not specifically detect $sPLA_2$ activity, but can also detect the activity of intracellular PLA_2s.

The above assays were used not only to follow the refolding and purification of recombinant $sPLA_2$s during the multiple steps, but they were also used to measure the specific activities of the final $sPLA_2$ product, in line with the above mass spectrometry and SDS-PAGE gel analyses. Detailed kinetic properties of all $sPLA_2$s can be found in Singer et al. (2002) along with calcium requirements, substrate and inhibitor specificity, sensitivity to thiol-reducing agents, and interfacial vesicle-binding properties.

REFERENCES

Abita, J. P., Lazdunski, M., Bonsen, P. P., Pieterson, W. A., & de Haas, G. H. (1972). Zymogen-enzyme transformations. On the mechanism of activation of prophospholipase A. *European Journal of Biochemistry*, *30*(1), 37–47.

Chen, J., Engle, S. J., Seilhamer, J. J., & Tischfield, J. A. (1994). Cloning, expression and partial characterization of a novel rat phospholipase A2. *Biochimica et Biophysica Acta*, *1215*(1–2), 115–120. http://doi.org/10.1016/0005-2760(94)90099-X.

Cupillard, L., Koumanov, K., Matt'ei, M. G., Lazdunski, M., & Lambeau, G. (1997). Cloning, chromosomal mapping, and expression of a novel human secretory phospholipase A2. *The Journal of Biological Chemistry*, *272*, 15745–15752.

Cupillard, L., Mulherkar, R., Gomez, N., Kadam, S., Valentin, E., Lazdunski, M., et al. (1999). Both group IB and group IIA secreted phospholipases A2 are natural ligands of the mouse 180-kDa M-type receptor. *The Journal of Biological Chemistry*, *274*, 7043–7051.

Degousee, N., Ghomashchi, F., Stefanski, E., Singer, A. G., Smart, B. P. B. N., Reithmeier, R., et al. (2001). Groups IV, V and X phospholipases A2 in human neutrophils. *The Journal of Biological Chemistry*, *277*, 5061–5073.

Dennis, E. A., Cao, J., Hsu, Y.-H., Magrioti, V., & Kokotos, G. (2011). Phospholipase A2 enzymes: Physical structure, biological function, disease implication, chemical inhibition, and therapeutic intervention. *Chemical Reviews*, *111*(10), 6130–6185. http://doi.org/10.1021/cr200085w.

Eerola, L. I., Surrel, F., Nevalainen, T. J., Gelb, M. H., Lambeau, G., & Laine, V. J. (2006). Analysis of expression of secreted phospholipases A2 in mouse tissues at protein and mRNA levels. *Biochimica et Biophysica Acta*, *1761*(7), 745–756.

Franson, R., Patriarca, P., & Elsbach, P. (1974). Phospholipid metabolism by phagocytic cells. Phospholipases A2 associated with rabbit polymorphonuclear leukocyte granules. *Journal of Lipid Research*, *15*(4), 380–388.

Gelb, M. H., Valentin, E., Ghomashchi, F., Lazdunski, M., & Lambeau, G. (2000). Cloning and recombinant expression of a structurally novel human secreted phospholipase A2. *The Journal of Biological Chemistry*, *275*, 39823–39826.

Hanasaki, K., Ono, T., Saiga, A., Morioka, Y., Ikeda, M., Kawamoto, K., et al. (1999). Purified group X secretory phospholipase A(2) induced prominent release of arachidonic acid from human myeloid leukemia cells. *The Journal of Biological Chemistry*, *274*, 34203–34211.

Hirel, P. H., Schmitter, M. J., Dessen, P., Fayat, G., & Blanquet, S. (1989). Extent of N-terminal methionine excision from Escherichia coli proteins is governed by the side-chain length of the penultimate amino acid. *Proceedings of the National Academy of Sciences of the United States of America*, *86*(21), 8247–8251.

Jemel, I., Ii, H., Oslund, R. C., Payré, C., Dabert-Gay, A.-S., Douguet, D., et al. (2011). Group X secreted phospholipase A2 proenzyme is matured by a furin-like proprotein convertase and releases arachidonic acid inside of human HEK293 cells. *The Journal of Biological Chemistry*, *286*(42), 36509–36521. http://doi.org/10.1074/jbc.M111.268540.

Kramer, R. M., Hession, C., Johansen, B., Hayes, G., McGray, P., Chow, E. P., et al. (1989). Structure and properties of a human non-pancreatic phospholipase A2. *The Journal of Biological Chemistry*, *264*, 5768–5775.

Lambeau, G., & Gelb, M. H. (2008). Biochemistry and physiology of mammalian secreted phospholipases A2. *Annual Review of Biochemistry*, *77*, 495–520.

Murakami, M., Masuda, S., Shimbara, S., Ishikawa, Y., Ishii, T., & Kudo, I. (2005). Cellular distribution, post-translational modification, and tumorigenic potential of human group III secreted phospholipase A2. *The Journal of Biological Chemistry*, *280*(26), 24987–24998.

Murakami, M., Sato, H., Miki, Y., Yamamoto, K., & Taketomi, Y. (2015). A new era of secreted phospholipase A_2. *Journal of Lipid Research*, *56*(7), 1248–1261. http://doi.org/10.1194/jlr.R058123.

Murakami, M., Taketomi, Y., Sato, H., & Yamamoto, K. (2011). Secreted phospholipase A2 revisited. *Journal of Biochemistry*, *150*(3), 233–255. http://doi.org/10.1093/jb/mvr088.

Nevalainen, T. J., Haapamaki, M. M., & Gronroos, J. M. (2000). Roles of secretory phospholipases A(2) in inflammatory diseases and trauma. *Biochimica et Biophysica Acta*, *1488*(1–2), 83–90.

Oslund, R. C., & Gelb, M. H. (2008). Highly specific and pan-specific inhibitors of mammalian secreted phospholipases A2. *Journal of Medicinal Chemistry*, *51*(15), 4708–4714.

Rouault, M., Bollinger, J. G., Lazdunski, M., Gelb, M. H., & Lambeau, G. (2003). Novel mammalian group XII secreted phospholipase A2 lacking enzymatic activity. *Biochemistry*, *42*, 11494–11503.

Rouault, M., Le Calvez, C., Boilard, E., Surrel, F., Singer, A., Ghomashchi, F., et al. (2007). Recombinant production and properties of binding of the full set of mouse secreted phospholipases A2 to the mouse M-type receptor. *Biochemistry*, *46*(6), 1647–1662.

Singer, A. G., Ghomashchi, F., Le Calvez, C., Bollinger, J., Bezzine, S., Rouault, M., et al. (2002). Interfacial kinetic and binding properties of complete set of human and mouse groups I, II, V, X, and XII secreted phospholipases A2. *The Journal of Biological Chemistry*, *277*, 48535–48549.

Smart, B. P., Pan, Y. H., Weeks, A. K., Bollinger, J. G., Bahnson, B. J., & Gelb, M. H. (2004). Inhibition of the complete set of mammalian secreted phospholipases A2 by indole analogs: A structure-guided study. *Bioorganic & Medicinal Chemistry*, *12*, 1737–1749.

Snitko, Y., Koduri, R. S., Han, S. K., Othman, R., Baker, S. F., Molini, B., et al. (1997). Mapping the interfacial binding surface of human secretory class IIa phospholipase A2. *Biochemistry*, *36*, 14325–14333.

Snitko, Y., Yoon, E. T., & Cho, W. (1997). High specificity of human secretory class II phospholipase A2 for phosphatidic acid. *The Biochemical Journal*, *321*, 737–741.

Takabatake, N., Sata, M., Inoue, S., Shibata, Y., Abe, S., Wada, T., et al. (2005). A novel polymorphism in secretory phospholipase A2-IID is associated with body weight loss in chronic obstructive pulmonary disease. *American Journal of Respiratory and Critical Care Medicine*, *172*(9), 1097–1104. http://doi.org/10.1164/rccm.200503-319OC.

Thannhauser, T. W., Konishi, Y., & Scheraga, H. A. (1984). Sensitive quantitative analysis of disulfide bonds in polypeptides and proteins. *Analytical Biochemistry*, *138*, 181–188.

Valentin, E., Ghomashchi, F., Gelb, M. H., Lazdunski, M., & Lambeau, G. (1999). On the diversity of secreted phospholipases A2: Cloning, tissue distribution, and functional expression of 2 novel mouse group II enzymes. *The Journal of Biological Chemistry*, *274*, 31195–31202.

Valentin, E., Ghomashchi, F., Gelb, M. H., Lazdunski, M., & Lambeau, G. (2000). Novel human secreted phospholipase A2 with homology to the group III bee venom enzyme. *The Journal of Biological Chemistry*, *275*, 7492–7496.

Valentin, E., Koduri, R. S., Scimeca, J. C., Carle, G., Gelb, M. H., Lazdunski, M., et al. (1999). Cloning and recombinant expression of a novel mouse-secreted phospholipase A2. *The Journal of Biological Chemistry*, *274*, 19152–19160.

Valentin, E., & Lambeau, G. (2000). Increasing molecular diversity of secreted phospholipases A2 and their receptors and binding proteins. *Biochimica et Biophysica Acta*, *1488*, 59–70.

Valentin, E., Singer, A. G., Ghomashchi, F., Lazdunski, M., Gelb, M. H., & Lambeau, G. (2000). Cloning and recombinant expression of human group IIF secreted phospholipase A2. *Biochemical and Biophysical Research Communications*, *279*, 223–228.

Valentinis, B., Romano, G., Peruzzi, F., Morrione, A., Prisco, M., Soddu, S., et al. (1999). Growth and differentiation signals by the insulin-like growth factor 1 receptor in hemopoietic cells are mediated through different pathways. *The Journal of Biological Chemistry*, *274*(18), 12423–12430.

Wang, Y., Kowalski, P. E., Thalmann, I., Ornitz, D. M., Mager, D. L., & Thalmann, R. (1998). Otoconin-90, the mammalian otoconial matrix protein, contains two domains of homology to secretory phospholipase A2. *Proceedings of the National Academy of Sciences of the United States of America*, *95*(26), 15345–15350.

Wijewickrama, G. T., Albanese, A., Kim, Y. J., Oh, Y. S., Murray, P. S., Takayanagi, R., et al. (2006). Unique membrane interaction mode of group IIF phospholipase A2. *The Journal of Biological Chemistry*, *281*(43), 32741–32754. http://doi.org/10.1074/jbc.M606311200.

CHAPTER FOUR

Cellular Assays for Evaluating Calcium-Dependent Translocation of cPLA$_2\alpha$ to Membrane

B. Yun*, C.C. Leslie*,†,[1]
*National Jewish Health, Denver, CO, United States
†University of Colorado Denver, Aurora, CO, United States
[1]Corresponding author: e-mail address: lesliec@njhealth.org

Contents

Abstract

The group IVA phospholipase A$_2$, commonly called cytosolic phospholipase A$_2\alpha$ (cPLA$_2\alpha$), is a widely expressed enzyme that hydrolyzes membrane phospholipid to produce arachidonic acid and lysophospholipids, which are precursors for a number of bioactive lipid mediators. Arachidonic acid is metabolized through the cyclooxygenase and lipoxygenase pathways for production of prostaglandins and leukotrienes that

Methods in Enzymology, Volume 583
ISSN 0076-6879
http://dx.doi.org/10.1016/bs.mie.2016.09.006

regulate normal physiological processes and contribute to disease pathogenesis. $cPLA_2\alpha$ is composed of an N-terminal C2 domain and a C-terminal catalytic domain that contains the Ser-Asp catalytic dyad. The catalytic domain contains phosphorylation sites and basic residues that regulate the catalytic activity of $cPLA_2\alpha$. In response to cell stimulation, $cPLA_2\alpha$ is rapidly activated by posttranslational mechanisms including increases in intracellular calcium and phosphorylation by mitogen-activated protein kinases. In resting cells, $cPLA_2\alpha$ is localized in the cytosol but translocates to membrane including the Golgi, endoplasmic reticulum, and the peri-nuclear membrane in response to increases in intracellular calcium. Calcium binds to the C2 domain, which promotes the interaction of $cPLA_2\alpha$ with membrane through hydrophobic interactions. In this chapter, we describe assays used to study the calcium-dependent translocation of $cPLA_2\alpha$ to membrane, a regulatory step necessary for access to phospholipid and release of arachidonic acid.

1. INTRODUCTION

Mammalian cells contain several types of phospholipase A_2 (PLA_2) enzymes that fall into six main classes: GIV PLA_2 (cytosolic PLA_2, $cPLA_2$), GVI PLA_2 (calcium-independent PLA_2, $iPLA_2$), GVII and GVIII PLA_2 (platelet-activating factor-acetylhydrolases, PAF-AHs), GXV PLA_2 (lysosomal PLA_2), GXVI PLA_2 (adipose PLA_2), and several groups of secreted PLA_2 ($sPLA_2$). The properties of these enzymes are described in a number of recent reviews (Dennis, Cao, Hsu, Magrioti, & Kokotos, 2011; Ghosh, Tucker, Burchett, & Leslie, 2006; Lambeau & Gelb, 2008; Shayman, Kelly, Kollmeyer, He, & Abe, 2011; Stafforini, 2009). Although several of PLA_2s can release arachidonic acid for eicosanoid production, only $cPLA_2\alpha$ exhibits selectivity for releasing *sn-2* arachidonic acid, and its role in initiating production of eicosanoids in cells stimulated with diverse agonists is well documented (Clark et al., 1991; Hanel, Schüttel, & Gelb, 1993; Leslie, 2004). $cPLA_2\alpha$ is widely expressed in cells and tissues and is evolutionarily conserved reflecting its functional importance for the production of bioactive lipid mediators. Eicosanoids are a diverse group of oxygenated metabolites of arachidonic acid that include prostaglandins and thromboxane produced from the cyclooxygenase pathways, and leukotrienes, lipoxins and oxoeicosanoids that are derived from 5-, 12-, and 15-lipoxygenases (Buckley, Gilroy, & Serhan, 2014; Funk, 2001; Peters-Golden & Henderson, 2007; Powell & Rokach, 2013). Cyclooxygenases are widely expressed reflecting the ability of most cells to produce prostanoids, which regulate physiological processes such as female reproduction, kidney function, and hemostasis (Nakahata, 2008; Smith & Langenbach, 2001).

They also promote acute and chronic inflammation, but can be immunosuppressive and participate in the resolution of inflammation (Aoki & Narumiya, 2012; Kalinski, 2012; Ricciotti & Fitzgerald, 2011; Tilley, Coffman, & Koller, 2001). Products of the lipoxygenase pathways are immune modulators that are either proinflammatory (leukotrienes, oxoeicosanoids), or act as antiinflammatory mediators (lipoxins) (Buckley et al., 2014; Peters-Golden & Henderson, 2007; Powell & Rokach, 2013). Eicosanoids are released by cells and act through cell type-specific G-protein coupled receptors to promote diverse biological responses (Back et al., 2014; Woodward, Jones, & Narumiya, 2011).

$cPLA_2\alpha$ is subject to complex mechanisms of regulation that are important for controlling the levels of free arachidonic acid available for metabolism through the cyclooxygenase and lipoxygenase pathways. Initial characterization of $cPLA_2\alpha$ showed that it required calcium for activity *in vitro* consistent with cellular studies showing that calcium-mobilizing agonists induce arachidonic acid release (Clark, Milona, & Knopf, 1990; Fernandez & Balsinde, 1991; Kramer, Roberts, Manetta, & Putnam, 1991; Leslie, Voelker, Channon, Wall, & Zelarney, 1988; Rittenhouse-Simmons & Deykin, 1977; Rzigalinski, Blackmore, & Rosenthal, 1996). Calcium was found to promote the interfacial binding of $cPLA_2\alpha$ to phospholipid vesicles *in vitro* and to cellular membranes (Channon & Leslie, 1990; Clark et al., 1991; Glover et al., 1995; Hirabayashi et al., 1999; Peters-Golden, Song, Marshall, & Brock, 1996; Schievella, Regier, Smith, & Lin, 1995). $cPLA_2\alpha$ is composed of two independently folded domains, an N-terminal calcium-dependent lipid-binding (C2) domain and a C-terminal catalytic domain that contains the Ser-Asp active site dyad (Dessen et al., 1999; Perisic, Fong, Lynch, Bycroft, & Williams, 1998). The C2 domain mediates the calcium-dependent translocation of $cPLA_2\alpha$ to intracellular membranes in response to increases in calcium in stimulated cells (Evans, Spencer, Zweifach, & Leslie, 2001; Gijón, Spencer, Kaiser, & Leslie, 1999; Perisic, Paterson, Mosedale, Lara-González, & Williams, 1999). In unstimulated cells, $cPLA_2\alpha$ is in the cytosol where it is unable to firmly bind membranes at levels of cytosolic calcium ($\sim$50 n*M*) found in resting cells (Evans et al., 2001). When intracellular calcium concentration $[Ca^{2+}]_i$ increases above 100–125 n*M*, $cPLA_2\alpha$ preferentially translocates to the Golgi apparatus, and at higher $[Ca^{2+}]_i$ (210–280 n*M*) to the endoplasmic reticulum (ER) and peri-nuclear membrane (PNM) (Evans et al., 2001). The localization of full-length $cPLA_2\alpha$ and the C2 domain to these intracellular membranes is identical indicating that the

C2 domain determines the targeting specificity (Evans, Gerber, Murray & Leslie, 2004; Evans & Leslie, 2004; Evans et al., 2001).

C2 domains were first identified in classical protein kinase C as the module responsible for mediating calcium-dependent membrane binding and have since been found in many proteins that function in processes at membranes such as trafficking and signaling (Corbalan-Garcia & Gomez-Fernandez, 2014; Nalefski & Falke, 1996). They are composed of ~130 amino acids folded into an antiparallel beta-sandwich structure composed of eight beta strands connected by loops (Sutton, Davletov, Berghuis, Sudhof, & Sprang, 1995). Two ions of calcium bind to acidic residues in the calcium-binding loops (CBL) at the membrane-binding face of the $cPLA_2\alpha$ C2 domain (Bittova, Sumandea, & Cho, 1999; Nalefski & Falke, 1998; Perisic et al., 1999). Once calcium neutralizes the negative electrostatic potential of the CBLs, the affinity of the $cPLA_2\alpha$ C2 domain for membrane increases that involves penetration of exposed hydrophobic residues in CBLs into membrane (Ball, Nielsen, Gelb, & Robinson, 1999; Davletov, Perisic, & Williams, 1998; Murray & Honig, 2002; Nalefski & Falke, 1998; Perisic et al., 1999). This mode of membrane binding is distinct from the "electrostatic switch" mechanism of the PKC C2 domain, which changes from electronegative to electropositive after binding calcium leading to binding of basic residues in the CBLs to anionic phospholipids (Evans et al., 2004; Nalefski, Slazas, & Falke, 1997; Nalefski et al., 2001). These differences in the properties of C2 domains influence the organelle targeting specificity in cells (Evans, Murray, Leslie, & Falke, 2006; Stahelin, Rafter, Das, & Cho, 2003; Verdaguer, Corbalan-Garcia, Ochoa, Fita, & Gomez-Fernandez, 1999). The $cPLA_2\alpha$ C2 domain preferentially binds to phosphatidylcholine and targets intracellular membranes through hydrophobic interactions, whereas the electropositive properties of PKC C2 domain promote targeting to the inner leaflet of the plasma membrane that is enriched in anionic phospholipid (Evans et al., 2004; Evans et al., 2006; Medkova & Cho, 1998; Nalefski et al., 1998).

The calcium-dependent translocation of $cPLA_2\alpha$ mediated by the C2 domain brings the catalytic domain in proximity with the membrane, which binds in a calcium-independent manner through hydrophobic interactions (Evans & Leslie, 2004). A comparison of calcium-induced translocation of fluorescent protein (FP)-tagged C2 domain (FP-C2 domain) and FP-$cPLA_2\alpha$ to Golgi showed rapid dissociation of the C2 domain from membrane as intracellular calcium decreased but full-length $cPLA_2\alpha$ remained associated with membrane (Evans & Leslie, 2004). A calcium increase of sufficient duration is necessary for stable, calcium-independent association of $cPLA_2\alpha$ catalytic domain with membrane (Hirabayashi et al., 1999; Tucker et al., 2009). The release of calcium from intracellular stores in cells stimulated with physiological agonists induces the rapid translocation of $cPLA_2\alpha$ to Golgi. However, the influx of extracellular calcium triggered by store depletion is required for stable binding of $cPLA_2\alpha$ and release of arachidonic acid (Tucker et al., 2009).

Translocation of $cPLA_2\alpha$ to membrane is necessary but not sufficient for $cPLA_2\alpha$ to release arachidonic acid (Evans, Fergus, & Leslie, 2002; Reed et al., 2011; Tucker et al., 2009). The catalytic activity of $cPLA_2\alpha$ is independently regulated by phosphorylation, and by binding of basic residues in the catalytic domain to anionic components in the membrane (Qiu, Gijón, de Carvalho, Spencer, & Leslie, 1998; Reed et al., 2011; Tucker et al., 2009). The phosphorylation of $cPLA_2\alpha$ on S505 by mitogen-activated protein kinases enhances the catalytic activity of $cPLA_2\alpha$ *in vitro* and its ability to release arachidonic acid in cells (Bayburt & Gelb, 1997; Börsch-Haubold et al., 1998; Das, Rafter, Kim, Gygi, & Cho, 2003; Ghosh et al., 2006; Hefner et al., 2000; Kramer et al., 1996; Lin et al., 1993; Lin et al., 1993; Nemenoff et al., 1993; Tucker et al., 2009). The activity of $cPLA_2\alpha$ assayed *in vitro* is also enhanced by polyphosphoinositides particularly phosphatidylinositol 4,5-bisphosphate (PIP_2) that act through basic residues in the catalytic domain (Das & Cho, 2002; Leslie & Channon, 1990; Mosior, Six, & Dennis, 1998; Six & Dennis, 2003). A comparison of wild type and mutant forms of $cPLA_2\alpha$ expressed in cells showed that phosphorylation of S505 and the basic residues in the catalytic domain act to regulate hydrolytic activity (arachidonic acid release) but do not regulate the ability of $cPLA_2\alpha$ to translocate to Golgi (Evans et al., 2002; Reed et al., 2011; Schievella et al., 1995; Tucker et al., 2009).

The focus of this chapter is to describe imaging techniques to evaluate $cPLA_2\alpha$ translocation to intracellular membranes in response to stimuli that increase calcium. These studies primarily involve the expression of FP-$cPLA_2\alpha$ and monitoring translocation by live-cell imaging. Expression of wild type and mutant forms of full-length $cPLA_2\alpha$ and the C2 domain are used to understand the role of calcium, phosphorylation, and basic residues in regulating translocation. Importantly we found that FP-$cPLA_2\alpha$ is fully functional and can reconstitute stimulus-dependent release of arachidonic acid when expressed in cells lacking endogenous $cPLA_2\alpha$. Using this model system, the regulation of $cPLA_2\alpha$ translocation can be directly correlated with its ability to release arachidonic acid (Reed et al., 2011; Tucker et al., 2009).

2. CONSTRUCTS, CELL TYPES, AND TRANSFECTION PROTOCOLS

2.1 Generation of Wild Type and Mutant Constructs of FP-cPLA$_2\alpha$ and the FP-C2 Domain

For imaging experiments, cDNA encoding human cPLA$_2\alpha$ (residues 1–749) (GenBank Accession # M72393) was amplified by polymerase chain reaction (PCR) using primers containing restriction enzyme sites, and the PCR product cloned into pEGFP-C3 (Clontech, Mountain View, CA, USA) to generate EGFP-cPLA$_2\alpha$ (Evans et al., 2001). Depending on the specific application, cPLA$_2\alpha$ with N-terminal ECFP and EYFP were also generated. The cPLA$_2\alpha$ C2 domain (residues 17–148) was amplified using PCR and cloned into pEYFP-C3. Plasmid DNA encoding FP-cPLA$_2\alpha$ or FP-C2 domain was isolated by Endo-Free Maxi Plasmid Isolation Kit (Qiagen, Valencia, CA, USA) for transfection. The FP vectors originally used are no longer available from Clontech but can be obtained from Addgene, Cambridge, MA, USA. Newer versions such as AcGFP1 (GFP), ZsYellow1 (YFP), and AmCyan1 (CFP) are available from Clontech. To generate constructs for expression using adenovirus, FP-cPLA$_2\alpha$ or -C2 domain including the FP coding region were amplified, and the PCR product cloned into pVQAd5CMVK-NpA shuttle plasmid (ViraQuest Inc., North Liberty, IA, USA). Adenoviruses were produced by ViraQuest Inc.

A number of mutated forms of FP-cPLA$_2\alpha$ were also generated. Point mutations were introduced by site-directed mutagenesis (Agilent Technologies, Inc., Santa Clara, CA, USA). These included mutations of (1) serine residues phosphorylated by MAPKs (S505A), by calcium calmodulin-dependent protein kinase II (S515A) and by MAPK-interacting kinase I (S727A); (2) an essential calcium-binding residue in the C2 domain (D43N); (3) basic residues in the catalytic domain implicated in activation by PIP$_2$ (K488N/K543N/K544N); (4) a tryptophan residue in the catalytic domain that increases membrane residency time (W464A, W464F); (5) mutation of the active site serine nucleophile (S228A); and (6) sites mutated in human cPLA$_2\alpha$ deficiency in the C2 domain (S111P) and in the catalytic domain (R485H) (Evans & Leslie, 2004; Evans et al., 2002; Gijón et al., 1999; Reed et al., 2011; Tucker et al., 2009). FP-constructs of the protein kinase Cα C2 domain (residues 158–286) (Accession # X53479) in which the CBLs were replaced with CBLs from human cPLA$_2\alpha$ C2 domain were assembled from overlapping PCR fragments as described (Evans et al., 2004).

2.2 Cell Types for Monitoring $cPLA_2\alpha$ Translocation

2.2.1 MDCK Cells

Many cell types can be used to study the localization of $cPLA_2\alpha$ to specific subcellular compartments using immunofluorescence in fixed cells or to study calcium-dependent translocation by live-cell imaging. In this section the cell types we used to study $cPLA_2\alpha$ translocation to membrane are described. MDCK cells (ATCC, Manassas, VA, USA) were used (1) to identify subcellular organelles targeted by $cPLA_2\alpha$, (2) to determine the $[Ca^{2+}]_i$ required for association of FP-$cPLA_2\alpha$ and FP-C2 domain with intracellular membranes using calcium-clamping techniques, (3) to determine the kinetics of translocation using live-cell imaging, and (4) to determine the residues and structural features of $cPLA_2\alpha$ in determining translocation to membrane (Evans & Leslie, 2004; Evans et al., 2001; Evans et al., 2002; Evans et al., 2004; Evans et al., 2006). After adherence in culture, MDCK spread out to relatively large, flat cells making them useful for microscopy. Studies have shown that $cPLA_2\alpha$-mediates arachidonic acid release from MDCK cells in response to ATP acting through purinergic receptors to transiently increase $[Ca^{2+}]_i$, and by ionomycin to elicit a transient or sustained, saturating $[Ca^{2+}]_i$ increase depending on the concentration (Shehnaz, Torres, Balboa, & Insel, 2000; Xing, Firestein, Shen, & Insel, 1997).

2.2.2 Mouse Lung Fibroblasts: Isolation and Immortalization With SV40

In order to directly correlate the translocation properties of wild type and mutant forms of FP-$cPLA_2\alpha$, and their ability to release arachidonic acid (see Section 4), we developed a lung fibroblast cell line lacking endogenous $cPLA_2\alpha$. Fibroblasts can be isolated from mouse lungs with relative ease by using an explant protocol in which fibroblasts grow out from small pieces of minced mouse lungs on tissue culture dishes (Ghosh et al., 2004). Since fibroblasts have a growth advantage relative to other cell types in the lung, a relatively pure population is obtained after approximately three passages. Lung fibroblasts were isolated from $cPLA_2\alpha$ knockout and wild type mice, and immortalized using SV40 to obtain continuously growing cell lines (Ghosh et al., 2004; Stewart, Ghosh, Spencer, & Leslie, 2002). A comparison of immortalized lung fibroblasts from wild type ($IMLF^{+/+}$) and $cPLA_2\alpha$ knockout mice ($IMLF^{-/-}$) stimulated with either serum or A23187/phorbol myristate acetate confirmed that arachidonic acid release is mediated by $cPLA_2\alpha$ (Ghosh et al., 2004;

Stewart et al., 2002). Expression of FP-cPLA$_2\alpha$ in IMLF$^{-/-}$ using adenovirus fully reconstituted stimulus-dependent arachidonic acid release confirming that FP-cPLA$_2\alpha$ is functional. The use of adenovirus for transfection is very useful since the amount of cPLA$_2\alpha$ expression can be carefully titrated to achieve levels similar to the amount of endogenous cPLA$_2\alpha$ found in wild type cells (Stewart et al., 2002; Tucker et al., 2009).

Isolation of Mouse Lung Fibroblasts

1. Fibroblasts were isolated from wild type and cPLA$_2\alpha^{-/-}$ mice, which were generated using 129 embryonic stem cells in a C57BL/6 strain as previously described (Bonventre et al., 1997). cPLA$_2\alpha^{-/-}$ breeder mice on a Balb/c background were provided by Dr. Joseph Bonventre. Balb/c mice were obtained from The National Cancer Institute and Charles River. The constitutive pla2g4a knockout (129/SvEv-C57BL/6) is available from Taconic Biosciences, Hudson, NY. To isolate fibroblasts, mouse lungs are placed in a glass petri dish containing ~5–10 mL of sterile phosphate-buffered saline (PBS) (Corning, Corning, NY, USA) then cut into small pieces (~1–2 mm) in the petri dish using sterile scalpels (#15, QuickMedical Corp., Issaquah, WA, USA).
2. The lung pieces are washed five times with sterile PBS (Corning, Corning, NY, USA) then transferred to 100-mm tissue culture dishes (two dishes/lung) containing Dulbecco's modified Eagle's medium (DMEM) (Corning) with 10% fetal bovine serum (FBS) (GE Healthcare Life Sciences, Logan, UT, USA), 1% streptomycin/penicillin solution 100× (Corning) (culture medium) additionally supplemented with 100 μg/mL Normocin (InvivoGen, San Diego, CA, USA), an antimicrobial reagent to prevent bacteria, mycoplasma, and fungal contamination. A low volume of medium (~3 mL) is used so that the lung pieces do not float but remain in contact with the dish. The medium is changed daily.
3. Fibroblasts start growing out of the tissue explants in ~3 days. After 5 days, the pieces of lung are removed from the dish and the fibroblasts are grown in fresh medium containing 100 μg/mL Normocin for 2 more days.
4. The fibroblasts are washed, trypsinized (Trypsin-EDTA, Corning) and all the cells from the original 100 mm dish are plated into a 100 mm dish and grown to ~80% confluence. The medium is refreshed every 3 days. The cells are passaged three times at a split ratio of 1:2.

SV40 Immortalization

1. The fibroblasts are infected with SV40 after the third passage. Fibroblasts are trypsinized and then centrifuged in a 15-mL falcon tube at $500 \times g$ for 5 min. The medium is removed and SV40 is added to 1×10^6 cells (multiplicity of infection 30) in 1 mL culture medium, and gently mixed. The mixture is incubated at 37°C for 1 h and then plated in a 100 mm tissue culture dish in culture medium. SV40 virus was provided by Dr. James L. Cook, Loyola University but it can also be purchased from ATCC.
2. At 3 days postinfection (the cells may be nearly confluent at this time), the medium is replaced with DMEM containing 2% FBS and the cells cultured for 10 days. After ~6–7 days in 2% FBS medium, colonies will be visible along with many dead cells.
3. On day 10 the cultures are washed with PBS to remove the dead cells, and the colonies are trypsinized. Clones are selected by serial dilution in 96 well tissue culture plates in 100 μL culture medium as described (Fuller, Takahashi, & Hurrell, 2001).
4. Alternatively, 10 days after adding SV40, trypsinized colonies can be plated into 100 mm tissue culture dishes at low cell density (~1×10^4 cells/100 mm dish) and the cells grown for ~3–4 days for colony formation. A single colony is isolated by placing a sterile cloning cylinder over the colony on the dish, trypsinizing the cells inside the cylinder and plating the cells again at low density to repeat the process of colony isolation to obtain clones.
5. Frozen stocks are prepared by dispensing 1×10^6 cells in 1 mL of DMEM containing 55% FBS and 10% DMSO (Sigma-Aldrich, St. Louis, MO, USA).

2.3 Transfection Protocols for Expression of FP-$cPLA_2\alpha$ and FP-C2 Domain

In this section, we describe procedures for transient transfection of MDCK cells and generation of MDCK cells stably expressing FP-tagged proteins. Expression protocols using the adenoviral vector system in IMLF cells are also described.

2.3.1 Procedure for Transient Transfection of Plasmid DNA in MDCK Cells

1. MDCK cells are plated on 35 mm glass bottom dishes (MatTek Corp., Ashland, MA, USA) at $1 \times 10^4/cm^2$ in 2 mL of culture medium. The MatTek dishes used for immunofluorescence (cat. # p35G-0-14-C) and

for live-cell imaging (p35G-1.5-14-C) contain a 14 mm glass coverslip. Cells are grown for 24 h to <50% confluence at the time of transfection.

2. Dilute 5 μL of FuGENE-6 (Promega, Madison, WI, USA) in 100 μL serum- and antibiotics-free DMEM in a 1.5-mL tube. FuGENE-6 is used since it has relatively low cytotoxicity in MDCK cells. Add 1 μg of endotoxin-free plasmid DNA to the diluted FuGENE-6, mix immediately, and incubate for 45 min at room temperature.
3. The FuGENE-6/plasmid mixture is added to the culture medium and the cells incubated in 5% CO_2 at 37°C for ~48 h for expression.

2.3.2 Procedure for Developing MDCK Cells Stably Expressing FP-cPLA$_2\alpha$

1. MDCK cells are plated into 6-well tissue culture plates ($1 \times 10^4/cm^2$) in 2 mL culture medium and grown for 24 h to <50% confluence.
2. Prepare and add DNA/FuGENE-6 complex to cells as described in (Section 2.3.1) then incubate (5% CO_2 at 37°C) for 48 h.
3. Trypsinize the cells and plate into 60 mm dishes in medium supplemented with 5 mg/mL antibiotic G418 (Thermo Fisher Scientific, Waltham, MA, USA). The medium is changed daily for 14 days. This step will select cells that stably express pEGFP-cPLA$_2\alpha$ since the construct renders them resistant to killing by the antibiotic G418.
4. After 14 days selection, cells are trypsinized, serially diluted, and then plated into 96-well plates in culture medium with 5 mg/mL G418 to select single clones (Fuller et al., 2001).
5. Western blotting is carried out to confirm that EGFP-cPLA$_2\alpha$ is stably expressed in the clones by using anti-GFP antibody (Abcam, Cambridge, MA, USA). Alternatively, western blotting with antibody to cPLA$_2\alpha$ (Cell Signaling Technology, Beverly, MA, USA) will show the extent of EGFP-cPLA$_2\alpha$ expression (m.w. ~125 kDa) compared to endogenous cPLA$_2\alpha$ (~100 kDa). In addition, microscopy is used to confirm EGFP-cPLA$_2\alpha$ fluorescence (Evans et al., 2001).
6. Once confirmed, cells are maintained in culture medium containing 3 mg/mL G418 and frozen stocks prepared. Expression can lessen over time in stable clones but cells with highest expression of EGFP-cPLA$_2\alpha$ can be enriched using fluorescence-activated cell sorting.

2.3.3 Procedure for Expressing FP-cPLA$_2\alpha$ in IMLF Using Adenovirus

IMLF can be transfected using Lipofectamine 3000 (Thermo Fisher Scientific, Waltham, MA, USA) but transfection efficiency is higher using recombinant adenovirus (described later), which gives very good expression in a short-time relative to other methods.

1. IMLF are plated (1×10^4) on the glass coverslip portion of 35 mm MatTek dishes in 200 μL culture medium and grown overnight. The dishes are carefully handled to avoid spilling the liquid on the glass insert into the entire dish.
2. The cells are washed twice with DMEM containing 0.1% bovine serum albumin (Sigma-Aldrich) (BSA medium), and then incubated for 2 h (5% CO_2, 37°C) in 150 μL BSA medium containing recombinant adenoviruses (multiplicity of infection ~4–60) (Stewart et al., 2002).
3. The dish is flooded with culture medium (1.5 mL) and the cells are incubated at 37°C in 5% CO_2 for 26 h, the time required for optimal expression of FP-cPLA$_2\alpha$ (Stewart et al., 2002).

3. PROTOCOLS TO STUDY THE ROLE OF CALCIUM IN REGULATING THE TRANSLOCATION OF FP-cPLA$_2\alpha$ AND FP-C2 DOMAIN

3.1 Identification of Organelles Targeted by FP-cPLA$_2\alpha$ and FP-C2 Domain in Response to Calcium Increase

Immunofluorescence microscopy and imaging of live cells were used to identify the organelles targeted by FP-cPLA$_2\alpha$ and FP-C2 domain in MDCK cells and IMLF stimulated with calcium-mobilizing agonists. In unstimulated cells FP-cPLA$_2\alpha$ is uniformly distributed throughout the cytosol. It translocates to Golgi, ER, and the PNM by increases in $[Ca^{2+}]_i$ but does not localize to the plasma membrane. In this section, we describe protocols to identify subcellular membranes targeted by FP-cPLA$_2\alpha$ in fixed cells using antibodies to organelle markers, and in live cells using fluorescent dyes that localize to specific organelles.

3.1.1 Procedure to Identify Organelles by Immunofluorescence Microscopy

1. MDCK cells-expressing FP-cPLA$_2\alpha$ or FP-C2 domain are cultured in MatTek dishes in BSA medium and incubated overnight to quiesce the cells.
2. Cells are washed with Hepes-buffered (25 m*M*, pH 7.4) Hank's balanced salts solution (Corning) (HHBSS) and treated with calcium-mobilizing agonists such as 10 μ*M* ionomycin (EMD Millipore,

Billerica, MA, USA) or 100 μ*M* ATP (Sigma-Aldrich, St. Louis, MO, USA) for 45 s to 2 min. Ionomycin at 10 μ*M* induces a saturating, sustained increase in $[Ca^{2+}]_i$, and ATP induces a transient increase in $[Ca^{2+}]_i$ due to the release of calcium from intracellular stores. Extra dishes of cells are included as controls that are not stimulated but carried through the fixation procedure.

3. Cells are washed with PBS twice and fixed with 3.2% paraformaldehyde (Electron Microscopy Sciences, Hatfield, PA, USA) in PBS containing 3% sucrose (Sigma-Aldrich) for 15 min then washed with PBS five times.
4. The cells are permeabilized with 0.1% Triton X-100 (Sigma-Aldrich) in PBS for 15 min, washed with PBS five times, and then incubated in blocking solution (10% FBS in PBS) for 30 min.
5. Primary antibodies against markers specific for Golgi (Golgin 97 from Thermo Fisher Scientific, Waltham, MA, USA; giantin from Abcam) and PNM (nucleoporin p62 from BD Bioscience, San Jose, CA, USA) are added to the fixed cells (1:100 in blocking buffer). The cells are incubated at room temperature for 1 h or at 4°C overnight then washed with PBS five times.
6. Cells are washed with blocking solution five times at 2 min intervals. A TRITC-conjugated secondary antibody (1:100, Jackson ImmunoResearch Laboratories Inc., West Grove, PA, USA) is added and the cells incubated at room temperature for 1 h. For dual imaging it is important to consider the degree of color separation in selecting the fluorophore for the secondary antibody. Information can be found on the Jackson ImmunoResearch Laboratories Inc. (https://www.jacksonimmuno.com/technical/products/conjugate-selection) and Thermo Fisher Scientific websites (https://www.thermofisher.com/us/en/home/life-science/cell-analysis/labeling-chemistry/fluorescence-spectraviewer.html). Fluorophores in the red range such as TRITC and Texas Red are useful for imaging with EGFP.
7. The cells are imaged using an inverted Nikon Diaphot microscope although any inverted microscope capable of fluorescent imaging can be used that has components needed for imaging acquisition such as a digital camera, stage controller, lamp, and computer.
8. A similar fixation protocol is used to study localization of FP-cPLA$_2\alpha$ to Golgi in IMLF, which are incubated in phenol red-free DMEM (Thermo Fisher Scientific, Waltham, MA, USA) containing 0.1% BSA then stimulated with 10% mouse serum (Atlanta Biologicals,

Norcross, GA, USA) to induce a transient increase in $[Ca^{2+}]_i$. The primary antibody to the Golgi marker giantin (1:200, Abcam) and Texas Red-conjugated secondary antibody (1:100, Jackson ImmunoResearch Laboratories) are used.

9. FITC and TRITC filters are used for excitation of EGFP-cPLA$_2\alpha$ and TRITC- or Texas Red, respectively, using an exposure time of ~500–1000 ms.
10. A drop of oil is placed on the oil immersion objective (40 × or 60 ×) and the dish containing fixed cells in PBS placed on the stage. A field of cells for imaging is selected quickly using bright field and then focused in the selected fluorescence ranges. Sequential imaging of Z-stacks can be taken to identify regions where pixels of each fluorophore are above background fluorescence. A useful discussion of methodologies used for determining fluorescence colocalization and quantitation using Pearson's correlation coefficient can be found in Pompey, Michaely, and Luby-Phelps (2013).

3.1.2 Procedure to Identify Organelles Using Live-Cell Imaging

1. Dyes such as ER Tracker Blue-White DPX (Thermo Fisher Scientific, Waltham, MA, USA) intended for live-cell imaging are also used to identify organelles targeted by cPLA$_2\alpha$ (Evans et al., 2001). MDCK cells-expressing EGFP-C2 domain or EGFP-cPLA$_2\alpha$ are incubated with 0.1–1.0 μM ER Tracker in HHBSS for 15–30 min at 37°C to stain ER. The lowest dye concentration as possible is used to minimize potential labeling artifacts.
2. Cells are stimulated with 1 μM ionomycin and time-lapse images recorded using FITC optics. ER tracker is imaged with a DAPI filter (374/430 nm for excitation/emission). Refer to Section 3.3.1 for details of live-cell imaging.
3. Organelles targeted by cPLA$_2\alpha$ can also be identified by expressing FP-tagged proteins that are known to localize to specific organelles. We used pECFP-GT (Clontech) encoding the N-terminal 82 residues of β-1,4-galactosyltransferase, which localizes to medial and *trans*-Golgi cisternae (Sciaky et al., 1997). MDCK coexpressing EYFP-cPLA$_2\alpha$ and ECFP-GT are stimulated with 5 μM ATP in HHBSS. Cells are imaged every 3 s using an Olympus inverted microscope (Olympus, Waltham, MA, USA) with a 60 ×, 1.25 NA oil immersion objective and ECFP (430 nm) and EYFP (510 nm) filters.

3.2 Calcium-Clamping Method to Determine the $[Ca^{2+}]_i$ Required for FP-cPLA$_2\alpha$ and FP-C2 Domain Translocation to Intracellular Membranes

Steady-state levels of $[Ca^{2+}]_i$ were established using a calcium-clamping protocol to determine the amount of $[Ca^{2+}]_i$ required for FP-cPLA$_2\alpha$ and FP-C2 domain to bind Golgi, ER, and PNM (Evans et al., 2001).

1. MDCK cells-expressing EGFP-cPLA$_2\alpha$ are plated on MatTek dishes (1×10^4 cells/cm^2) in culture medium and grown overnight. The cells are washed with and incubated in BSA medium overnight to quiesce the cells.
2. The cells are washed with and incubated in HHBSS containing 1 m*M* probenecid (Thermo Fisher Scientific, Waltham, MA, USA), 10 μ*M* Fura2-AM (Thermo Fisher Scientific, Waltham, MA, USA), and 1% DMSO for 1.5 h at 37°C. Cells are then washed with and incubated in HHBSS containing 1 m*M* probenecid for 30 min for deesterification of Fura2-AM.
3. MDCK cells are incubated in HHBSS containing 3 m*M* EGTA for at least 45 min to deplete internal calcium stores. The cells are then treated with 10 μ*M* ionomycin to make cell membranes permeable to $[Ca^{2+}]_i$. Once the intracellular Ca^{2+}-stores are depleted they cannot be refilled since the cells are in medium containing EGTA (Sigma-Aldrich) (Albert & Tashjian, 1984; Drummond, Lee, Resendez, & Steinhardt, 1987; Hoth & Penner, 1992).
4. $CaCl_2$ (Sigma-Aldrich) (~20–23 μL of 100 m*M*) is added to 1 mL of the HHBSS/EGTA/ionomycin medium covering the cells. The medium is carefully mixed and the cells incubated for >5 min for the $[Ca^{2+}]_i$ to equilibrate before imaging. The process of $CaCl_2$ addition, in increments of 2–3 μL, mixing, equilibration, and imaging are repeated after each addition of extracellular calcium. This results in incremental increases in $[Ca^{2+}]_i$ in the physiological range of ~10–600 n*M* that remains stable for >10 min. It is not possible to establish sustained $[Ca^{2+}]_i$ to predetermined levels by manipulating $[Ca^{2+}]_e$ since $[Ca^{2+}]_i$ is lower than $[Ca^{2+}]_e$ at equilibrium. Therefore, the procedure requires that $[Ca^{2+}]_i$ be determined empirically using ratiometric measurements of Fura2 fluorescence.
5. After each addition of $[Ca^{2+}]_e$, images of Fura2 fluorescence at 340 and 380 nm, and EGFP fluorescence at 480 nm are taken. The $[Ca^{2+}]_i$ is determined using the equation, $[Ca^{2+}]_i = K_D \times \beta \times (R - R_{min})/R_{max} - R)$ (Grynkiewicz, Poenie, & Tsien, 1985; Roe, Lemasters, & Herman,

1990). K_D is the dissociation constant for Ca^{2+} of Fura2 (224 n*M*), *R* is the ratio of background-corrected fluorescence from the cytoplasmic area of individual cells at 340 and 380 nm; R_{max} and R_{min} are determined by *in situ* calibration after imaging at each level of $[Ca^{2+}]_i$. R_{max} is the ratio of Fura2 fluorescence in the presence of saturating calcium, and R_{min} is the ratio at zero calcium. β is the ratio of fluorescence at 340 and 380 nm of Fura2 in zero and saturating calcium. Methods used for calibration and other techniques for measuring calcium dynamics are thoroughly detailed in a review by Palmer and Tsien (2006).

6. In addition to imaging Fura2 in the cytosol after each addition of calcium to the medium, the EGFP-cPLA$_2\alpha$ fluorescence signal is monitored and regions of interest (ROI) corresponding to the Golgi, ER, and PNM are defined. Having established that EGFP-cPLA$_2\alpha$ localizes to these intracellular membranes by using organelle-specific antibodies (as described in Section 3.1.1), identifying these regions in cells exhibiting FP-cPLA$_2\alpha$ translocation is straightforward due to their characteristic morphology and cellular distribution, and allows ROI to be selected. Representative EGFP-cPLA$_2\alpha$ images at Golgi, ER, and PNM at steady-state levels of $[Ca^{2+}]_i$ is shown in Fig. 1. Results from several calcium clamp experiments established that ~100–250 n*M* $[Ca^{2+}]_i$ is required for EGFP-cPLA$_2\alpha$ to bind Golgi, and 210–280 n*M* $[Ca^{2+}]$ to bind ER and PNM (Evans et al., 2001).
7. Single-cell imaging is performed on a Nikon inverted microscope using a 40 ×, 1.3 NA oil immersion objective, Fura2 and FITC filter sets, and a CCD camera. Image acquisition and analysis are performed with SlideBook software (Intelligent Imaging Innovations, Inc., Denver, CO, USA).
8. Having determined the $[Ca^{2+}]_i$ required for FP-cPLA$_2\alpha$ to bind to Golgi, ER, and PNM, the calcium-clamping protocol is then used to directly compare the membrane association of ECFP-cPLA$_2\alpha$ and EYFP-C2 domain when coexpressed in MDCK cells (Evans & Leslie, 2004). In order to directly compare the translocation properties of ECFP-cPLA$_2\alpha$ and EYFP-C2 domain, it is important that they are coexpressed in the same cell since this controls for cell-to-cell differences in the rate and amplitude of $[Ca^{2+}]_i$ changes. At resting levels of $[Ca^{2+}]_i$ (<50 n*M*), both constructs are distributed throughout the cytoplasm but the EYFP-C2 domain is also in the nucleoplasm, whereas full-length

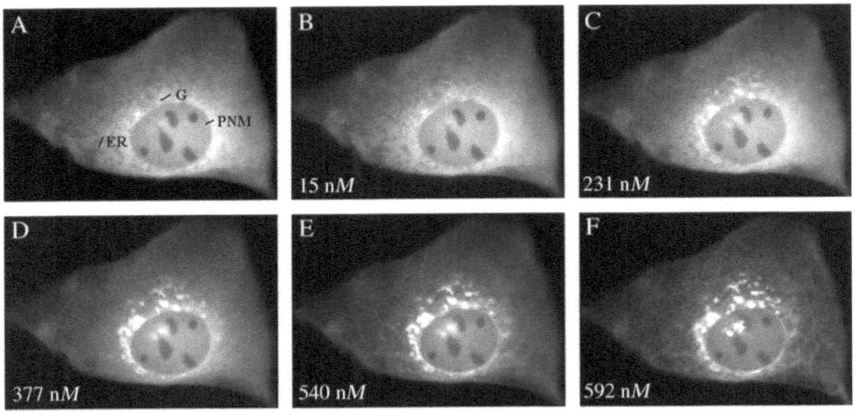

Fig. 1 Calcium clamp method to determine $[Ca^{2+}]_i$ required for association of EGFP-cPLA$_2\alpha$ to intracellular membranes. MDCK cells-expressing EGFP-cPLA$_2\alpha$ and loaded with Fura2 were imaged following step increases in the $[Ca^{2+}]_e$ of an EGTA-buffered bath solution that resulted in step increases in $[Ca^{2+}]_i$ of the cells (calcium clamp condition). (A–F) Fluorescence images of EGFP-cPLA$_2\alpha$ are shown at zero (A), and at the indicated (B–F) $[Ca^{2+}]_i$ values. The EGFP fluorescence at endoplasmic reticulum (ER), Golgi (G), and peri-nuclear membrane (PNM) (marked inside the cell in A) is shown at the indicated $[Ca^{2+}]_i$. *Adapted from Figure 4 of Evans, J. H., Spencer, D. M., Zweifach, A., & Leslie, C. C. (2001). Intracellular calcium signals regulating cytosolic phospholipase A$_2$ translocation to internal membranes.* Journal of Biological Chemistry, 276*(32), 30150–30160.*

ECFP-cPLA$_2\alpha$ is excluded because of its larger size. After each increase in $[Ca^{2+}]_e$, similar fractions of ECFP-cPLA$_2\alpha$ and EYFP-C2 domain colocalize with the same intracellular membranes establishing that the C2 domain provides targeting specificity and that they have the same calcium sensitivities.

3.3 Protocol for Time-Lapse Imaging to Correlate the Kinetics of FP-cPLA$_2\alpha$ and FP-C2 Domain Translocation With $[Ca^{2+}]_i$

3.3.1 Experiments Using MDCK Cells

1. MDCK cells stably expressing EGFP-cPLA$_2\alpha$ or EGFP-C2 domain grown on 35 mm MatTek dishes are loaded with Fura2 as described in Section 3.2 (Evans et al., 2001).
2. The cells in 1 mL of HHBSS are placed on the microscope stage and focused. It is useful to have ~3 cells in the field of view. A portion of the medium (200 μL) is removed and then imaging is initiated. Stimuli (5 ×) are added into the 200 μL of medium, which is then added back to the cells. The medium is quickly mixed with a Pipetman using care not

to change the focus. The stimuli ATP (100 μ*M*), thapsigargin (2 μ*M*) (Thermo Fisher Scientific, Waltham, MA, USA), or ionomycin (1 μ*M*) induce transient increases in $[Ca^{2+}]_i$.

3. Cells are imaged with an inverted Nikon microscope using a 40×, 1.3 NA oil immersion objective, Fura2 and FITC filter sets, and a CCD camera. Images of Fura2 are taken at 340 and 380 nm, and EGFP at 480 nm, at 4–11 s intervals for ~30 min with an exposure time of ~500–1000 ms.
4. The $[Ca^{2+}]_i$ is determined according to Grynkiewicz et al. (1985) as described in Section 3.2 (#5) using 140 n*M* as the K_D for Fura2.
5. EGFP fluorescence is determined for different ROI corresponding to Golgi, ER, and PNM by calculating the F_t/F_0, where F_t is the background- and bleach-corrected EGFP fluorescence at time *t*, and F_0 is the background-corrected fluorescence at time 0. Background fluorescence is the average pixel value of fluorescence from an area of the culture dish without cells and this is subtracted from the average pixel values from the Golgi or other cellular ROI at each time point. Bleach-correction factors are determined at each time point by determining the background-corrected average pixel values for an ROI corresponding to the entire cell and then normalizing those values to the initial value at zero time. The background and bleach-corrected fluorescence, F_t, is calculated by dividing the background-corrected average value for each ROI at each time point by the bleach-correction factor at the corresponding time point. The normalized fluorescence intensity (F_t/F_0) at Golgi is calculated by dividing F_t by the fluorescence value at time zero, F_0. Live time-lapse images can be converted to a movie by exporting the data as a movie file format using the microscope software such as SlideBook 6, and the movie can be played using QuickTime player. At least 15–30 cells are analyzed from several independent experiments. A representative experiment depicting translocation of EGFP-C2 domain to Golgi, ER, and PNM and increases in $[Ca^{2+}]_i$ in response to ATP is shown in Fig. 2.

3.3.2 Experiments Using IMLF

1. IMLF-expressing EGFP-$cPLA_2\alpha$ are grown on 35 mm MatTek dishes, washed and incubated overnight in phenol red-free BSA medium to quiesce the cells.
2. They are washed and incubated in stimulation medium containing 2 m*M* probenecid (Thermo Fisher Scientific, Waltham, MA, USA)

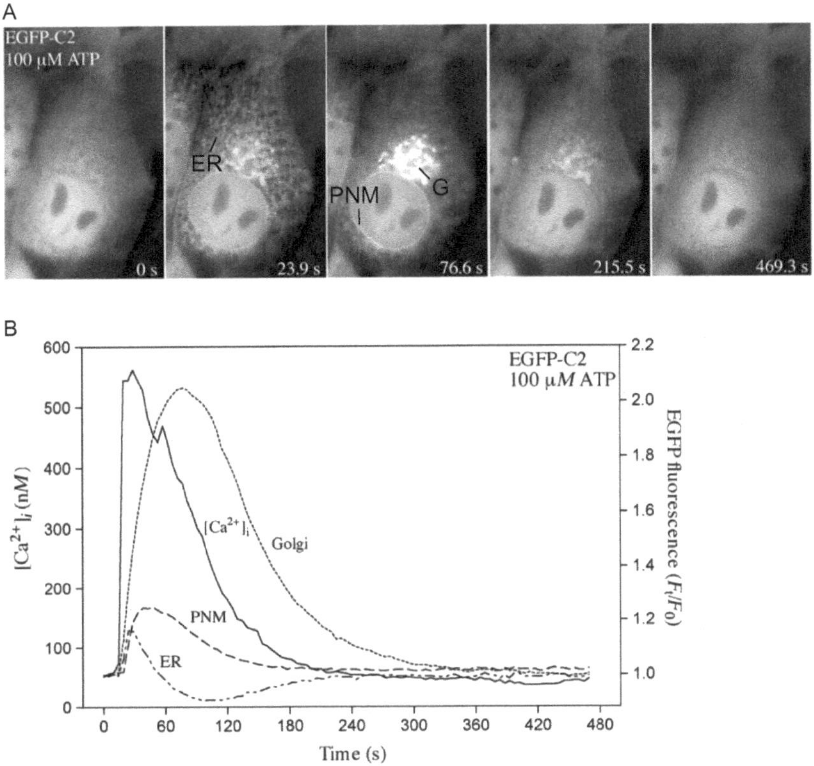

Fig. 2 Correlation of EGFP-cPLA$_2$α C2 domain (EGFP-C2) translocation with $[Ca^{2+}]_i$ in response to ATP stimulation. MDCK cells-expressing EGFP-C2 were loaded with Fura2 and then stimulated with 100 μ*M* ATP. (A) Time-lapse images of EGFP-C2 and Fura2 fluorescence were taken over time as indicated after ATP application. (B) The relative EGFP fluorescence at ROI corresponding to the ER, Golgi, and PNM, and the $[Ca^{2+}]_i$ determined from each image, were measured and depicted graphically. *Adapted from Figure 5 of Evans, J. H., Spencer, D. M., Zweifach, A., & Leslie, C. C. (2001). Intracellular calcium signals regulating cytosolic phospholipase A$_2$ translocation to internal membranes.* Journal of Biological Chemistry, *276(32), 30150–30160.*

and then loaded with 5 μ*M* Fura Red-AM (Thermo Fisher Scientific, Waltham, MA, USA) for 30 min at room temperature in the presence of 0.02% pluronic acid. The cells are washed with and maintained in HHBSS containing 2 m*M* probenecid for imaging. Probenecid is an anion transporter inhibitor that helps prevent the extrusion of calcium indicators from cells. Although many of our early experiments in MDCK cells used Fura2 as the calcium indicator for wide-field microscopy we more recently have used Fura Red. Fura2 is excited at ultraviolet wavelength, which may have adverse effects on cell function. For example, in experiments studying the role of calcium in regulating the translocation of EGFP-$cPLA_2\alpha$ in Fura2-loaded macrophages during phagocytosis of zymosan, we found that UV excitation to image Fura2-inhibited phagocytosis (Girotti, Evans, Burke & Leslie, 2004). Fura Red is structurally and chemically similar to Fura2 and can be used for ratiometric $[Ca^{2+}]_i$ measurements (Kurebayashi, Harkins, & Baylor, 1993; Takahashi, Camacho, Lechleiter, & Herman, 1999). Since the excitation/emission spectra of Fura Red are shifted to the red range, there is less interference from autofluorescence in most cells.

3. IMLF are stimulated with 10% mouse serum and imaged using a 40 ×, oil immersion objective on an inverted Zeiss 200 M microscope. Images are collected every 3 s using a FITC filter for EGFP fluorescence, and F403 and F470 filters for Fura Red fluorescence (Tucker et al., 2009).
4. The Zeiss 200 M microscopy system has a heat controlled environmental chamber for imaging at 37°C. However, live-cell imaging can be carried out at room temperature, the conditions used for experiments with MDCK cells.
5. Data are acquired and analyzed using Slidebook software. Translocation data are calculated based on average fluorescence intensity of EGFP-$cPLA_2\alpha$ on the Golgi in each cell. Values are corrected by subtracting background fluorescence, and for differential bleaching at each wavelength through the duration of imaging and expressed relative to time 0 (F_t/F_0). Ratios of bound to unbound calcium fluorescence intensities (F403/F470) are calculated and corrected for background fluorescence and expressed relative to time zero (R_t/R_0).

3.4 Dual Time-Lapse Imaging of FP-$cPLA_2\alpha$ and FP-C2 Domain Translocation

3.4.1 *Procedure to Establish That the $cPLA_2\alpha$ C2 Domain Functions as a $[Ca^{2+}]_i$ Sensor*

In order to directly compare the translocation properties of FP-$cPLA_2\alpha$ and FP-C2 domain in response to calcium transients, we first established that the $cPLA_2\alpha$ C2 domain could be used as a $[Ca^{2+}]_i$ sensor to track relative changes in $[Ca^{2+}]_i$ (Evans & Leslie, 2004).

1. MDCK cells-expressing EGFP-$cPLA_2\alpha$ C2 are loaded with Fura2-AM as described in Section 3.2.
2. The cells are incubated with EGTA in HHBSS to deplete intracellular Ca^{2+} stores, and then made permeable to $[Ca^{2+}]_i$ by adding ionomycin as described in Section 3.2.

3. Cells are imaged using an Olympus inverted microscope with a 40 ×, 1.35 NA oil immersion objective and 340/380 nm and 465 nm excitation for Fura2 and EGFP, respectively.
4. At ~7–8 s after the start of imaging, the medium is exchanged with fresh HHBSS/EGTA solution (1 mL) containing 10 μ*M* ionomycin and 2.5 m*M* $CaCl_2$ to increase $[Ca^{2+}]_i$. Alternatively these experiments can be done by maintaining the cells in HHBSS without EGTA, so that the internal stores are not initially depleted. $[Ca^{2+}]_i$ is then increased by adding 100 μL of 10 μ*M* ionomycin (in EGTA/HHBSS buffer) to 900 μL of EGTA/HHBSS buffer, which causes release of calcium from intracellular stores. This results in faster, higher amplitude $[Ca^{2+}]_i$ increase but both procedures give similar results (Evans & Leslie, 2004).
5. At ~3–4 min after adding extracellular calcium, 1 mL of EGTA/HHBSS buffer is added to the cells during imaging to rapidly decrease $[Ca^{2+}]_i$.
6. As shown in Fig. 3, EGFP-cPLA$_2$α C2 domain translocation to Golgi began after the $[Ca^{2+}]_i$ exceeded the threshold for membrane association and increased at a similar rate as increase in $[Ca^{2+}]_i$. Upon addition of extracellular EGTA, $[Ca^{2+}]_i$ and EGFP-C2 domain fluorescence on the Golgi decreased at a similar rate. The results show that the C2 domain acts as a $[Ca^{2+}]_i$ sensor since its association with Golgi is proportional to $[Ca^{2+}]_i$.

3.4.2 Procedure Comparing FP-cPLA$_2$α and FP-C2 Domain Translocation

To study the contribution of the C2 and catalytic domains in regulating translocation to Golgi, we directly compared the translocation properties of FP-cPLA$_2$α and FP-C2 domain coexpressed in MDCK.

1. MDCK cells coexpressing ECFP-cPLA$_2$α and EYFP-C2 domain are incubated under calcium-clamping conditions resulting in depletion of intracellular calcium stores and treated with ionomycin as described in Section 3.2.
2. $[Ca^{2+}]_i$ is increased by adding extracellular calcium then decreased by adding EGTA at ~240 s as described in Section 3.4.1.
3. As shown in Fig. 4, the rate of EYFP-C2 domain translocation to Golgi is greater than ECFP-cPLA$_2$α. The EYFP-C2 domain fluorescence on the Golgi decreased to baseline values after adding EGTA reflecting the decrease in $[Ca^{2+}]_i$. However, ECFP-cPLA$_2$α fluorescence on the Golgi remained elevated for several minutes indicating that the catalytic

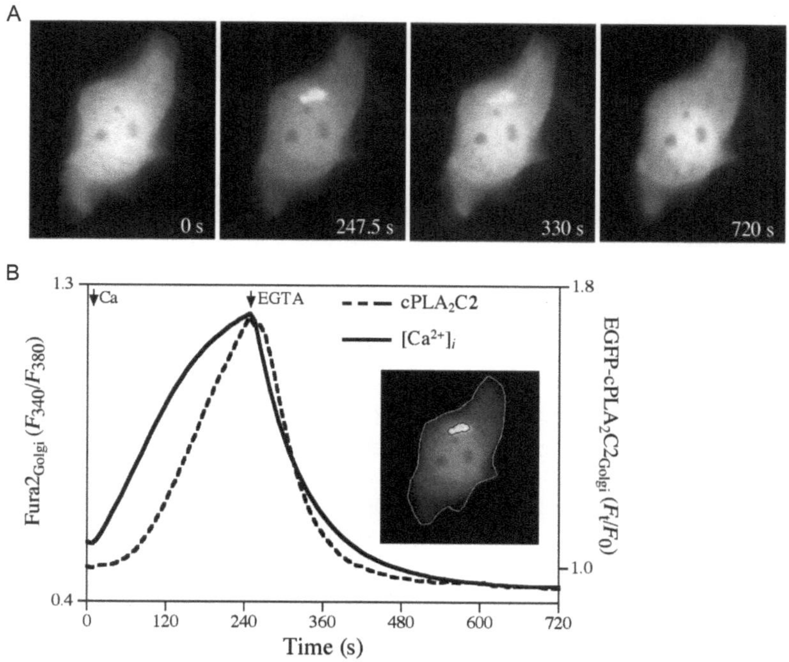

Fig. 3 cPLA$_2\alpha$ C2 domain functions as a $[Ca^{2+}]_i$ sensor. (A) $CaCl_2$ was added 7.5 s after initiation of imaging and EGTA at 247.5 s to MDCK cells-expressing EGFP-cPLA$_2\alpha$ C2 domain and loaded with Fura2 under calcium clamp conditions. Fluorescence images of EGFP-cPLA$_2\alpha$ C2 domain (cPLA$_2$C2) at the indicated times are shown. (B) Graphical analysis of Fura2 and EGFP fluorescence from the area of Golgi (Fura2$_{Golgi}$ and EGFP-cPLA$_2$C2$_{Golgi}$, inset, black line) is shown. *Adapted from Figure 2 of Evans, J. H., & Leslie, C. C. (2004). The cytosolic phospholipase A$_2$ catalytic domain modulates association and residence time at Golgi membranes.* Journal of Biological Chemistry, *279(7), 6005–6016.*

domain increases the residence time on the membrane in a calcium-independent manner. Similar results are observed in control experiments in which the fluorescent tags on cPLA$_2\alpha$ and the C2 domain are switched.

4. A similar protocol can also be used to determine the basis for the differences in translocation properties of full-length cPLA$_2\alpha$ and the C2 domain by coexpressing mutant forms of FP-cPLA$_2\alpha$ and FP-C2 domain. The increased residence time of full-length cPLA$_2\alpha$ on the Golgi is mediated by a tryptophan residue in the catalytic domain (Evans & Leslie, 2004). Using the same protocol as described earlier, MDCK cells coexpressing ECFP-cPLA$_2\alpha$W464G (or W464A) and EYFP-C2 domain showed that both proteins associated with Golgi

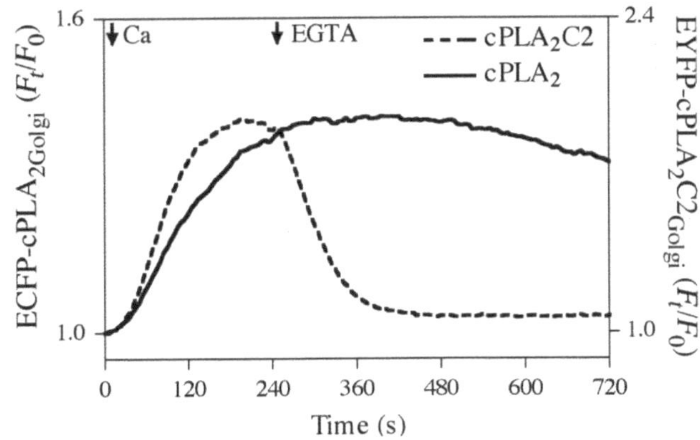

Fig. 4 Rate of membrane association and dissociation of ECFP-cPLA$_2\alpha$ and EYFP-C2 domain as a function of $[Ca^{2+}]_i$. $CaCl_2$ was added at 7.5 s after initiation of imaging and EGTA at 247.5 s to MDCK cells coexpressing ECFP-cPLA$_2\alpha$ and EYFP-cPLA$_2\alpha$ C2 domain under calcium clamp conditions. Graphical analysis of ECFP and EYFP fluorescence from the area of Golgi (ECFP-cPLA$_{2Golgi}$ and EYFP-cPLA$_2$C2$_{Golgi}$) is shown. *Adapted from Figure 4 of Evans, J. H., & Leslie, C. C. (2004). The cytosolic phospholipase A$_2$ catalytic domain modulates association and residence time at Golgi membranes.* Journal of Biological Chemistry, *279(7), 6005–6016.*

when $[Ca^{2+}]_i$ is increased. However, unlike wild type, the ECFP-cPLA$_2\alpha$W464G mutant dissociates from the Golgi when EGTA was added to decrease $[Ca^{2+}]_i$ as observed for the C2 domain (Evans & Leslie, 2004).

4. PROTOCOL FOR CORRELATING TRANSLOCATION OF FP-cPLA$_2\alpha$ AND ARACHIDONIC ACID RELEASE

By reconstituting wild type and mutant forms of cPLA$_2\alpha$ in IMLF$^{-/-}$ lacking endogenous cPLA$_2\alpha$, the functional role of phosphorylation sites and basic residues in the catalytic domain on calcium-dependent translocation and arachidonic acid release can be directly compared (Tucker et al., 2009).

4.1 Dual Time-Lapse Imaging of Wild Type and Mutant FP-cPLA$_2\alpha$

1. IMLF$^{-/-}$ cultured in MatTek dishes are coinfected with recombinant adenoviruses for 26 h to coexpress wild type ECFP-cPLA$_2\alpha$ and various mutant forms of EYFP-cPLA$_2\alpha$ as described in Section 2.3.3.

2. Imaging is initiated and then the cells are stimulated with 10% serum. Sequential images are collected using CFP and YFP filters every 3 s after serum stimulation using a 40 × oil immersion objective. Results are corrected for background fluorescence and differential bleaching and expressed relative to time zero (F_t/F_0).

4.2 Protocol to Determine Arachidonic Acid Release of Wild Type and Mutant FP-cPLA$_2\alpha$

1. To compare the ability of wild type and mutant forms of FP-cPLA$_2\alpha$ to release arachidonic acid, IMLF$^{-/-}$ are plated in 48-well plates (1.25×10^4 cells/cm^2 in 250 μL of culture medium).
2. After 18 h incubation in 5% CO_2 at 37°C, the cells are washed twice with and incubated in 100 μL BSA medium containing adenoviruses. Cells in separate wells are infected with adenoviruses for expression of wild type FP-cPLA$_2\alpha$ and mutant forms of FP-cPLA$_2\alpha$.
3. After incubation for 1.5 h, 150 μL of BSA medium containing 0.2 μC/mL [^{3}H]arachidonic acid [5,6,8,9,11,12,14,15-^{3}H]arachidonic acid (100 Ci/mmol) (PerkinElmer, Waltham, MA, USA) is added to each well. The cells are incubated for 26 h for adenovirus-mediated expression and incorporation of [^{3}H]arachidonic acid into membrane phospholipid.
4. The cells are washed twice with and incubated in fresh BSA medium, then stimulated with 10% serum. The culture medium is collected at various times after stimulation (5–30 min), centrifuged for 10 min at 15,000 rpm in 1.5-mL microfuge tubes to pellet any floating cells and radioactivity determined in the supernatant by scintillation counting.
5. Cells remaining in the wells are scraped into 50 μL of 0.1% Triton X-100 containing protease inhibitors, and an aliquot of the solution is used to determine the total cellular radioactivity. The amount of radioactivity released into the culture medium is calculated as a percentage of the total radioactivity (cells plus medium) in each well.
6. The remainder of the cell lysate is used to determine the relative levels of expression of wild type and mutant forms of FP-cPLA$_2\alpha$ in each well. It is necessary that wild type and mutant FP-cPLA$_2\alpha$ are expressed in the separate wells at similar levels in order to compare their ability to release arachidonic acid. This is achieved by infecting neighboring wells of IMLF$^{-/-}$ with three different amounts of adenoviruses that result in a range of expression to ensure that at least one well of wild type and one well of mutant cPLA$_2\alpha$ exhibit comparable expression.

Western blotting using primary antibody to cPLA$_2\alpha$ is used to determine the relative expression levels of wild type and mutant forms of FP-cPLA$_2\alpha$.

Expression of wild type FP-cPLA$_2\alpha$ in cells lacking endogenous cPLA$_2\alpha$ reconstitutes arachidonic acid release confirming that it is fully functional providing a model to directly determine the role of phosphorylation sites, calcium-dependent binding to the C2 domain, and basic residues in the catalytic domain that bind PIP$_2$ in regulating translocation and arachidonic acid release. Results demonstrate that phosphorylation and the PIP$_2$-binding site regulate catalytic activity but not calcium-dependent translocation (Reed et al., 2011; Tucker et al., 2009). The results are also supported by findings studying the interfacial kinetics and binding of cPLA$_2\alpha$ *in vitro*. A description of the methods for the expression of wild type and mutant forms of cPLA$_2\alpha$ in insect cells using baculovirus, purification, vesicle-binding assays, and kinetic studies are beyond the scope of this chapter but are described in detail elsewhere (Reed et al., 2011; Tucker et al., 2009).

ACKNOWLEDGMENT

The work carried out in the author's laboratory was supported by National Institutes of Health Grants HL34303 and HL61378.

REFERENCES

Albert, P. R., & Tashjian, A. H., Jr. (1984). Relationship of thyrotropin-releasing hormone-induced spike and plateau phases in cytosolic free Ca2+ concentrations to hormone secretion. Selective blockade using ionomycin and nifedipine. *Journal of Biological Chemistry*, *259*(24), 15350–15363.

Aoki, T., & Narumiya, S. (2012). Prostaglandins and chronic inflammation. *Trends in Pharmacological Sciences*, *33*(6), 304–311.

Back, M., Powell, W. S., Dahlen, S. E., Drazen, J. M., Evans, J. F., Serhan, C. N., … Rovati, G. E. (2014). Update on leukotriene, lipoxin and oxoeicosanoid receptors: IUPHAR Review 7. *British Journal of Pharmacology*, *171*(15), 3551–3574.

Ball, A., Nielsen, R., Gelb, M. H., & Robinson, B. H. (1999). Interfacial membrane docking of cytosolic phospholipase A2 C2 domain using electrostatic potential-modulated spin relaxation magnetic resonance. *Proceedings of the National Academy of Sciences of the United States of America*, *96*(12), 6637–6642.

Bayburt, T., & Gelb, M. H. (1997). Interfacial catalysis by human 85 kDa cytosolic phospholipase A2 on anionic vesicles in the scooting mode. *Biochemistry*, *36*(11), 3216–3231.

Bittova, L., Sumandea, M., & Cho, W. (1999). A structure-function study of the C2 domain of cytosolic phospholipase A$_2$. *Journal of Biological Chemistry*, *274*(14), 9665–9672.

Bonventre, J. V., Huang, Z., Taheri, M. R., O'Leary, E., Li, E., Moskowitz, M. A., & Sapirstein, A. (1997). Reduced fertility and postischaemic brain injury in mice deficient in cytosolic phospholipase A$_2$. *Nature*, *390*(6660), 622–625.

Börsch-Haubold, A. G., Bartoli, F., Asselin, J., Dudler, T., Kramer, R. M., Apitz-Castro, R., … Gelb, M. H. (1998). Identification of the phosphorylation sites of cytosolic phospholipase A$_2$ in agonist-stimulated human platelets and HeLa cells. *Journal of Biological Chemistry*, *273*(8), 4449–4458.

Buckley, C. D., Gilroy, D. W., & Serhan, C. N. (2014). Proresolving lipid mediators and mechanisms in the resolution of acute inflammation. *Immunity*, *40*(3), 315–327.

Channon, J., & Leslie, C. C. (1990). A calcium-dependent mechanism for associating a soluble arachidonoyl-hydrolyzing phospholipase A_2 with membrane in the macrophage cell line, RAW 264.7. *Journal of Biological Chemistry*, *265*(10), 5409–5413.

Clark, J. D., Lin, L.-L., Kriz, R. W., Ramesha, C. S., Sultzman, L. A., Lin, A. Y., … Knopf, J. L. (1991). A novel arachidonic acid-selective cytosolic PLA_2 contains a Ca^{2+}-dependent translocation domain with homology to PKC and GAP. *Cell*, *65*(6), 1043–1051.

Clark, J. D., Milona, N., & Knopf, J. L. (1990). Purification of a 110-kilodalton cytosolic phospholipase A_2 from the human monocytic cell line U937. *Proceedings of the National Academy of Sciences of the United States of America*, *87*(19), 7708–7712.

Corbalan-Garcia, S., & Gomez-Fernandez, J. C. (2014). Signaling through C2 domains: More than one lipid target. *Biochimica et Biophysica Acta*, *1838*(6), 1536–1547.

Das, S., & Cho, W. (2002). Roles of catalytic domain residues in interfacial binding and activation of group IV cytosolic phospholipase A2. *Journal of Biological Chemistry*, *277*(26), 23838–23846.

Das, S., Rafter, J. D., Kim, K. P., Gygi, S. P., & Cho, W. (2003). Mechanism of group IVA cytosolic phospholipase A_2 activation by phosphorylation. *Journal of Biological Chemistry*, *278*(42), 41431–41442.

Davletov, B., Perisic, O., & Williams, R. L. (1998). Calcium-dependent membrane penetration is a hallmark of the C2 domain of cytosolic phospholipase A_2 whereas the C2A domain of synaptotagmin binds membranes electrostatically. *Journal of Biological Chemistry*, *273*(30), 19093–19096.

Dennis, E. A., Cao, J., Hsu, Y. H., Magrioti, V., & Kokotos, G. (2011). Phospholipase A2 enzymes: Physical structure, biological function, disease implication, chemical inhibition, and therapeutic intervention. *Chemical Reviews*, *111*(10), 6130–6185.

Dessen, A., Tang, J., Schmidt, H., Stahl, M., Clark, J. D., Seehra, J., & Somers, W. S. (1999). Crystal structure of human cytosolic phospholipase A_2 reveals a novel topology and catalytic mechanism. *Cell*, *97*(3), 349–360.

Drummond, I. A., Lee, A. S., Resendez, E., Jr., & Steinhardt, R. A. (1987). Depletion of intracellular calcium stores by calcium ionophore A23187 induces the genes for glucose-regulated proteins in hamster fibroblasts. *Journal of Biological Chemistry*, *262*(26), 12801–12805.

Evans, J. H., Fergus, D. J., & Leslie, C. C. (2002). Inhibition of the MEK1/ERK pathway reduces arachidonic acid release independently of $cPLA_2$ phosphorylation and translocation. *BMC Biochemistry [Computer File]*, *3*(1), 30.

Evans, J. H., Gerber, S. H., Murray, D., & Leslie, C. C. (2004). The calcium binding loops of the cytosolic phospholipase A_2 C2 domain specify targeting to Golgi and ER in live cells. *Molecular Biology of the Cell*, *15*(1), 371–383.

Evans, J. H., & Leslie, C. C. (2004). The cytosolic phospholipase A_2 catalytic domain modulates association and residence time at Golgi membranes. *Journal of Biological Chemistry*, *279*(7), 6005–6016.

Evans, J. H., Murray, D., Leslie, C. C., & Falke, J. J. (2006). Specific translocation of protein kinase Cα to the plasma membrane requires both Ca^{2+} and PIP_2 recognition by its C2 domain. *Molecular Biology of the Cell*, *17*(1), 56–66.

Evans, J. H., Spencer, D. M., Zweifach, A., & Leslie, C. C. (2001). Intracellular calcium signals regulating cytosolic phospholipase A_2 translocation to internal membranes. *Journal of Biological Chemistry*, *276*(32), 30150–30160.

Fernandez, B., & Balsinde, J. (1991). Receptor-mediated activation of arachidonic acid release in mouse peritoneal macrophages is linked to extracellular calcium influx. *Biochemical and Biophysical Research Communications*, *180*(2), 1036–1040.

Fuller, S. A., Takahashi, M., & Hurrell, J. G. (2001). Cloning of hybridoma cell lines by limiting dilution. *Current Protocols in Molecular Biology*1:II:11.8:11.8.1–11.8.2.

Funk, C. D. (2001). Prostaglandins and leukotrienes: Advances in eicosanoid biology. *Science, 294*(5548), 1871–1875.

Ghosh, M., Stewart, A., Tucker, D. E., Bonventre, J. V., Murphy, R. C., & Leslie, C. C. (2004). Role of cytosolic phospholipase A_2 in prostaglandin E_2 production by lung fibroblasts. *American Journal of Respiratory Cell and Molecular Biology, 30*(1), 91–100.

Ghosh, M., Tucker, D. E., Burchett, S. A., & Leslie, C. C. (2006). Properties of the group IV phospholipase A_2 family. *Progress in Lipid Research, 45*(6), 487–510.

Gijón, M. A., Spencer, D. M., Kaiser, A. L., & Leslie, C. C. (1999). Role of phosphorylation sites and the C2 domain in regulation of cytosolic phospholipase A_2. *Journal of Cell Biology, 145*(6), 1219–1232.

Girotti, M., Evans, J. H., Burke, D., & Leslie, C. C. (2004). Cytosolic phospholipase A_2 translocates to forming phagosomes during phagocytosis of zymosan in macrophages. *Journal of Biological Chemistry, 279*(18), 19113–19121.

Glover, S., de Carvalho, M. S., Bayburt, T., Jonas, M., Chi, E., Leslie, C. C., & Gelb, M. H. (1995). Translocation of the 85-kDa phospholipase A_2 from cytosol to the nuclear envelope in rat basophilic leukemia cells stimulated with calcium ionophore or IgE/antigen. *Journal of Biological Chemistry, 270*(25), 15359–15367.

Grynkiewicz, G., Poenie, M., & Tsien, R. Y. (1985). A new generation of Ca^{2+} indicators with greatly improved fluorescence properties. *Journal of Biological Chemistry, 260*(6), 3440–3450.

Hanel, A. M., Schüttel, S., & Gelb, M. H. (1993). Processive interfacial catalysis by mammalian 85-kilodalton phospholipase A_2 enzymes on product-containing vesicles: Application to the determination of substrate preferences. *Biochemistry, 32*(23), 5949–5958.

Hefner, Y., Borsch-Haubold, A. G., Murakami, M., Wilde, J. I., Pasquet, S., Schieltz, D., ... Gelb, M. H. (2000). Serine 727 phosphorylation and activation of cytosolic phospholipase A_2 by MNK1-related protein kinases. *Journal of Biological Chemistry, 275*(48), 37542–37551.

Hirabayashi, T., Kume, K., Hirose, K., Yokomizo, T., Iino, M., Itoh, H., & Shimizu, T. (1999). Critical duration of intracellular Ca2+ response required for continuous translocation and activation of cytosolic phospholipase A2. *Journal of Biological Chemistry, 274*(8), 5163–5169.

Hoth, M., & Penner, R. (1992). Depletion of intracellular calcium stores activates a calcium current in mast cells. *Nature, 355*(6358), 353–356.

Kalinski, P. (2012). Regulation of immune responses by prostaglandin E_2. *Journal of Immunology, 188*(1), 21–28.

Kramer, R. M., Roberts, E. F., Manetta, J., & Putnam, J. E. (1991). The Ca^{2+}-sensitive cytosolic phospholipase A_2 is a 100-kDa protein in human monoblast U937 cells. *Journal of Biological Chemistry, 266*(8), 5268–5272.

Kramer, R. M., Roberts, E. F., Um, S. L., Börsch-Haubold, A. G., Watson, S. P., Fisher, M. J., & Jakubowski, J. A. (1996). p38 mitogen-activated protein kinase phosphorylates cytosolic phospholipase A_2 ($cPLA_2$) in thrombin-stimulated platelets. *Journal of Biological Chemistry, 271*(44), 27723–27729.

Kurebayashi, N., Harkins, A. B., & Baylor, S. M. (1993). Use of fura red as an intracellular calcium indicator in frog skeletal muscle fibers. *Biophysical Journal, 64*(6), 1934–1960.

Lambeau, G., & Gelb, M. H. (2008). Biochemistry and physiology of mammalian secreted phopholipases A2. *Annual Review of Biochemistry*, 77, 495–520.

Leslie, C. C. (2004). Regulation of the specific release of arachidonic acid by cytosolic phospholipase A_2. *Prostaglandins, Leukotrienes, and Essential Fatty Acids, 70*(4), 373–376.

Leslie, C. C., & Channon, J. Y. (1990). Anionic phospholipids stimulate an arachidonoyl-hydrolyzing phospholipase A_2 from macrophages and reduce the calcium requirement for activity. *Biochimica et Biophysica Acta, 1045*(3), 261–270.

Leslie, C. C., Voelker, D. R., Channon, J. Y., Wall, M. M., & Zelarney, P. T. (1988). Properties and purification of an arachidonoyl-hydrolyzing phospholipase A_2 from a macrophage cell line, RAW 264.7. *Biochimica et Biophysica Acta, 963*(3), 476–492.

Lin, L.-L., Wartmann, M., Lin, A. Y., Knopf, J. L., Seth, A., & Davis, R. J. (1993). cPLA$_2$ is phosphorylated and activated by MAP kinase. *Cell, 72*(2), 269–278.

Medkova, M., & Cho, W. (1998). Mutagenesis of the C2 domain of protein kinase C-alpha. Differential roles of Ca2+ ligands and membrane binding residues. *Journal of Biological Chemistry, 273*(28), 17544–17552.

Mosior, M., Six, D. A., & Dennis, E. A. (1998). Group IV cytosolic phospholipase A_2 binds with high affinity and specificity to phosphatidylinositol 4,5-bisphosphate resulting in dramatic increases in activity. *Journal of Biological Chemistry, 273*(4), 2184–2191.

Murray, D., & Honig, B. (2002). Electrostatic control of the membrane targeting of C2 domains. *Molecular Cell, 9*(1), 145–154.

Nakahata, N. (2008). Thromboxane A2: Physiology/pathophysiology, cellular signal transduction and pharmacology. *Pharmacology & Therapeutics, 118*(1), 18–35.

Nalefski, E. A., & Falke, J. J. (1996). The C2 domain calcium-binding motif: Structural and functional diversity. *Protein Science, 5*(12), 2375–2390.

Nalefski, E. A., & Falke, J. J. (1998). Location of the membrane-docking face on the Ca^{2+}-activated C2 domain of cytosolic phospholipase A_2. *Biochemistry, 37*(51), 17642–17650.

Nalefski, E. A., McDonagh, T., Somers, W., Seehra, J., Falke, J. J., & Clark, J. D. (1998). Independent folding and ligand specificity of the C2 calcium-dependent lipid binding domain of cytosolic phospholipase A_2. *Journal of Biological Chemistry, 273*(3), 1365–1372.

Nalefski, E. A., Slazas, M. M., & Falke, J. J. (1997). Ca^{2+}-signaling cycle of a membrane-docking C2 domain. *Biochemistry, 36*(40), 12011–12018.

Nalefski, E. A., Wisner, M. A., Chen, J. Z., Sprang, S. R., Fukuda, M., Mikoshiba, K., & Falke, J. J. (2001). C2 domains from different Ca2 + signaling pathways display functional and mechanistic diversity. *Biochemistry, 40*(10), 3089–3100.

Nemenoff, R. A., Winitz, S., Qian, N.-X., Van Putten, V., Johnson, G. L., & Heasley, L. E. (1993). Phosphorylation and activation of a high molecular weight form of phospholipase A_2 by p42 microtubule-associated protein 2 kinase and protein kinase C. *Journal of Biological Chemistry, 268*(3), 1960–1964.

Palmer, A. E., & Tsien, R. Y. (2006). Measuring calcium signaling using genetically targetable fluorescent indicators. *Nature Protocols, 1*(3), 1057–1065.

Perisic, O., Fong, S., Lynch, D. E., Bycroft, M., & Williams, R. L. (1998). Crystal structure of a calcium-phospholipid binding domain from cytosolic phospholipase A_2. *Journal of Biological Chemistry, 273*(3), 1596–1604.

Perisic, O., Paterson, H. F., Mosedale, G., Lara-González, S., & Williams, R. L. (1999). Mapping the phospholipid-binding surface and translocation determinants of the C2 domain from cytosolic phospholipase A_2. *Journal of Biological Chemistry, 274*(21), 14979–14987.

Peters-Golden, M., & Henderson, W. R., Jr. (2007). Leukotrienes. *New England Journal of Medicine, 357*(18), 1841–1854.

Peters-Golden, M., Song, K., Marshall, T., & Brock, T. (1996). Translocation of cytosolic phospholipase A_2 to the nuclear envelope elicits topographically localized phospholipid hydrolysis. *Biochemical Journal, 318*(3), 797–803.

Pompey, S. N., Michaely, P., & Luby-Phelps, K. (2013). Quantitative fluorescence co-localization to study protein-receptor complexes. *Methods in Molecular Biology, 1008*, 439–453.

Powell, W. S., & Rokach, J. (2013). The eosinophil chemoattractant 5-oxo-ETE and the OXE receptor. *Progress in Lipid Research*, *52*(4), 651–665.

Qiu, Z.-H., Gijón, M. A., de Carvalho, M. S., Spencer, D. M., & Leslie, C. C. (1998). The role of calcium and phosphorylation of cytosolic phospholipase A_2 in regulating arachidonic acid release in macrophages. *Journal of Biological Chemistry*, *273*(14), 8203–8211.

Reed, K. A., Tucker, D. E., Aloulou, A., Adler, D., Ghomashchi, F., Gelb, M. H., … Boutaud, O. (2011). Functional characterization of mutations in inherited human cPLA(2) deficiency. *Biochemistry*, *50*(10), 1731–1738.

Ricciotti, E., & Fitzgerald, G. A. (2011). Prostaglandins and inflammation. *Arteriosclerosis, Thrombosis, and Vascular Biology*, *31*(5), 986–1000.

Rittenhouse-Simmons, S., & Deykin, D. (1977). The mobilization of arachidonic acid in platelets exposed to thrombin or ionophore A23187. Effects of adenosine triphosphate deprivation. *Journal of Clinical Investigation*, *60*(2), 495–498.

Roe, M. W., Lemasters, J. J., & Herman, B. (1990). Assessment of Fura-2 for measurements of cytosolic free calcium. *Cell Calcium*, *11*(2–3), 63–73.

Rzigalinski, B. A., Blackmore, P. F., & Rosenthal, M. D. (1996). Arachidonate mobilization is coupled to depletion of intracellular calcium stores and influx of extracellular calcium in differentiated U937 cells. *Biochimica et Biophysica Acta*, *1299*(3), 342–352.

Schievella, A. R., Regier, M. K., Smith, W. L., & Lin, L.-L. (1995). Calcium-mediated translocation of cytosolic phospholipase A_2 to the nuclear envelope and endoplasmic reticulum. *Journal of Biological Chemistry*, *270*(51), 30749–30754.

Sciaky, N., Presley, J., Smith, C., Zaal, K. J., Cole, N., Moreira, J. E., … Lippincott-Schwartz, J. (1997). Golgi tubule traffic and the effects of brefeldin a visualized in living cells. *Journal of Cell Biology*, *139*(5), 1137–1155.

Shayman, J. A., Kelly, R., Kollmeyer, J., He, Y., & Abe, A. (2011). Group XV phospholipase A(2), a lysosomal phospholipase A(2). *Progress in Lipid Research*, *50*(1), 1–13.

Shehnaz, D., Torres, B., Balboa, M. A., & Insel, P. A. (2000). Pyridoxal-phosphate-6-azophenyl-2′,4′-disulfonate (PPADS), a putative P2Y(1) receptor antagonist, blocks signaling at a site distal to the receptor in Madin-Darby canine kidney-D(1) cells. *Journal of Pharmacology and Experimental Therapeutics*, *292*(1), 346–350.

Six, D. A., & Dennis, E. A. (2003). Essential Ca^{2+}-independent role of the group IVA cytosolic phospholipase A_2 C2 domain for interfacial activity. *Journal of Biological Chemistry*, *278*(26), 23842–23850.

Smith, W. L., & Langenbach, R. (2001). Why there are two cyclooxygenase isozymes. *Journal of Clinical Investigation*, *107*(12), 1491–1495.

Stafforini, D. M. (2009). Biology of platelet-activating factor acetylhydrolase (PAF-AH, lipoprotein associated phospholipase A2). *Cardiovascular Drugs and Therapy*, *23*(1), 73–83.

Stahelin, R. V., Rafter, J. D., Das, S., & Cho, W. (2003). The molecular basis of differential subcellular localization of C2 domains of protein kinase C-alpha and group IVa cytosolic phospholipase A_2. *Journal of Biological Chemistry*, *278*(14), 12452–12460.

Stewart, A., Ghosh, M., Spencer, D. M., & Leslie, C. C. (2002). Enzymatic properties of human cytosolic phospholipase A_2g. *Journal of Biological Chemistry*, *277*(33), 29526–29536.

Sutton, R. B., Davletov, B. A., Berghuis, A. M., Sudhof, T. C., & Sprang, S. R. (1995). Structure of the first C2 domain of synaptotagmin I: A novel Ca^{2+}/phospholipid-binding fold. *Cell*, *80*(6), 929–938.

Takahashi, A., Camacho, P., Lechleiter, J. D., & Herman, B. (1999). Measurement of intracellular calcium. *Physiological Reviews*, *79*(4), 1089–1125.

Tilley, S. L., Coffman, T. M., & Koller, B. H. (2001). Mixed messages: Modulation of inflammation and immune responses by prostaglandins and thromboxanes. *Journal of Clinical Investigation*, *108*(1), 15–23.

Tucker, D. E., Ghosh, M., Ghomashchi, F., Loper, R., Suram, S., St. John, B., ... Leslie, C. C. (2009). Role of phosphorylation and basic residues in the catalytic domain of cytosolic phospholipase $A_2\alpha$ in regulating interfacial kinetics and binding and cellular function. *Journal of Biological Chemistry*, *284*(14), 9596–9611.

Verdaguer, N., Corbalan-Garcia, S., Ochoa, W. F., Fita, I., & Gomez-Fernandez, J. C. (1999). Ca(2+) bridges the C2 membrane-binding domain of protein kinase Calpha directly to phosphatidylserine. *EMBO Journal*, *18*(22), 6329–6338.

Woodward, D. F., Jones, R. L., & Narumiya, S. (2011). International Union of Basic and Clinical Pharmacology. LXXXIII: Classification of prostanoid receptors, updating 15 years of progress. *Pharmocological Reviews*, *63*(3), 471–538.

Xing, M., Firestein, B. L., Shen, G. H., & Insel, P. A. (1997). Dual role of protein kinase C in the regulation of $cPLA_2$-mediated arachidonic acid release by P_{2U} receptors in MDCK-D_1 cells: Involvement of MAP kinase-dependent and -independent pathways. *Journal of Clinical Investigation*, *99*(4), 805–814.

CHAPTER FIVE

Secreted Phospholipase A_2 Specificity on Natural Membrane Phospholipids

K. Yamamoto[*,†,‡], Y. Miki[*], H. Sato[*], R. Murase[*], Y. Taketomi[*], M. Murakami[*,§,1]

[*]Lipid Metabolism Project, Tokyo Metropolitan Institute of Medical Science, Tokyo, Japan
[†]Faculty of Bioscience and Bioindustry, Tokushima University, Tokushima, Japan
[‡]PRIME, Japan Agency for Medical Research and Development, Tokyo, Japan
[§]AMED-CREST, Japan Agency for Medical Research and Development, Tokyo, Japan
[1]Corresponding author: e-mail address: murakami-mk@igakuken.or.jp

Contents

Abstract

The secreted phospholipase A_2 ($sPLA_2$) family contains 10 catalytically active isoforms. Current *in vitro* biochemical studies have shown that individual $sPLA_2$s have distinct substrate selectivity in terms of the polar head groups or sn-2 fatty acids of their substrate phospholipids. Importantly, transgenic or knockout mice for distinct $sPLA_2$s display non-overlapping phenotypes, arguing that they do act on different phospholipid substrates and mobilize unique lipid metabolites *in vivo*. In an effort to comprehensively understand lipid metabolism driven by individual $sPLA_2$s under pathophysiological conditions, we took advantages of mass spectrometric lipidomics technology to monitor the spatiotemporal changes in phospholipids (substrates) and products (fatty acids,

Methods in Enzymology, Volume 583
ISSN 0076-6879
http://dx.doi.org/10.1016/bs.mie.2016.09.007

lysophospholipids, and their metabolites) in tissues or cells of $sPLA_2$-transgenic or knockout mice. The *in vivo* lipidomic data were compared with the *in vitro* activity of recombinant $sPLA_2$s toward phospholipid mixtures extracted from the target tissues, cells, or extracellular membrane components on which $sPLA_2$s may intrinsically act. These approaches reveal that the overall tendency in *in vitro* assays using natural membranes is recapitulated in several *in vivo* systems, often with even more selective patterns of hydrolysis. In this chapter, we will summarize current understanding of the *in vivo* substrate specificity of $sPLA_2$s toward natural membrane phospholipids.

1. INTRODUCTION

Phospholipase A_2 (PLA_2) is a group of lipolytic enzymes that hydrolyze the sn-2 position of glycerophospholipids to yield fatty acids and lysophospholipids, which often act as precursors for a wide variety of lipid mediators (Fig. 1A). In contrast to group IVA cytosolic PLA_2 ($cPLA_2\alpha$), which plays a central role in arachidonic acid (AA) metabolism

A

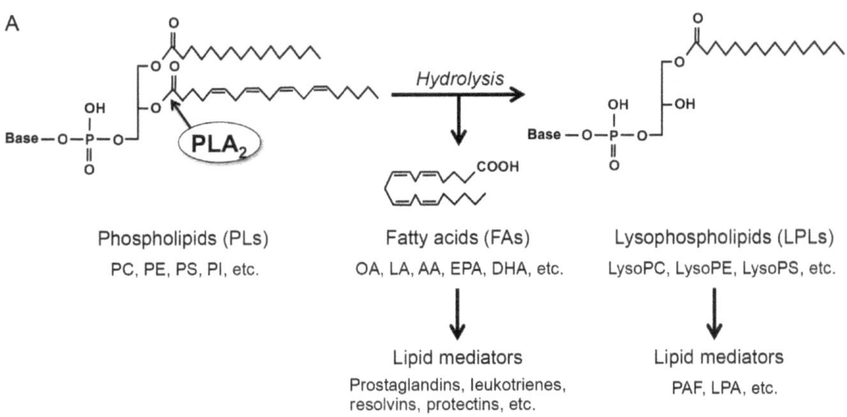

B

Mouse lines	Expected alterations			Advantages	Disadvantages
	PLs	FAs	LPLs		
TG mice	Decrease	Increase	Increase	Easy to monitor due to large changes	Not always reflecting physiological functions
KO mice	Increase	Decrease	Decrease	Often difficult to monitor due to small changes masked by high backgrounds	Likely reflecting physiological functions

Fig. 1 PLA_2 reaction and lipid dynamics in PLA_2 TG or KO mice. (A) PLA_2 reaction. (B) Expected alterations of substrates (phospholipids) and products (fatty acids, lysophospholipids, and their metabolites) in $sPLA_2$ TG or KO mice. *OA*, oleic acid; *LA*, linoleic acid; *PI*, phosphatidylinositol.

(Leslie, 1997; Ueno et al., 2011; Uozumi et al., 1997), the physiological roles and underlying lipid metabolisms driven by the secreted PLA_2 ($sPLA_2$) family, which comprises 10 catalytically active isoforms in mammals (Lambeau & Gelb, 2008; Murakami, Sato, Miki, Yamamoto, & Taketomi, 2015; Murakami et al., 2011), have remained obscure over the last few decades. In order to precisely understand the biological roles of $sPLA_2$s *in vivo*, it is important to consider when and where different $sPLA_2$s are expressed, which isoforms are involved in specific types of pathophysiology, and how the $sPLA_2$s exhibit their unique functions by driving specific types of lipid metabolism. Recent studies using mice gene-manipulated for $sPLA_2$s, in combination with lipidomics technology, have revealed that the enzymes have diverse functions, driving unique lipid pathways in response to specific extracellular microenvironmental cues (Duchez et al., 2015; Escoffier et al., 2010; Henderson et al., 2007; Labonte et al., 2006; Miki et al., 2013; Murase et al., 2016; Ohtsuki et al., 2006; Sato et al., 2011, 2010, 2014; Taketomi et al., 2013; Vijay et al., 2015; Yamamoto et al., 2015). With an absolute requirement for millimolar concentrations of Ca^{2+} for catalysis, $sPLA_2$s principally target phospholipids in extracellular spaces, such as adjacent cell membranes, microparticles, surfactant, lipoproteins, microbial membranes, and dietary lipids *in vivo* (Murakami et al., 2015, 2011). Here, we overview the methodology to determine the *in vivo* substrate specificity of $sPLA_2$s toward natural membrane phospholipids.

2. KEY FACTORS DETERMINING THE SUBSTRATE SPECIFICITY OF $sPLA_2$s *IN VIVO*

The *in vitro* enzyme activity of $sPLA_2$s could be influenced by the assay conditions employed, such as the composition of the substrate phospholipids, concentrations of $sPLA_2$s, presence of detergents, and pH. Hence, the enzymatic properties of individual $sPLA_2$s determined in different studies are not entirely identical. Since natural biological membranes contain numerous phospholipid species, results obtained using artificial phospholipid vesicles comprising one or a few phospholipid species may not always reflect the true enzymatic properties of a given $sPLA_2$. The mobilization of distinct lipids by $sPLA_2$s *in vivo* appears to rely on not only their intrinsic enzymatic properties, but also tissue- or disease-specific contexts such as the lipid composition of target membranes or the spatial and temporal availability of downstream lipid-metabolizing enzymes, which may account for why

$sPLA_2s$ exert distinct functions with different lipid mediator profiles in distinct settings. Theoretically, lipidomics analyses of the affected tissues in $sPLA_2$-knockout (KO) mice in comparison with wild-type (WT) mice could identify the specific substrates and products of individual $sPLA_2s$. Although this strategy would seem to be simple and straightforward, several issues need to be considered carefully.

First, it is necessary to discriminate whether the observed difference in lipid profiles between WT and KO mice is "direct" or "indirect," that is, whether the observed changes in lipid profiles arise due to the direct action of a given $sPLA_2$ or whether they merely reflect changes in the compensatory expression or function of other PLA_2s or lipases (particularly $cPLA_2\alpha$ in the case of eicosanoid profiles) in the ongoing process of pathology. Second, the high background levels or rapid turnover rates of lipids in the whole tissue may mask subtle and spatiotemporal changes in $sPLA_2$-driven lipid metabolism occurring in a local niche. Indeed, oxygenated fatty acid metabolites can be rapidly degraded or further converted to other metabolites due to their intrinsic instability, and lysophospholipids can be rapidly reacylated back to phospholipids. These issues often hinder the precise assignment of $sPLA_2$-dependent lipid metabolism *in situ*. Third, as a result of secondary effects, some phenotypes of $sPLA_2$ KO mice may be manifested in tissues where the enzyme does not exist. In such cases, lipidomics using the affected tissues does not necessarily mirror the true nature of the $sPLA_2$ enzymatic action, and the analysis needs to be performed using distal tissues where the enzyme is intrinsically expressed. Fourth, since $sPLA_2s$ often act on extracellular phospholipids supplied from tissue microenvironments, the action of $sPLA_2s$ cannot be evaluated adequately in culture using a single cell type. This approach may be applicable when the target phospholipids reside in $sPLA_2$-secreting cells themselves or in microvesicles released from them, but in cases where target membranes reside in, or are supplied by neighboring cells or other sources, a proper coculture system using $sPLA_2$-secreting cells and phospholipid-supplying cells is needed to evaluate the actions of $sPLA_2$ *in vitro*.

Application of lipidomics analysis to $sPLA_2$-overexpressing transgenic (TG) mice often provides informative insights into the potential substrate specificity of $sPLA_2s$ *in vivo* (Murase et al., 2016; Ohtsuki et al., 2006; Sato et al., 2014; Taketomi et al., 2013; Yamamoto et al., 2015). The overall phenotypes of TG mice for different $sPLA_2s$ are not entirely identical, implying that individual $sPLA_2s$ have distinct enzymatic properties, acting on different phospholipids and mobilizing different lipid metabolites *in vivo*.

In this situation, TG mice are useful for screening potential substrates and products driven by any given $sPLA_2$ in specific tissues, since lipid mobilization in $sPLA_2$-TG mice is typically large and easy to monitor using lipidomics approaches. However, the results obtained from TG mice should be interpreted with caution, as super-physiological levels of $sPLA_2$ could cause artificial phenotypes even in tissues or cells where the enzyme is not expressed endogenously. Nonetheless, if TG mice for a certain $sPLA_2$ display a particular phenotype opposite to that observed in KO mice for the same $sPLA_2$, we might conclude that this phenotype does reflect the intrinsic role of this specific $sPLA_2$ (Murase et al., 2016; Sato et al., 2014; Taketomi et al., 2013; Yamamoto et al., 2015). It is thus possible to identify $sPLA_2$-dependent lipid pathways by searching in KO mice for decreased lipid metabolites (fatty acids, lysophospholipids, or their derivatives) whose levels are reciprocally increased in TG mice (Fig. 1B).

As an alternative approach, examination of the *in vitro* hydrolysis of natural membrane phospholipids extracted from the affected tissues by recombinant $sPLA_2$, particularly at a low (reasonably physiologically relevant) concentration of the enzyme, could provide an additional clue to the *in vivo* substrates and products of this $sPLA_2$ (Miki et al., 2013; Yamamoto et al., 2015). The profiles of fatty acids and lysophospholipids released by a given $sPLA_2$ from natural membrane phospholipids can be influenced by the intrinsic substrate specificity of the enzyme as well as the tissue-specific phospholipid composition. In the latter context, for instance, if the proportion of AA relative to other fatty acids in phospholipids is larger, it could be expected that the release of AA by $sPLA_2$ would be apparently greater, whereas the release of eicosapentaenoic acid (EPA) would be generally small, as the EPA content of phospholipids is low in most tissues. Nevertheless, although the substrate specificity of $sPLA_2$s differs according to the *in vitro* assay conditions employed, particularly when excess amounts of the enzymes are used (where many phospholipid species, even non-physiological substrates, can be hydrolyzed), the overall tendency in *in vitro* assays using natural membranes is recapitulated in several *in vivo* systems, often with even more selective patterns of hydrolysis that are relevant to the results of studies using KO and/or TG mice.

3. METHODOLOGIES OF LIPIDOMICS ANALYSIS

Herein, we describe the procedure for lipidomics analysis using electrospray ionization-mass spectrometry (ESI-MS) to determine the lipid-metabolic pathways driven by $sPLA_2$s *in vivo*. Lipidomics is an approach suitable for obtaining structural information about individual phospholipid molecular species and for acquiring quantitative profiles using a combination of liquid chromatography (LC) and MS. The amounts of samples used for LC-MS depend on the estimated contents of lipids in target tissues or cells, as well as the sensitivity of the mass spectrometer. Phospholipids are generally the most abundant lipids in cells, whereas the amounts of fatty acid-oxygenated metabolites in tissues and cells are very low. To assess all lipids from a single sample, the lipids are extracted from given tissues or cells using different methods and analyzed under three distinct LC-MS conditions. The entire process is illustrated in Fig. 2.

3.1 Sample Preparation

Although most phospholipid and lysophospholipid species can be extracted from biological aqueous samples with chloroform/methanol, the yields of acidic lysophospholipids, such as lysophosphatidic acid, lysophosphatidylserine, and lysophosphatidylinositol, are low unless the aqueous phase is acidified. However, acidification can cause undesirable degradation of plasmalogen-type phospholipids. To avoid this decomposition problem during the extraction process, we use a simple multistep extraction procedure that allows rapid analysis of phospholipid and lysophospholipid molecular species. Animal tissues or cells should be

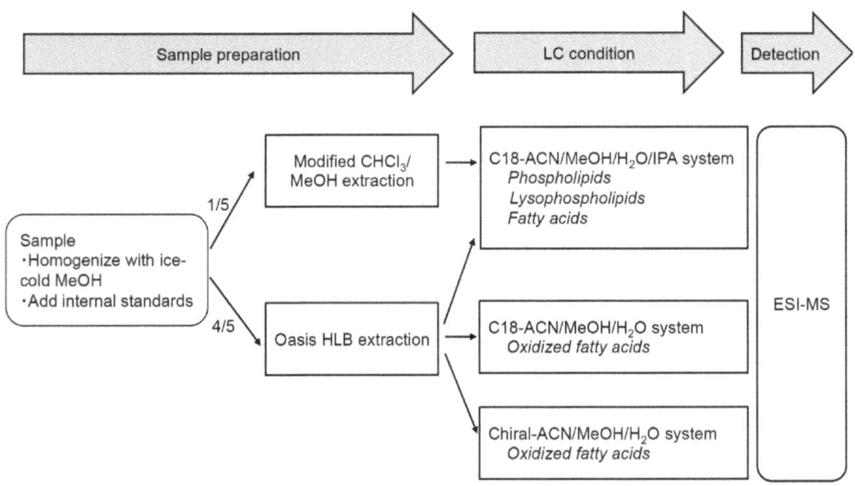

Fig. 2 Schematic flowchart of ESI-MS lipidomics. For details, please see the text. *ACN*, acetonitrile; *MeOH*, methanol; *$CHCl_3$*, chloroform; *IPA*, 2-propanol.

quickly frozen in liquid N_2 and stored at −80°C to avoid oxidation and degradation. Tissues are soaked in 10 volumes of ice-cold methanol and then homogenized with a Polytron homogenizer or powerful beads crusher. Cell pellets are thawed on ice in 1 mL of ice-cold methanol. To calibrate the recovery of each lipid throughout the procedure, a mixture of internal standards, such as d5-labeled EPA, d4-labeled prostaglandin (PG) E_2, and lysophosphatidylcholine (LysoPC) with a 17:0 fatty acyl chain [LysoPC(17:0)], is added to each sample just before lipid extraction. The amounts of internal standards are determined according to the sensitivity and calibration range of the system.

For extraction of phospholipids and lysophospholipids, one-fifth of the homogenized sample is added to 10 volumes of 20 m*M* Tris–HCl (pH 7.4). Lipids are extracted by adding four volumes of chloroform/methanol (2/1; v/v), followed by two volumes of 20 m*M* Tris–HCl (pH 7.4)-saturated chloroform/methanol (2/1; v/v). Each extraction step is followed by vortexing for 60 s and centrifugation for 5 min at 2000 rpm at 4°C, to resolve the organic and aqueous phases. The lower organic phases from the first and second neutral extraction steps are pooled and dried up under a stream of N_2 gas. The remaining aqueous phase is acidified with ice-cold 0.3 M citric acid solution to pH 3–4 (checked by spotting a small aliquot onto a pH paper) and then extracted twice with ice-cold chloroform/methanol (2/1; v/v) saturated with 20 m*M* Tris–HCl (pH 7.4). After vigorous vortexing, the organic phases are pooled and neutralized to pH 6–7 (checked using a pH paper as above) with 10% (v/v) triethylamine in chloroform/methanol (2/1; v/v) on ice. The neutralized organic phase from the acidic extraction is combined with the initial neutral-extracted fraction and dried up under N_2 gas. The resulting lipid film is dissolved in chloroform/methanol (2/1; v/v) and stored at −20°C until use.

For extraction of fatty acid metabolites, ice-cold 0.1% (v/v) hydrochloric acid in water is added to the remaining four-fifths of the sample to give a final methanol concentration of 10% (v/v). The sample in 10% methanol is applied to an Oasis HLB cartridge (Waters, Milford, MA), washed with hexane, eluted with methyl formate, dried up under N_2 gas, and stored at −80°C until use.

3.2 Liquid Chromatography

Reverse-phase LC with a C18 column (Kinetex C18, 2.1 × 150 mm, 1.7 μm, Phenomenex, Torrance, CA) is used for separation of fatty acids, fatty acid-oxygenated metabolites, lysophospholipids, and phospholipids.

Due to the diverse hydrophobicity of these lipids, gradient elution from the LC is required for comprehensive analysis. Two solvent types, acetonitrile/methanol/water/2-propanol and acetonitrile/methanol/water, can be used, depending on the target panel of lipids. For analyses of fatty acids, lysophospholipids, and phospholipids, the samples are applied to the column and separated by a step gradient including mobile phase A [acetonitrile/methanol/water = 1/1/1 (v/v/v) containing 5 μ*M* phosphoric acid and 1 m*M* ammonium formate] and mobile phase B (2-propanol containing 5 μ*M* phosphoric acid and 1 m*M* ammonium formate) at a flow rate of 0.2 mL/min at 50°C. For analyses of fatty acid metabolites, the samples are applied to the column and separated by a step gradient including mobile phase C (water containing 0.1% acetic acid) and mobile phase D (acetonitrile/methanol = 4/1; v/v) at a flow rate of 0.2 mL/min at 45°C. To distinguish the chirality (R or S) of the oxidized lipids, chiral HPLC analysis is carried out using a Chiralpak IA-3 column (4.6 × 150 mm i.d., 3-μm particles, Daicel, Tokyo, Japan) with mobile phase E [acetonitrile/methanol/water = 68/17/15 (v/v/v) containing 0.1% acetic acid] at a flow rate of 0.2 mL/min at 25°C.

3.3 Detection of Lipids by ESI-MS

Output from the LC is directly connected to ESI-MS, which is considered to be a soft ionization method widely used for lipidomics. The MS/MS fragmentation patterns of individual lipids can be used for their selective detection and identification (Fig. 3A). For example, the collision-induced dissociation spectrum for phosphatidylethanolamine (PE) with stearic acid and docosahexaenoic acid (DHA) [PE(18:0/22:6)] provides a deprotonated ion of $m/z = 790.5$. The ion of $m/z = 327.2$ represents a specific fatty acid (DHA; C22:6) for selection, which is suitable for obtaining structural information on each phospholipid molecular species. To monitor and quantify the levels of targeted lipids, a multiple reaction monitoring method (MRM) (also known as selected reaction monitoring) was devised using signature ion fragments for each molecule. This method is capable of monitoring hundreds of compounds in a single run. Parameters for compound detection including declistering potential and collision energy, major factors that affect the sensitivity of MRM detection are optimized using authentic compounds. Typical chromatograms obtained using our system with a 4000Q-TRAP quadrupole-linear ion trap hybrid mass spectrometer (AB Sciex, Framingham, MA) are shown in Fig. 3B, in which phospholipid,

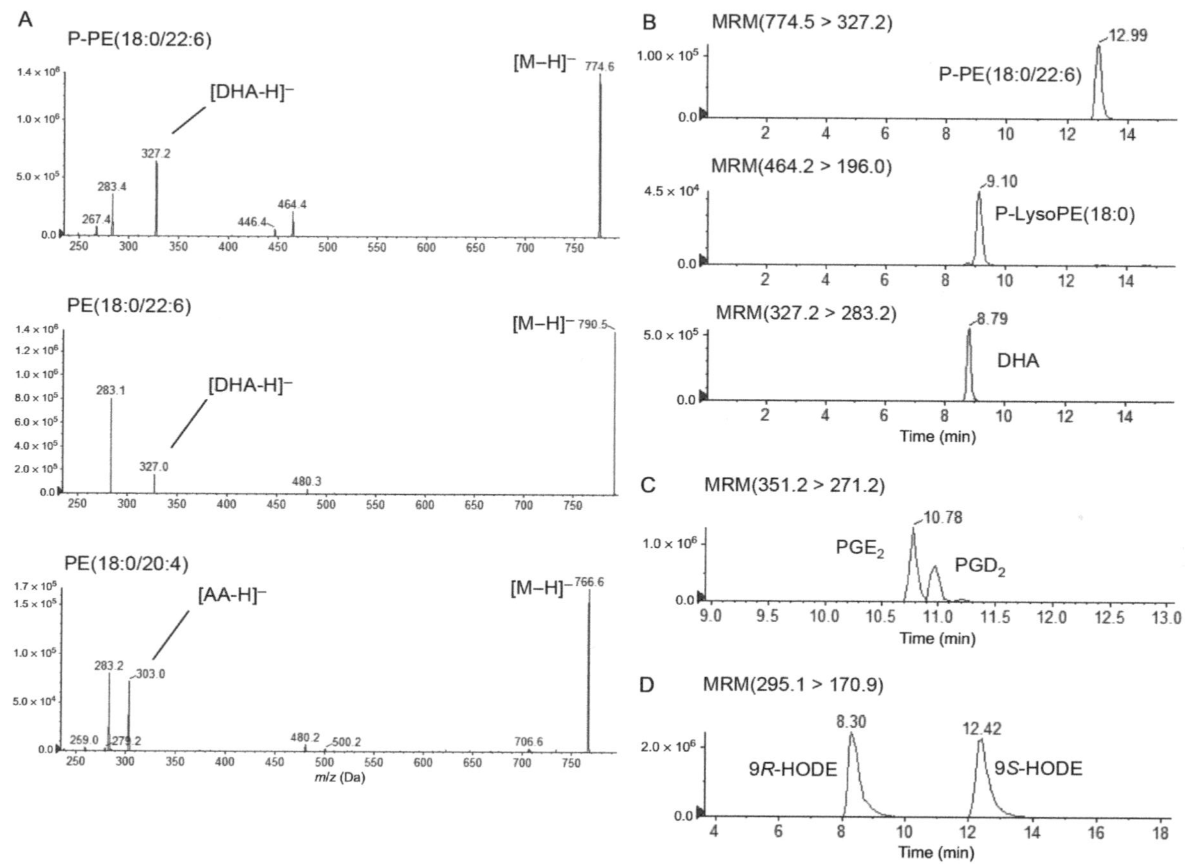

Fig. 3 LC-MS detection. (A) MS/MS spectra for P-PE(18:0/22:6), PE(18:0/22:6), and PE(18:0/20:4). The fragment peak of a specific fatty acid is used for identification by MRM. (B) Chromatographic profiles of P-PE(18:0/22:6), P-LysoPE(18:0), and DHA in a C18-acetonitrile/methanol/water/2-propanol system. (C) Chromatographic resolution of PGE_2 and PGD_2 in a C18-acetonitrile/methanol/water system. (D) Chromatograms for 9R-HODE and 9S-HODE obtained by chiral-phase chromatography. For details, see text.

lysophospholipid, and fatty acid are separated in a single run. PGD_2 and PGE_2 ($m/z = 351.2$ for both in the negative ESI mode) do not give specific MS/MS fragments to distinguish, requiring chromatographic separation for sorting (Fig. 3C). The deprotonated ions for 9R-hydroxyoctadecaenoic acid (HODE) and 9S-HODE share the same fragment (m/z of $295.1 > 170.9$), but show different chromatographic retention times according to their chiral configurations (Fig. 3D). Deuterated fatty acids or eicosanoids and lysophospholipids with an odd-numbered fatty acid are commercially available and can be used as internal standards for normalization. The lipid contents of samples are calibrated using authentic lipids as standards.

3.4 PLA_2 Reaction

Total lipids are extracted from mouse tissues by the method described earlier and further purified by straight-phase chromatography. The samples extracted in chloroform are applied to a Sep-Pak Silica Cartridge (Waters), washed sequentially with acetone and chloroform/methanol (9/1; v/v), eluted with chloroform/methanol (3/1; v/v), and dried up under N_2 gas. The amounts of total phospholipids in the samples are determined by the inorganic phosphorus assay (Eaton & Dennis, 1976). The membrane mimic composed of tissue-extracted lipids (typically 1–10 μ*M*) is sonicated for 5 min in 100 m*M* Tris–HCl (pH 7.4) containing 4 m*M* $CaCl_2$, and then incubated for appropriate periods with various concentrations of recombinant $sPLA_2$ (typically 10–100 ng) at 37°C. After incubation, the lipids were mixed with internal standards, extracted, and subjected to LC-MS for detection of phospholipids (substrate), fatty acids, and lysophospholipids (products) as noted earlier.

4. EXAMPLES FOR DETERMINATION OF THE $sPLA_2$-INTRINSIC LIPID PATHWAYS

We now describe several examples of the unique lipid pathways directed by $sPLA_2$s *in vivo*, as revealed by the approaches described earlier using $sPLA_2$ gene-manipulated mice in combination with lipidomics analysis.

4.1 $sPLA_2$-IIF Generates Lysoplasmalogen in the Epidermis

$sPLA_2$-IIF is expressed in terminally differentiated keratinocytes in the suprabasal epidermis and markedly increased in psoriatic skin (Yamamoto et al., 2015). $sPLA_2$-IIF TG mice spontaneously developed psoriasis-like epidermal hyperplasia and alopecia, with preferential hydrolysis of

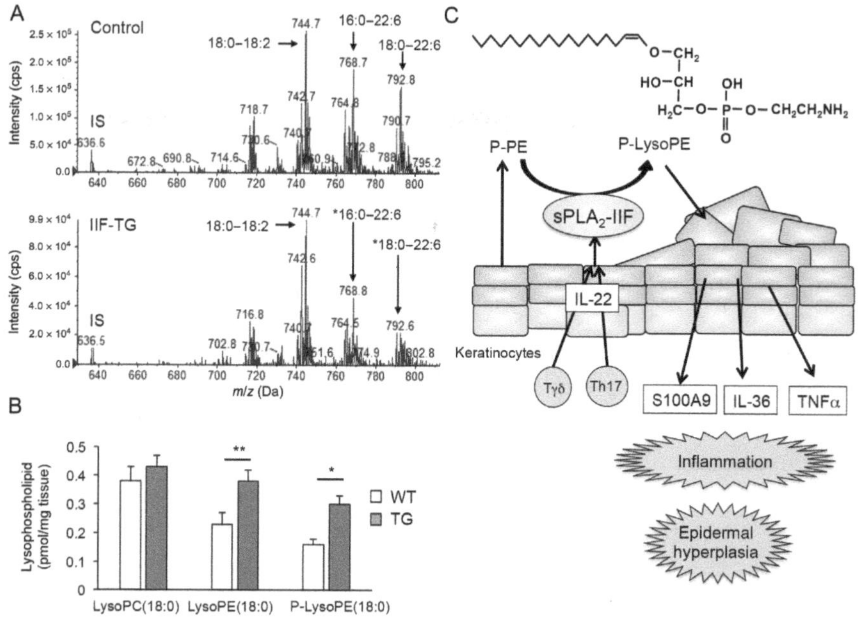

Fig. 4 sPLA$_2$-IIF releases P-LysoPE in hyperplastic epidermis. (A) Representative ESI-MS profiles of PE molecular species in the skins of sPLA$_2$-IIF TG and WT mice. Peaks for DHA (22:6)-containing PE species are selectively decreased in sPLA$_2$-IIF TG mice. (B) Quantification of lysophospholipids in the skins of sPLA$_2$-IIF TG and WT mice. LysoPE and P-LysoPE, but not LysoPC, are increased in TG mice (mean ± SEM, $n = 5$, $^*p < 0.05$, and $^{**}p < 0.01$). (C) A schematic diagram of the sPLA$_2$-IIF-driven lipid pathway in psoriasis.

DHA-containing PE to give rise to acyl and plasmalogen (P-) forms of lysophosphatidylethanolamine (LysoPE) molecular species in addition to DHA (Fig. 4A and B). Conversely, sPLA$_2$-IIF-deficient mice were protected from epidermal hyperplasia in psoriasis, contact dermatitis, and skin cancer, where P-LysoPE (ethanolamine lysoplasmalogen) was selectively reduced in the KO skin relative to WT skin. Moreover, when the phospholipid mixture extracted from mouse skin was used as a substrate, recombinant sPLA$_2$-IIF selectively hydrolyzed DHA-bearing P-PE (ethanolamine plasmalogen) to yield DHA and P-LysoPE at a physiological concentration (Yamamoto et al., 2015). Importantly, topical application of lysoplasmalogen, but not DHA-derived protectin D1, restored the defective epidermal hyperplasia in the skin of sPLA$_2$-IIF-deficient mice. In summary, sPLA$_2$-IIF preferentially hydrolyzes plasmalogen secreted from keratinocytes to produce lysoplasmalogen, which in turn promotes the hyperproliferation and activation of keratinocytes, leading to aggravation of epidermal-hyperplasic disorders (Fig. 4C).

4.2 $sPLA_2$-IID Mobilizes ω3 PUFA-Derived Lipid Mediators in the Lymph Nodes

$sPLA_2$-IID is preferentially and abundantly expressed in dendritic cells in lymphoid organs (Miki et al., 2013), suggesting its role in adaptive immunity. In the elicitation phase of contact hypersensitivity, a useful model of Th1-driven inflammation, the resolution, but not propagation, of skin inflammation was delayed in $sPLA_2$-IID-deficient mice. Although the phenotype became apparent in the skin, the $sPLA_2$-IID-driven lipid pathway responsible for this phenotype, as revealed by lipidomics, was assigned to the lymph nodes in which $sPLA_2$-IID is abundantly expressed, rather than to the skin in which its expression is very low. In draining lymph nodes, the levels of PUFAs, including AA (ω6), EPA, and DHA (ω3), as well as their oxygenated antiinflammatory metabolites, such as 15-deoxy-prostaglandin J_2 (15d-PGJ_2), 18-hydroxypentaenoic acid (18-HEPE), and resolvin D1 (RvD1), were reduced in $sPLA_2$-IID-deficient mice relative to WT mice, the reduction of the ω3 lipid products being more prominent (Fig. 5A). When natural membrane phospholipids extracted from the lymph nodes were used as substrates, recombinant $sPLA_2$-IID selectively hydrolyzed PE with AA and DHA, with a tendency to prefer the latter, in preference to other phospholipid species (Miki et al., 2013). Together, $sPLA_2$-IID preferentially hydrolyzes AA- or DHA-containing PE in lymph nodes, thereby mobilizing AA- and particularly DHA-derived antiinflammatory lipid mediators that attenuate the Th1-mediated immunity (Fig. 5B). Structures of representative EPA/DHA-derived antiinflammatory lipid mediators coupled with $sPLA_2$-IID are illustrated in Fig. 5C.

4.3 $sPLA_2$-X Releases ω3 PUFAs in the Colon and Spermatozoa

$sPLA_2$-X is a relatively well-studied $sPLA_2$ isoform. In most studies, however, the mechanistic action of $sPLA_2$-X had been attributed only to AA metabolism, whereas its ability to release ω3 PUFAs remained poorly understood. Physiologically, $sPLA_2$-X is abundantly expressed in the colorectal epithelium and sperm acrosomes (Escoffier et al., 2010; Murase et al., 2016; Sato et al., 2011). Our study using $sPLA_2$-X-deficient mice in combination with lipidomics revealed that $sPLA_2$-X did release ω3 PUFAs *in vivo* (Fig. 6A), thereby protecting against dextran sulfate-induced colitis and promoting sperm fertility (Murase et al., 2016). In colitis, $sPLA_2$-X-driven ω3 PUFAs attenuated the Th17 immune response partly through the fatty acid receptor GPR120, and intrarectal administration of ω3 PUFAs (EPA and DHA)

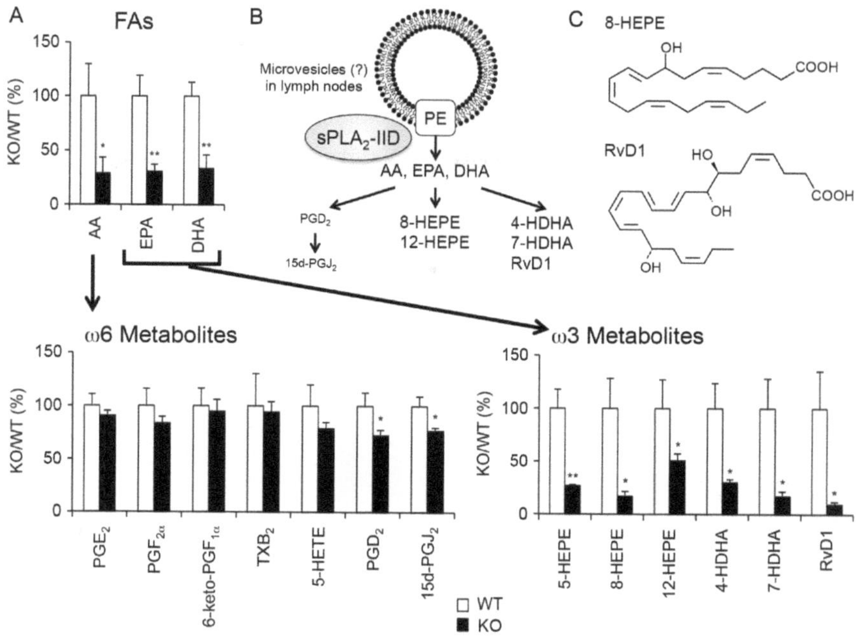

Fig. 5 sPLA$_2$-IID is preferentially coupled with ω3 PUFA metabolites. (A) In the elicitation phase of contact hypersensitivity, the lymph node levels of PUFAs (AA, EPA, and DHA) are markedly lower in sPLA$_2$-IID KO mice (*solid bars*) than in WT mice (*clear bars*). Among the PUFA metabolites, ω6 AA-derived PGD$_2$ and its downstream metabolite 15d-PGJ$_2$ are slightly and selectively decreased, whereas ω3 EPA/DHA-derived metabolites are markedly and globally reduced, in KO mice relative to WT mice (mean ± SEM, $n = 10$, $*p < 0.05$, and $**p < 0.01$). (B) A schematic diagram of the sPLA$_2$-IID-driven lipid pathway in lymph nodes. (C) Structures of 18-HEPE (hydroxyeicosapentaenoic acid) and RvD1.

rescued sPLA$_2$-X-deficient mice from the colitis. Conversely, sPLA$_2$-X TG mice displayed systemic immunosuppressive phenotypes, accompanied by marked elevation of free ω3 PUFAs and their metabolites (Fig. 6B). Moreover, sPLA$_2$-X secreted from sperm acrosomes selectively hydrolyzed DHA- or docosapentaenoic acid (DPA)-containing PC in the sperm membrane to release DHA, DPA, and LysoPC, and the defective fertilization ability of sPLA$_2$-X KO sperm was corrected by these lipids (DPA in particular), suggesting the importance of the sPLA$_2$-X-driven DPA for successful fertilization (Murase et al., 2016). Overall, these observations establish a previously unrecognized role of sPLA$_2$-X as another ω3 PUFA mobilizer (in addition to sPLA$_2$-IID; see earlier) *in vivo* (Fig. 6C). Distinct lines of evidence suggest that sPLA$_2$-X mobilizes ω6 AA-derived eicosanoids in the asthmatic lung (Henderson et al., 2007), suggesting that the mobilization of ω3 vs ω6 PUFAs by sPLA$_2$-X occurs in tissue- or disease-specific contexts.

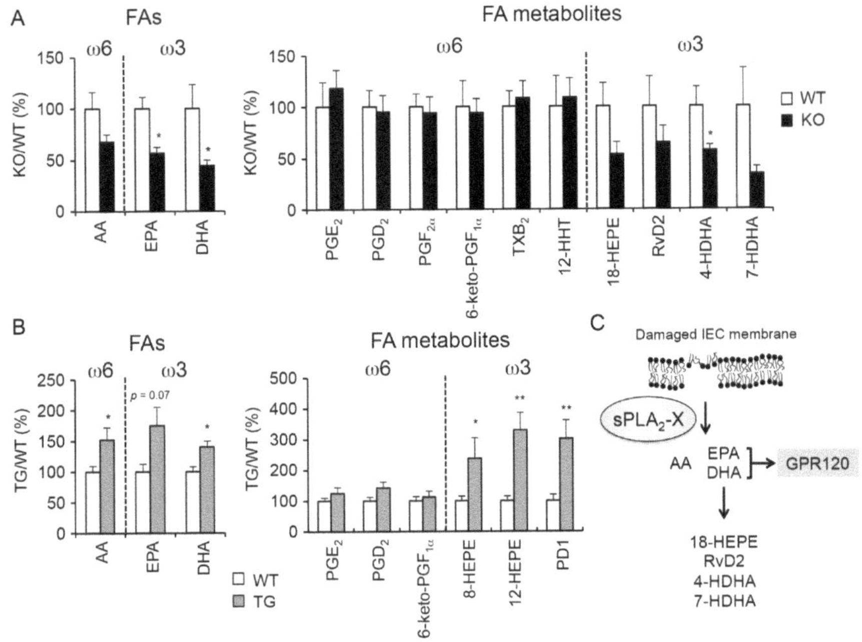

Fig. 6 sPLA$_2$-X is preferentially coupled with ω3 PUFA metabolites. (A) In colitis, the colon levels of PUFAs (AA, EPA, and DHA) are substantially lower in sPLA$_2$-X KO mice (*solid bars*) than in WT mice (*clear bars*). Among the PUFA metabolites, several ω3 EPA/DHA-derived metabolites, rather than ω6 AA-derived metabolites, are consistently reduced in KO mice relative to WT mice. (B) The splenic levels of AA, EPA, and DHA are all elevated in sPLA$_2$-X TG mice (*gray bars*) over WT mice (*clear bars*). Among the metabolites, ω3 metabolites are increased in preference to ω6 eicosanoids in KO mice relative to WT mice. (C) A schematic diagram of the sPLA$_2$-X-driven lipid pathway in colitis. *IEC*, intestinal epithelial cells. Values are mean ± SEM ($n = 5$, $^{*}p < 0.05$, and $^{**}p < 0.01$) in (A and B).

4.4 sPLA$_2$-IIE Hydrolyzes Lipoprotein Phospholipids in Obesity

In mice, sPLA$_2$-IIE is robustly induced in adipocytes during diet-induced obesity (Sato et al., 2014). When fed a high-fat diet, sPLA$_2$-IIE-deficient mice displayed lower weight gain, adiposity, and fatty liver, accompanied by alterations in plasma lipoprotein profiles, suggesting that adipocyte-inducible sPLA$_2$-IIE targets phospholipids in lipoproteins. In lipoprotein particles, PC is a dominant phospholipid component (>90%), whereas other phospholipids such as PE and phosphatidylserine (PS) are present only at trace levels. Lipidomics of phospholipids extracted from lipoproteins in sPLA$_2$-IIE-deficient mice in comparison with WT mice revealed that sPLA$_2$-IIE preferentially hydrolyzed PE and PS, with no fatty acid selectivity (Fig. 7A and B). Thus, the hydrolysis of minor lipoprotein

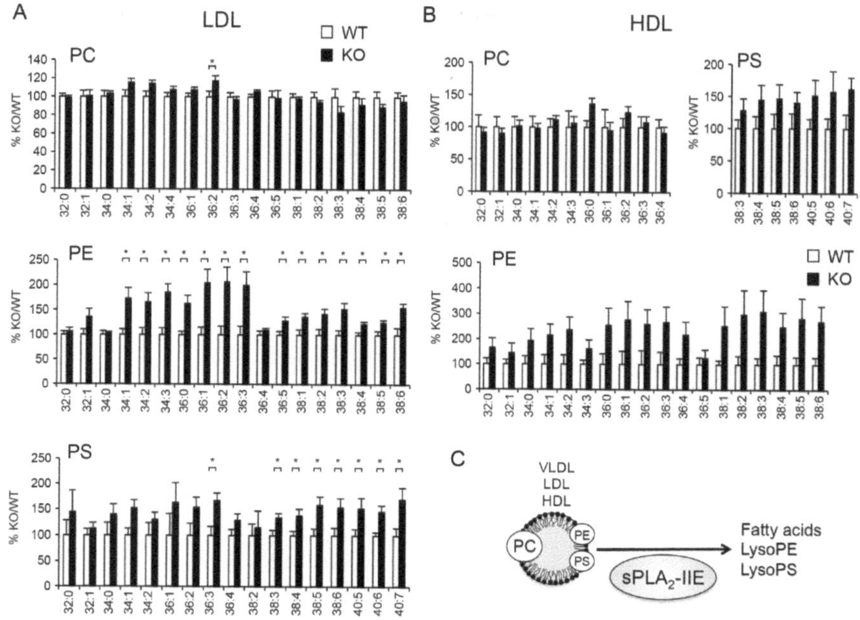

Fig. 7 sPLA$_2$-IIE preferentially hydrolyzes PE and PS in lipoproteins. (A and B) Phospholipid (PC, PE, and PS) profiles in LDL (A) and HDL (B) of sPLA$_2$-IIE KO and WT mice ($n=5$, mean ± SEM, $*p<0.05$). PE and PS molecular species in both lipoprotein particles are increased in sPLA$_2$-IIE KO mice relative to WT mice. (C) A schematic diagram of the sPLA$_2$-IIE-driven lipid pathway in lipoprotein metabolism during obesity.

phospholipids, PE and PS, by sPLA$_2$-IIE secreted from adipocytes may alter lipoprotein-mediated lipid delivery to adipose tissue and liver, thereby affecting diet-induced metabolic syndrome (Fig. 7C). It is also possible that the sPLA$_2$-IIE-driven LysoPE or LysoPS in adipose tissue may participate in the regulation of systemic metabolic states as potential lipid mediators, although this possibility awaits future studies.

5. CONCLUDING REMARKS

Recent advances in lipidomics technology, together with accumulated data obtained from sPLA$_2$ gene-manipulated mice, have allowed us to identify the lipid-metabolic pathways regulated by sPLA$_2$s in given tissue microenvironments. Individual sPLA$_2$s hydrolyze different phospholipids and mobilize distinct fatty acid or lysophospholipid metabolites in specific tissues. In the future, it will be important to translate the information obtained from mouse studies to humans. Hopefully, understanding the

$sPLA_2$-driven lipid pathways in distinct pathophysiological situations will be helpful for development of novel drugs that target each $sPLA_2$ or underlying lipid products.

ACKNOWLEDGMENTS

This work was supported by grants-in aid for Scientific Research from the Ministry of Education, Culture, Sports, Science and Technology of Japan 15K14957, 16H02613 (to M.M.), 26461671 (to K.Y.), 15K07959 (to H.S.), 25460087 (to Y.T.), and 26860051 (to Y.M.), and by AMED-CREST (to M.M.) and PRIME (to K.Y.) from the Agency for Medical Research and Development.

REFERENCES

Duchez, A. C., Boudreau, L. H., Naika, G. S., Bollinger, J., Belleannee, C., Cloutier, N., et al. (2015). Platelet microparticles are internalized in neutrophils via the concerted activity of 12-lipoxygenase and secreted phospholipase A_2-IIA. *Proceedings of the National Academy of Sciences of the United States of America*, *112*, E3564–E3573.

Eaton, B. R., & Dennis, E. A. (1976). Analysis of phospholipase C (*Bacillus cereus*) action toward mixed micelles of phospholipid and surfactant. *Archives of Biochemistry and Biophysics*, *176*, 604–609.

Escoffier, J., Jemel, I., Tanemoto, A., Taketomi, Y., Payre, C., Coatrieux, C., et al. (2010). Group X phospholipase A_2 is released during sperm acrosome reaction and controls fertility outcome in mice. *Journal of Clinical Investigation*, *120*, 1415–1428.

Henderson, W. R., Jr., Chi, E. Y., Bollinger, J. G., Tien, Y. T., Ye, X., Castelli, L., et al. (2007). Importance of group X-secreted phospholipase A_2 in allergen-induced airway inflammation and remodeling in a mouse asthma model. *Journal of Experimental Medicine*, *204*, 865–877.

Labonte, E. D., Kirby, R. J., Schildmeyer, N. M., Cannon, A. M., Huggins, K. W., & Hui, D. Y. (2006). Group 1B phospholipase A_2-mediated lysophospholipid absorption directly contributes to postprandial hyperglycemia. *Diabetes*, *55*, 935–941.

Lambeau, G., & Gelb, M. H. (2008). Biochemistry and physiology of mammalian secreted phospholipases A_2. *Annual Review of Biochemistry*, *77*, 495–520.

Leslie, C. C. (1997). Properties and regulation of cytosolic phospholipase A_2. *Journal of Biological Chemistry*, *272*, 16709–16712.

Miki, Y., Yamamoto, K., Taketomi, Y., Sato, H., Shimo, K., Kobayashi, T., et al. (2013). Lymphoid tissue phospholipase A_2 group IID resolves contact hypersensitivity by driving antiinflammatory lipid mediators. *Journal of Experimental Medicine*, *210*, 1217–1234.

Murakami, M., Sato, H., Miki, Y., Yamamoto, K., & Taketomi, Y. (2015). A new era of secreted phospholipase A_2. *Journal of Lipid Research*, *56*, 1248–1261.

Murakami, M., Taketomi, Y., Miki, Y., Sato, H., Hirabayashi, T., & Yamamoto, K. (2011). Recent progress in phospholipase A_2 research: From cells to animals to humans. *Progress in Lipid Research*, *50*, 152–192.

Murase, R., Sato, H., Yamamoto, K., Ushida, A., Nishito, Y., Ikeda, K., et al. (2016). Group X secreted phospholipase A_2 releases ω3 polyunsaturated fatty acids, suppresses colitis, and promotes sperm fertility. *Journal of Biological Chemistry*, *291*, 6895–6911.

Ohtsuki, M., Taketomi, Y., Arata, S., Masuda, S., Ishikawa, Y., Ishii, T., et al. (2006). Transgenic expression of group V, but not group X, secreted phospholipase A_2 in mice leads to neonatal lethality because of lung dysfunction. *Journal of Biological Chemistry*, *281*, 36420–36433.

Sato, H., Isogai, Y., Masuda, S., Taketomi, Y., Miki, Y., Kamei, D., et al. (2011). Physiological roles of group X-secreted phospholipase A_2 in reproduction, gastrointestinal phospholipid digestion, and neuronal function. *Journal of Biological Chemistry*, *286*, 11632–11648.

Sato, H., Taketomi, Y., Isogai, Y., Miki, Y., Yamamoto, K., Masuda, S., et al. (2010). Group III secreted phospholipase A_2 regulates epididymal sperm maturation and fertility in mice. *Journal of Clinical Investigation*, *120*, 1400–1414.

Sato, H., Taketomi, Y., Ushida, A., Isogai, Y., Kojima, T., Hirabayashi, T., et al. (2014). The adipocyte-inducible secreted phospholipases PLA2G5 and PLA2G2E play distinct roles in obesity. *Cell Metabolism*, *20*, 119–132.

Taketomi, Y., Ueno, N., Kojima, T., Sato, H., Murase, R., Yamamoto, K., et al. (2013). Mast cell maturation is driven via a group III phospholipase A_2-prostaglandin D_2-DP1 receptor paracrine axis. *Nature Immunology*, *14*, 554–563.

Ueno, N., Taketomi, Y., Yamamoto, K., Hirabayashi, T., Kamei, D., Kita, Y., et al. (2011). Analysis of two major intracellular phospholipases A_2 (PLA_2) in mast cells reveals crucial contribution of cytosolic $PLA_2\alpha$, not Ca^{2+}-independent $PLA_2\beta$, to lipid mobilization in proximal mast cells and distal fibroblasts. *Journal of Biological Chemistry*, *286*, 37249–37263.

Uozumi, N., Kume, K., Nagase, T., Nakatani, N., Ishii, S., Tashiro, F., et al. (1997). Role of cytosolic phospholipase A_2 in allergic response and parturition. *Nature*, *390*, 618–622.

Vijay, R., Hua, X., Meyerholz, D. K., Miki, Y., Yamamoto, K., Gelb, M. H., et al. (2015). Critical role of phospholipase A_2 group IID in age-related susceptibility to severe acute respiratory syndrome-CoV infection. *Journal of Experimental Medicine*, *212*, 1851–1868.

Yamamoto, K., Miki, Y., Sato, M., Taketomi, Y., Nishito, Y., Taya, C., et al. (2015). The role of group IIF-secreted phospholipase A_2 in epidermal homeostasis and hyperplasia. *Journal of Experimental Medicine*, *212*, 1901–1919.

CHAPTER SIX

Analyses of Calcium-Independent Phospholipase A_2beta (iPLA$_2\beta$) in Biological Systems

S.E. Barbour*, S. Ramanadham†,1
*University of Georgia at Athens, Athens, GA, United States
†Comprehensive Diabetes Center, University of Alabama at Birmingham, Birmingham, AL, United States
[1]Corresponding author: e-mail address: sramvem@uab.edu

Contents

Abstract

The Ca^{2+}-independent phospholipases A_2 (iPLA$_2$s) are part of a diverse family of PLA$_2$s, manifest activity in the absence of Ca^{2+}, are ubiquitous, and participate in a variety of biological processes. Among the iPLA$_2$s, the cytosolic iPLA$_2\beta$ has received considerable attention and ongoing studies from various laboratories suggest that dysregulation of iPLA$_2\beta$ can have a profound impact on the onset and/or progression of many diseases (e.g., cardiovascular, neurological, metabolic, autoimmune). Therefore, appropriate approaches are warranted to gain a better understanding of the role of iPLA$_2\beta$ *in vivo* and its contribution to pathophysiology. Given that iPLA$_2\beta$ is very labile, its basal expression is low in a number of cell systems, and that crystal structure of iPLA$_2\beta$ is not yet available, careful and efficient protocols are needed to appropriately assess iPLA$_2\beta$ biochemistry, dynamics, and membrane association. Here, step-by-step details are provided to (a) measure iPLA$_2\beta$-specific activity in cell lines or tissue preparations (using a simple radiolabel-based assay) and assess the impact of stimuli and inhibitors on resting- and disease-state iPLA$_2\beta$ activity, (b) purify the iPLA$_2\beta$ to near homogeneity (via sequential chromatography) from cell line or tissue preparations, enabling concentration of the enzyme for subsequent analyses (e.g., proteomics), and (c) employ hydrogen/deuterium exchange mass spectrometry analyses to probe both the structure of iPLA$_2\beta$ and dynamics of its association with the membranes, substrates, and inhibitors.

Methods in Enzymology, Volume 583
ISSN 0076-6879
http://dx.doi.org/10.1016/bs.mie.2016.09.001

1. INTRODUCTION

The Ca^{2+}-independent phospholipases A_2 (iPLA$_2$s) are members of a diverse family of PLA$_2$s that hydrolyze the *sn*-2 substituent from membrane phospholipids to release a free fatty acid and a lysophospholipid (Dennis, Cao, Hsu, Magrioti, & Kokotos, 2011; Gijon & Leslie, 1997). The iPLA$_2$s comprise Group VI PLA$_2$s and, in contrast to s(ecretory)PLA$_2$s and c(ytosolic)PLA$_2$s, do not require Ca^{2+} for either translocation to membrane or activity. Between 1994 and 2016, the Group VI PLA$_2$s have expanded to seven members: iPLA$_2\beta$ (VIA-1 and 2), iPLA$_2\gamma$ (VIB), iPLA$_2\delta$ (VIC), iPLA$_2\varepsilon$ (VID), iPLA$_2\zeta$ (VIE), and iPLA$_2\eta$ (VIF). Due to their shared homology with patatin, the iPLA$_2$s are included in the patatin-like protein family and also referred to as PNPLAs. The iPLA$_2$s also share a consensus GXSXG catalytic motif contained within a patatin-like lipase domain. The iPLA$_2$s manifest a variety of activities in addition to PLA$_2$, are ubiquitously expressed, and participate in a multitude of biological processes including fat catabolism, cell differentiation, maintenance of mitochondrial integrity, phospholipid remodeling, cell proliferation, signal transduction, and cell death. As such, dysregulation of iPLA$_2$s can influence the metabolic state, CNS function, cardiovascular performance, and cell survival and therefore have a profound impact on onset/progression of many diseases. To date, the best-characterized iPLA$_2$s are iPLA$_2\beta$, localized predominantly in the cytosol, and membrane-associated iPLA$_2\gamma$ (Mancuso, Jenkins, & Gross, 2000).

The iPLA$_2\beta$ (PNPLA9) is the most widely described of the iPLA$_2$s and expression of its activity was first described in P388D1 macrophage-like cells in 1994 (Ackermann, Kempner, & Dennis, 1994) and later shown to be the same enzyme (Balboa, Balsinde, Jones, & Dennis, 1997) as that cloned from Chinese hamster ovary cells in 1997 (Balboa et al., 1997; Jones et al., 1996; Tang et al., 1997). Unlike cPLA$_2$, which exhibits preference for hydrolysis of arachidonic acid from the *sn*-2 position of glycerophospholipids (Ghosh, Tucker, Burchett, & Leslie, 2006), the iPLA$_2$s do not demonstrate *sn-2* substrate specificity. The iPLA$_2$s manifest PLA$_2$/PLA$_1$, lysophospholipase (Lio & Dennis, 1998; Wolf & Gross, 1996), transacylase (Jenkins et al., 2004; Lio & Dennis, 1998), and thioesterase (Carper, Zhang, Turk, & Ramanadham, 2008; Jenkins, Yan, Mancuso, & Gross, 2006) activities.

1.1 iPLA$_2\beta$ Characteristics

The iPLA$_2\beta$ is an 84–88 kDa cytosolic protein with a serine lipase consensus sequence (GTSGT) in its catalytic domain that is preceded by eight

N-terminal ankyrin repeats (Gross, Ramanadham, Kruszka, Han, & Turk, 1993; Ma et al., 1997; Tang et al., 1997). The 88 kDa isoform is a product of a mRNA species that arises from an exon-skipping mechanism of alternative pre-mRNA splicing (Larsson, Claesson, & Kennedy, 1998) and contains a 54-amino acid sequence that interrupts the eighth ankyrin repeat. The iPLA$_2$β protein contains a caspase-3 cleavage site (DVTD), a putative bipartite nuclear localization sequence (KREFGEHTKMTDVKKPK), and, upon stimulation, can associate with multiple subcellular-localized proteins and mobilize into various subcellular organelles (Golgi, endoplasmic reticulum [ER], mitochondria, and nucleus).

To date, there are two recognized catalytic activators of iPLA$_2$β (ATP and calmodulin kinase IIβ) and one of its transcription (sterol regulatory element-binding protein). Inhibitors of iPLA$_2$β include arachidonyl trifluoromethyl ketone (AACOCF$_3$), methyl arachidonyl fluorophosphonate (MAFP), and palmitoyl trifluoromethyl ketone (PACOCF$_3$), which are sometimes used for "selective" inhibition of cPLA$_2$. In contrast, the bromoenol lactone (BEL) suicide inhibitor has been demonstrated to be a more selective inhibitor of iPLA$_2$ with little or no effect on cPLA$_2$ or sPLA$_2$ (Hazen, Zupan, Weiss, Getman, & Gross, 1991; Jenkins et al., 2004; Ma, Ramanadham, Hu, & Turk, 1998). Further, the *S*- and *R*-enantiomers of BEL exhibit selective potency for iPLA$_2$β and iPLA$_2$γ, respectively (Jenkins, Han, Mancuso, & Gross, 2002), and they have been used to distinguish biological processes impacted by the two isoforms. Fluoroketone- and oxadiazole-based compounds currently under development are proving to be just as potent as BEL, while being more specific for iPLA$_2$β and exhibiting reversible inhibition, without discernible toxicity (Ali et al., 2013; Dennis et al., 2011; Kalyvas et al., 2009; Kokotos et al., 2010; Li et al., 2011; Lopez-Vales et al., 2011, 2008; Mouchlis, Limnios, et al., 2016; Ong, Farooqui, Kokotos, & Farooqui, 2015).

2. EXPERIMENTAL PROCEDURES

2.1 iPLA$_2$β Activity Assay

Described is a facile assay that will provide rapid, selective, and quantifiable measurement of iPLA$_2$β-specific activity in a given preparation, without requiring purification of the enzyme.

Solutions needed

1. Homogenization buffer (HB, 250 m*M* sucrose, 40 m*M* Tris–HCl, pH 7.1 @ 4°C)
2. Assay buffer (AB, 200 m*M* Tris–HCl, pH 7.5 @ 37°C, 10 m*M* EGTA)
3. EGTA (10 m*M*)

4. ATP (Sigma, St. Louis, MO, A-5394; 1 or 10 m*M*)
5. *S*-BEL inhibitor (Cayman 10006801, prepare 10 m*M* stock in DMSO and store at −20°C in small aliquots. Do not reuse thawed aliquot. Final concentrations [0.10–10.0 μ*M*] should be contained in less than 1% by volume of DMSO.)
6. Radiolabeled substrate Perkin Elmer (NEC765010UC, L-α-1-palmitoyl-2-arachidonyl-*sn*-phosphatidylcholine [arachidonyl-1-^{14}C], or 16:0a/20:4-PC or PAPC, 2.5–5 μ*M* in 5 μL EtOH)
7. Oleate standard (Sigma 143-19-1; 5 mg/mL $CHCl_3$)
8. TLC solvent (petroleum ether/ethyl ether/acetic acid, 80/20/1 by volume). This solvent composition resolves fatty acids (R_f 0.58) from monoacylglycerol (R_f 0.24) and diacylglycerol (R_f 0.21)
9. 100% ethanol
10. 1-Butanol
11. Scintillation cocktail
12. Glycerol
13. β-Mercaptoethanol

Specific supplies/equipment needed

1. Centrifuge
2. Cold microfuge
3. Shaking water bath
4. Scintillation counter
5. Vortex
6. Stopwatch
7. Compressed nitrogen tank
8. Hair dryer
9. 10 × 75 mm glass assay tubes (best for viewing separation of organic and aqueous phases)
10. TLC tank
11. Silica gel G channeled/scored TLC plates (Analtech Uniplate 31711)
12. Iodine (Sigma 376558)
13. Glass Hamilton syringes (50 μL for substrate stock, 10 μL for diluted substrate for assay, 50 μL for TLC spotting; keep these separate and do not interchange.)
14. Repeater pipet for reagent additions (i.e., HB, AB, EGTA, butanol)
15. Scraper blade
16. 3″ × 5″ weighing paper
17. Scintillation vials (7 mL)
18. Scintillation cocktail (RPI 3a70B or equivalent)

Basic assay concept. iPLA$_2\beta$ in a given sample will hydrolyze the *sn*-2 (radiolabeled) fatty acid substituent from the provided substrate under zero-Ca^{2+} (no added Ca^{2+} + EGTA) conditions. The released fatty acid is then resolved by TLC, recovered for scintillation counting, and dpm used to determine specific enzyme activity.

Assay protocol (Fig. 1)

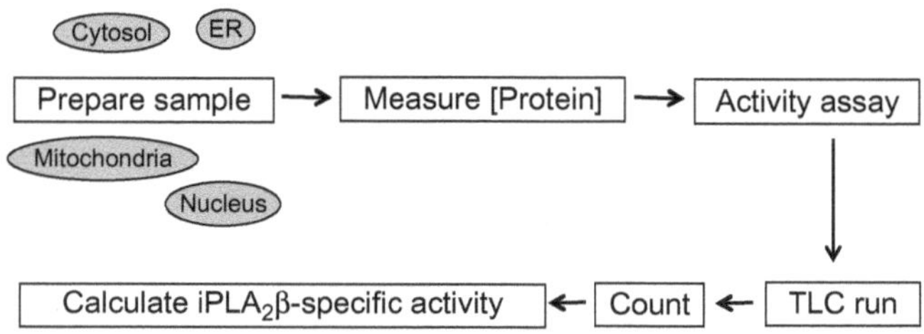

Fig. 1 iPLA$_2\beta$ activity assay flow.

1. Using your established methods, prepare sample (cytosolic, mitochondrial, ER, or nuclear fractions) from cells or tissues in HB . If only Ca^{2+}-independent PLA$_2$ activity is to be measured, EGTA can be included in the HB. Otherwise, it can be added to activity assay tubes. At all times, keep the sample preparation, assay reagents, and substrate on ice.
2. Measure protein concentration in the sample. Approximately 25–50 μg protein in aliquots of less than 50 μL should be the target and the volume of HB used in Step 1 volume should be titrated accordingly. For cells in 100 mm dish, use 200 μL HB for harvesting. For tissues, use sufficient volume (<1 mL) to allow efficient sonication and adjust the volume as needed during assay development. It is best to prepare a concentrated sample, which can be diluted subsequent to [protein] determination.
3. During sample preparation
   a. *Water shaking bath.* Turn on and equilibrate to 37°C.
   b. *Set up the activity assay tubes.* An initial protocol to establish the presence of iPLA$_2\beta$ activity in a sample will include assays in the absence and presence of ATP (1 or 10 m*M*, activator of iPLA$_2\beta$) or *S*-BEL (0.1–10.0 μ*M*, selective irreversible inhibitor of iPLA$_2\beta$). A typical assay template is as follows (tubes 2–5 are singlicates for a given sample; 6 on are for additional sample sets):
   c. *Prepare the substrate fresh each time.* PAPC is a common choice for the radiolabeled substrate, although 1-palmitoyl-2-palmitoyl-*sn*-

Tube #	Protein (μL)	HB (μL)	dH_2O (μL)	AB (μL)	EGTA (μL)	ATP (μL)	*S*-BEL (μL)	Substrate (μL)	TV (μL)
1 (blank)	0	100	35	50	10	0	0	5	200
2	Total = 100		35	50	10	0	0	5	200
3	Total = 100		-ATP (μL)	50	10	10	0	5	200
4	Total = 100		-BEL (μL)	50	10	0	10	5	200
5	Total = 100		-ATP -BEL (μL)	50	10	10	10	5	200
6	Total = 100		35	50	10	0	0	5	200
.	Total = 100		35	50	10	.	.	.	200
.	Total = 100		35	50	10	.	.	.	200
.	Total = 100		35	50	10	.	.	.	200
N (blank)	0	100	35	50	10	0	0	5	200

glycero-3-phosphocholine (DPPC), 1-palmitoyl-2-lauroyl-*sn*-glycero-3-phosphocholine (PLPC), and 1-*O*-(*Z*)-hexadec-1′-enyl-2-[9,10-$^3H_2$]octadec-9′-enoyl-*sn*-glycero-3-phosphocholine [(16:0p/18:1)-PC] are also suitable. The latter substrate is a plasmalogen phospholipid, containing a vinyl ether linkage in the *sn*-1 position, and is an extremely favored substrate by iPLA$_2$β. However, it can be more expensive (to buy or synthesize) and very labile. In initial experiments, one can use the final substrate concentration of 2.5–10 μ*M*. For a [^{14}C]-labeled substrate, 100,000 dpm in 5 μL is added to the assay tube and this is typically contained in 1 μL of the stock substrate. After determining the number (*N*) of assay tubes (including replicates), transfer ($N+2$) μL of stock substrate into a clean 12 × 75 mm glass tube using the Hamilton syringe designated for stock (highest counts). Commercial radiolabeled substrates are usually prepared in chloroform or methanol. Using a nitrogen stream, evaporate the solvent in the tube and then resuspend the lipid in $5 \times (N+2)$ μL of ethanol. This will provide an extra 10 μL of substrate. Count 2 μL prior to assay to ensure that it contains ~40,000 dpm (=100,000 in 5 μL) using a second Hamilton syringe (second most counts). At the end of the assay, during sample counting, count 5 μL of the substrate to confirm total added dpm to assay tubes and for calculation of % substrate hydrolyzed by iPLA$_2$β in the sample aliquot.

d. *Scintillation vials*. Set aside the required number (*N*) of scintillation vials with 5 mL scintillation cocktail. Number the caps. Scintillation cocktail splashing will wipe off markings on the tube!!

e. *TLC plates*. Number the TLC plates at the top of each lane with a pencil (Fig. 2, "x" at top of lanes). If the assay requires more than one plate, make sure to include a blank for each plate. If less than one plate is sufficient, snap off extra lanes and save.

f. *Oleate standard*. Place a drop (~10 μL) of oleic acid solution into the middle of the loading area in each lane (Fig. 2, lane 2). Set aside and let air-dry. Free fatty acid will comigrate with the oleic acid standard, which will be visible upon exposure to iodine vapor.

g. *TLC tank*. In a fume hood, prepare the TLC tank by inserting a U-shaped double-folded Whatman paper into the tank. Place solvent (100 mL) at bottom, cover the container, and apply a weight on top. Monitor solvent migration upward on the Whatman filter paper. This preparation is good for multiple runs (four plates) on the same day as long as the tank is kept air-tight. You can run TLC plates back to back or by using paper click separators, up to four plates at one time.

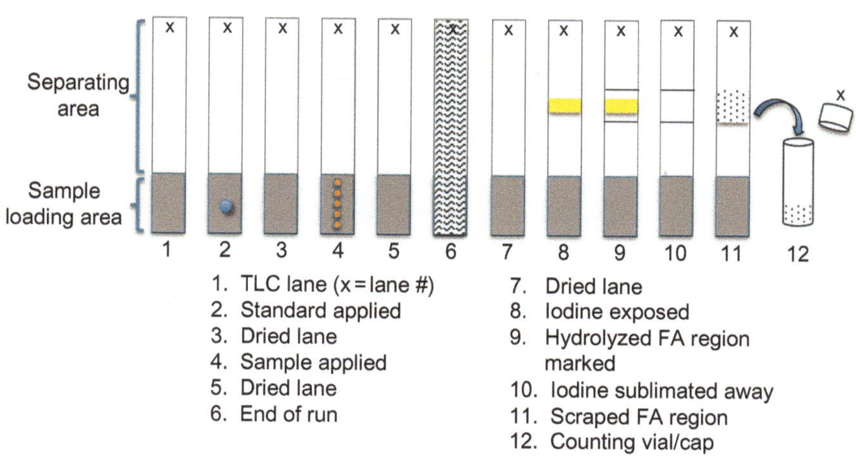

Fig. 2 Recovery of hydrolyzed fatty acid for quantitation.

   h. *Iodine vapor tank.* In a fume hood, add a few crystals of iodine to a separate tank and cover. This tank can be used repeatedly for the same purpose until the iodine is exhausted.

4. *Preparing the assay tubes* (done at room temperature [RT], unless otherwise noted)
   a. To all tubes, first add AB and EGTA.
   b. Second, add HB to all tubes. To the blank tubes, add 100 μL. For sample tubes, HB = 100 μL minus protein volume. Initial experiments should determine optimal protein amount that allows for linear accumulation of product during the course of the assay (to be contained in less than 50 μL).
      ⇒ Subsequent steps need to be performed rapidly.
   c. Third, add source of protein in HB.
   d. If testing ATP, add prior to substrate.
   e. If testing *S*-BEL, add after Step c and incubate at RT for 3 min. Include this in the time frame described in *Step 6*.
   f. When adding ATP and/or *S*-BEL, reduce the dH_2O volume by same amount.
5. *Assay.* Start the assay by adding substrate to tube. Prior to first addition, remove dead space in the second Hamilton syringe by flushing it with substrate two to three times (do not discard the substrate, empty it back into tube). Now, take up 5 μL into the syringe, immerse it to the bottom of assay tube, and deliver the substrate. Quickly vortex the tube and add it to the shaking water bath. Dip the syringe in a clean tube of EtOH to wash off the tip (do not flush the syringe again), prior to taking up the next aliquot of substrate.
6. *Stopping the reaction.* At a predetermined time (*Step 6a*), add 100 μL butanol (RT) to the tube, remove it from water bath. Vortex for 5 s and set aside (RT) in a test tube rack.
   a. Initial experiments should test the optimal assay time (from 1 to 10 min), during which substrate hydrolysis is linear with respect to time. The iPLA$_2$β-catalyzed hydrolysis typically occurs rapidly, between 3 and 5 min.
   b. The time elapsed between substrate and butanol additions should equal the chosen assay time. For example, if a 3-min assay duration is chosen, start with substrate addition to tube 1 at 0 min, tube 2 at 15 s, tube 3 at 30 s, …, and tube 12 at 2:45 min. Starting at 3 min, stop reaction at 15 s intervals. This interval can be lengthened or

shortened depending on the individual. Keep in mind to include at least one blank to each separate set of assay runs.

7. *Separating the hydrolyzed FA.* After all assay tubes have been collected, centrifuge them (RT, 2000 × *g*, 5 min) to separate the organic (upper) from the aqueous phase. The interface will be a thin protein band. Continue with the next step without delay.
8. *Spotting TLC lanes.* Tilt the tube toward you and using the third Hamilton syringe (lowest counts) and without piercing the interface, carefully remove 25 μL (a fourth) of the upper layer. Gently apply this aliquot onto the loading area (Fig. 2, lane 3). Dry the lanes with the hair dryer and place in the TLC tank.
9. *Identifying hydrolyzed FA region.* After the solvent migrates to the top of the TLC plate, remove the plate from tank and allow residual solvent to evaporate in a fume hood. Place the TLC plate in the iodine tank. When the oleic acid standard is visualized (by appearance of a yellow streak across the plate, Fig. 2, lane 8), remove the plate from tank and mark the boundaries of the standard (Fig. 2, lane 9). Allow the iodine to sublime under hood using a hair dryer. At this point, the plate can be set aside for later processing.
10. *Harvesting the hydrolyzed FA.* Using a blade, scrape the marked off region (Fig. 2, lane 11) into a counting vial (Fig. 2, lane 12). Add scintillation cocktail to vial, cap, and mix vigorously by hand.
11. *Counting.* Allow an O/N quenching period and then count the samples in a ^{14}C channel and record decays/minute (dpm). If only counts/minute (cpm) are reported, calculate dpms using efficiency of the counter.
12. *Activity calculation.* Subtract averaged blank dpm from spotted sample (25% of total) dpm and use that value to determine Specific Enzyme Activity:

$$\frac{4 \times \text{DPM}}{\text{SA label}} \times \frac{1}{\text{mg prt} \times \text{assay time(min)}} \times 1000 = \text{pmol/mg protein} \times \text{min}$$

Keys to achieving optimal results

1. iPLA$_2$β is very labile and sensitive to increases in temperature for prolonged periods. Therefore, any preparation in which iPLA$_2$β activity is to be measured needs to be maintained on ice.

2. When using a sonicator to prepare protein sample, use 1 s ON and 0.5 s OFF pulse for 15–20 s (~20% power, once for cells/2–3 times for tissue homogenate). Keep the sample tube immersed in ice during sonication.
3. Due to lability of iPLA$_2$β, it is best to perform the assay soon after the sample has been generated.
4. Utilizing samples prepared from frozen cells or tissues will give activity measurements, but they will tend to be much lower than freshly prepared samples.
5. iPLA$_2$β is stabilized by ATP, even at 37°C. Thus, storing a sample with ATP addition will preserve the activity. However, ATP is an activator of iPLA$_2$β and the specific activity measured will be higher than in the absence of ATP. This will preclude comparisons between samples prepared and assayed right away in the absence of ATP.
6. Once the activity assay procedure has commenced, it is best to carry through until the TLC plate has been run. The scraping and counting can be done later.

2.2 iPLA$_2$β Purification

(A) Below are details on how to purify iPLA$_2$β starting from tissue (i.e., brain, skeletal muscle, heart, >25–50 g) or cultured cells (15–20 × 175 cm^2 flasks; Carper et al., 2008; Ramanadham, Wolf, Jett, Gross, & Turk, 1994; Wolf & Gross, 1996) when a homogeneous concentrated enzyme preparation is needed for subsequent analyses.

Buffers needed **(adjust pH of buffers first before adding DTT):**

1. HB (250 m*M* sucrose, 10 m*M* imidazole, 10 m*M* KCl, pH 7.8 @ 4°C).
2. Pre-DEAE (anion exchange) dialysis buffer (15 m*M* imidazole, 5 m*M* K_2HPO_4, 10% glycerol, pH 7.8 @ 4°C).
3. DEAE column elution buffer (10 m*M* imidazole, 10 m*M* KCl, 100 m*M* NaCl, 10% glycerol, 1 m*M* DTT, pH 8.0 @ 4°C).
4. Pre-CF (chromatofocusing) dialysis buffer (10 m*M* imidazole, 10 m*M* KCl, 25% glycerol, 1 m*M* DTT, pH 8.4 @ 4°C).
5. CF column elution buffer (10% polybuffer 96 [GE Healthcare Life Sciences, Pittsburgh, PA, 17-0714-01], 5% polybuffer 74 [BOC Sciences, Shirley, NY, 82658-85-3], 25% glycerol, 1 m*M* DTT [Calbiochem, San Diego, CA, 578517], pH 6.9 @ 4°C).
6. ATP column elution buffer (10 m*M* imidazole, 25% glycerol, 1 m*M* DTT, pH 8.3 @ 4°C).
7. Calmodulin column elution buffer (10 m*M* HEPES, 1 m*M* EDTA, 340 m*M* sucrose, 1 m*M* DTT, 1 m*M* Triton X-100, pH 7.5).

Equipment/supplies needed

1. Cold room
2. Buffer pump
3. UV monitor
4. Chromatography columns and reservoirs (GE Healthcare)
5. DEAE Sephacel (GE Healthcare 17-0500-01) and PBE-94 CF (Axsyn, Chesham, UK, AX-9NNB) resins
6. ATP Agarose (Sigma A-9264) and Calmodulin-Sepharose (GE Healthcare 17-0529-01) column resins
7. Glass wool (Fisher, Hampton, NH, 22–289336)
8. Spectra/Por 1 dialysis tube (Spectrum Laboratories Inc., Rancho Dominguez, CA, 23 mm × 100 ft [132650], 40 mm × 100 ft [132660], and 50 mm × 100 ft [132665]; 6–8 kDa molecular weight cutoff)
9. Filters for buffers (Millipore, Darmstadt, Germany, GSWP04700)
10. Nunc freezing vials (ThermoFisher Scientific, Waltham, NH, 374080 [0.50 mL] and 374081 [1 mL])
11. Activity assay needs, as described in Section 2.1.

Columns preparation prior to sample loading

1. *DEAE-Sephacel column*
   - **(i)** On Day 1, place resin (160 mL for a 5 × 8 cm column) in a Buchner funnel with filter and wash with 1 *M* NaCl, 100 m*M* Na Acetate, pH 4 (1 L).
   - **(ii)** Wash the resin with dH_2O (1 L).
   - **(iii)** Wash with *Buffer 2* (1.5 L).
     ⇒ During each wash/rinse, gently swirl the resin using a glass rod; apply gentle vacuum to keep filter in place, allow resin to settle for 5 min; apply vacuum to remove wash but do not let the resin dry.
   - **(iv)** Place washed resin in 500 mL *Buffer 2* in a filtering flask with side-arm. Degas and allow resin to settle at 4°C.
   - **(v)** Remove floating particles by aspiration and pour resin mixture into column in the cold room and equilibrate with *Buffer 2* (1.5–2 L, O/N, 1.5–2 mL/min).
     ⇒ While loading the DEAE column, prepare remaining buffers.
     ⇒ Cannot reuse this column, prepare fresh each time.
2. *CF Column* (can be prepared at the same time and similarly to the DEAE column)
   - **(i)** Wash CF resin (60 mL) with 1 *M* NaCl, 100 m*M* Na Acetate, pH 4 (500 mL).

(ii) Rinse resin with dH_2O (1 L).
(iii) Rinse resin with *Buffer 4* (1 L).
(iv) Degas resin in *Buffer 4* (200 mL) and allow resin to settle.
(v) Aspirate floating particles and load slurry into column.
(vi) Level the column and apply air pressure to generate flow rate of 1.6 mL/min.
(vii) After column has packed, pump *Buffer 4* at 1.6 mL/min for 15 min to pack the column and then at 1.0 mL/mL O/N through Day 2.
(viii) Allow *Buffer 4* containing 1 m*M* DTT to flow through column (2 L @ 1.6 mL/min O/N into Day 3).
⇒ Can reuse the column after washing with buffer containing 1 *M* NaCl, 100 m*M* Na Acetate, pH 4 (2 L, O/N). Then, wash with dH_2O and store column in 30% EtOH.

3. *ATP column*
(i) Apply ATP-Agarose affinity resin to column (1 × 1 cm).
(ii) Rinse with dH_2O to remove EtOH.
(iii) Wash with 1 *M* NaCl (30–50 mL, 4°C).
(iv) Rinse column from top to bottom exhaustively with dH_2O.
(v) Equilibrate with *Buffer 6* (20 mL) and plug bottom until sample loading. Make sure to leave some buffer in the column so the resin does not dry.
⇒ Can reuse column after washing with 1 *M* NaCl (30 mL) followed by dH_2O. Store column in 30% EtOH.

4. *Calmodulin column*
(i) Apply to column 1 mL of the Calmodulin-Sepharose slurry.
(ii) Wash column with 25 mL of 25 m*M* imidazole (pH 8.0).
(iii) Wash column with 25 mL of 25 m*M* imidazole (pH 8.0) containing 5 m*M* $CaCl_2$.
⇒ Do not reuse this column—prepare it fresh each time.

Purification protocol **(all purification columns to be performed at 4°C):**

Day 1

1. Prepare crude cytosolic fraction in *Buffer 1*, using your established protocol.
2. Dialyze the fraction overnight against *Buffer 2*, using dialysis tubing previously equilibrated in the same buffer at 4°C.

Day 2

3. Load dialyzed fraction onto DEAE column (1.5 × 8 cm). Pass the eluent by gravity through a UV monitor to record the absorbance.
4. Wash DEAE columns with *Buffer 2* containing 1 m*M* DTT, until the A280 returns to baseline.
5. Elute iPLA$_2$β activity (in 3–5 mL fractions) using *Buffer 3*.
6. Measure iPLA$_2$β activity, as a reflection of activity present in each fraction. This can be achieved by immediate counting (@ 0.5 min) and will allow proceeding to the next step without delay. The fractions can be recounted after overnight quenching.
7. Pool fractions with activity and dialyze overnight against *Buffer 4*, at 4°C.

Day 3

8. Apply the dialyzed pool to PBE-94 CF column (1 × 8 cm, 1.8 mL/min).
9. Wash the CF column with *Buffer 4*.
10. Elute iPLA$_2$β activity (in 3–5 mL fractions) using *Buffer 4*.
11. Measure iPLA$_2$β activity in the fractions and pool fractions with activity.
12. Immediately load pool onto a 1 × 1 cm N6-[(6-aminohexyl) carbamomethyl]-ATP-agarose column. Reapply void one additional time.
13. Wash column with *Buffer 6* (25–50 mL).
14. Wash the column with *Buffer 6*, *Buffer 6* containing 10 m*M* AMP (Sigma A-2252; 5 × 5 mL fractions).
15. Wash the column exhaustively with *Buffer 6* to remove the AMP.
16. Elute iPLA$_2$β with *Buffer 6* containing 1 m*M* ATP (Sigma A-5394; 0.5–1.0 mL fractions).
17. The AMP and ATP fractions can be saved at 4°C and assayed for iPLA$_2$β activity the following day.

Day 4

18. Pool the fractions with activity and increase [Ca^{2+}] in pool to 5 m*M*.
19. Load pool onto a Calmodulin-Sepharose column (1 × 1 cm).
20. Collect the void and reload onto the column.
21. Wash column with *Buffer 7* containing 0.5 m*M* $CaCl_2$.
22. Elute iPLA$_2$β with *Buffer 7* containing 8 m*M* EGTA (0.5–1.0 mL fractions).

23. Measure activity and pool fractions with activity.
24. Concentrate pool to dryness in the presence of 10% SDS by lyophilization.
25. Resuspend in small volume (based on subsequent assay protocols) of buffer containing 25 m*M* Tris–HCl (pH 6.8), 5% glycerol, 0.5% β-mercaptoethanol.
26. Store sample in Nunc cryotubes at −80°C in small aliquots and before use, thaw aliquots on ice. Use aliquot only once; do not refreeze.

(B) An abbreviated alterative protocol can be employed if less starting material is available or if low basal activity in the sample is predicted, as follows:

1. Sonicate sample in HB (10 m*M* HEPES, 1 *M* EDTA, 340 m*M* sucrose, 1 m*M* DTT, pH 7.5).
2. Centrifuge sonicate (1000 × *g*, 10 min, 4°C).
3. Precipitate protein in the supernatant with the addition of ammonium sulfate (50%).
4. Centrifuge mixture (20,000 × *g*, 30 min, 4°C).
5. Resuspend pellet in *Buffer* 7.
6. Proceed with ATP and Calmodulin Sepharose column elution steps as earlier.
7. Concentrate and store sample as described earlier.

Calculations

1. Prior to loading a column, save a small aliquot of the pool to determine total loaded protein/activity.
2. Record each column load volume.
3. Keep track of the fraction volumes collected.
4. These will allow subsequent determination of specific enzyme activity and fold purification following each step.

Keys to optimal purification yields

1. Perform all purification steps at 4°C (or cold room).
2. Do not let the columns dry.
3. Make sure to add DTT to elute activity from the DEAE column. Use fresh resin for each preparation.
4. If the DEAE Sephacel resin is mishandled, cracks will develop when poured into column or after sample is loaded. In that case, gently swirl the column section with crack.

5. The CF column is unforgiving and if the resin dries at any point after protein loading, fresh column and sample will need to be prepared!!
6. Do not delay in proceeding from the CF to ATP column steps.
7. Eluant from the ATP column is stable and can be assayed the following day. Store at 4°C until assay.
8. Do not use peristaltic pump to elute protein out of any column as iPLA$_2$β will stick to pump tubing and can be denatured by the pump action.

2.3 Membrane Association of iPLA$_2$β

Like many other members of the PLA$_2$ family, iPLA$_2$β is a water-soluble protein that must associate with lipid membranes to bind substrate. Indeed, the iPLA$_2$β catalytic cycle has been described as consisting of four steps: (1) membrane association, (2) extraction of a substrate molecule into the active site, (3) catalysis, and (4) diffusion of reaction products away from the active site, setting the stage for another round of catalysis (Batchu, Hokynar, Jeltsch, Mattonet, & Somerharju, 2015; Mouchlis, Bucher, McCammon, & Dennis, 2015; Mouchlis & Dennis, 2016). Although iPLA$_2$β-selective inhibitors have been identified (Ali et al., 2013; Dennis et al., 2011; Hazen et al., 1991; Jenkins et al., 2004; Kalyvas et al., 2009; Kokotos et al., 2010; Li et al., 2011; Lopez-Vales et al., 2011, 2008; Ma et al., 1998; Mouchlis, Limnios, et al., 2016; Ong et al., 2015), a more thorough understanding of the molecular mechanisms underlying the membrane association, substrate binding, and even catalytic mechanism could lead to development of more efficacious and potentially clinically relevant inhibitors.

Investigation of these mechanisms has been hampered because a high-resolution structure of iPLA$_2$β is not yet available. However, Dennis and coworkers have developed an elegant approach to gain insight into the structure of iPLA$_2$β and the molecular mechanisms underlying its association with membranes, substrates, and inhibitors that relies on the coordinated use of three approaches: selective enzyme assays (like the one described earlier), molecular dynamics, and hydrogen/deuterium exchange mass spectrometry (DXMS; Cao, Burke, & Dennis, 2013; Hamuro et al., 2004; Hsu et al., 2013; Hsu, Burke, Li, Woods, & Dennis, 2009; Mouchlis et al., 2015; Mouchlis & Dennis, 2016; Mouchlis, Limnios, et al., 2016; Mouchlis, Morisseau, et al., 2016). DXMS has been broadly used to study protein dynamics, protein–protein, and protein–ligand interactions (Hamuro et al., 2004; Percy, Rey, Burns, & Schriemer, 2012). In principle,

DXMS measures solvent accessibility of amide protons by quantifying their exchange for deuterium when a protein is incubated in D_2O. Although there are other potential explanations for hydrogen/deuterium exchange (i.e., changes in hydrogen bonding, oligomerization, distal changes in protein conformation, and interaction with ligand), DXMS can be informative about local changes in protein conformation when coupled with other structural information (Cao et al., 2013; Gallagher & Hudgens, 2016; Percy et al., 2012). This approach has led to mapping of conformational changes in the group IVA phospholipase A_2 (cPLA$_2\alpha$) and group IA phospholipase A_2 upon binding to membranes, substrates, inhibitors, and divalent cations (Burke et al., 2009, 2008; Cao et al., 2013; Hsu et al., 2008; Mouchlis et al., 2015; Mouchlis & Dennis, 2016). In all of these reports, the DXMS data were interpreted and confirmed through molecular dynamics simulations and comparisons to the high-resolution 3D structures of each enzyme.

As noted earlier, such data are not yet available for iPLA$_2\beta$. To circumvent this, homology modeling approaches were used to predict structures of the ankyrin repeat (residues 88–474) and catalytic (residues 475–806) domains of human iPLA$_2\beta$ (Hsu et al., 2009). A BLAST homology search revealed that the ankyrin domain of iPLA$_2\beta$ is 51% homologous to human ankyrin-R (PDB 1N11) and the catalytic domain is 34% homologous to the potato acyl hydrolase patatin (PDB 1OXW; Hsu et al., 2009; Mouchlis & Dennis, 2016). After applying a variety of modeling approaches (Bennett-Lovsey, Herbert, Sternberg, & Kelley, 2008; Hsu et al., 2009; Rost, Yachdav, & Liu, 2004), homology models were developed that exhibited template modeling (TM, a measure of the similarity of two protein structures) scores of 0.77 (ankyrin domain) and 0.76 (catalytic domain; Hsu et al., 2009). The predicted structures were validated with DXMS, performed as described later.

2.3.1 Hydrogen/DXMS

The DXMS protocol essentially consists of four phases: (1) incubation of enzyme (±substrate, inhibitor, source of "membrane" or a combination) in D_2O buffer for various periods of time (10–10,000 s); (2) quenching of the exchange (reduce to pH 2.5 and temperature to 0°C) and denaturation of the enzyme (guanidine hydrochloride); (3) online pepsin digestion followed by C18 reverse-phase HPLC of the resultant peptides; and (4) analyses of deuterated peptides through electrospray ionization tandem mass spectrometry (ESI-MS/MS). For additional detail on the theory, implementation, and optimization of DXMS, the reader is referred to two excellent reviews (Gallagher & Hudgens, 2016; Percy et al., 2012). The optimized protocol for analysis of iPLA$_2$β is described in detail later:

Phase 1. *Preparation of deuterated samples.* DXMS has been used to assess changes in iPLA$_2$β conformation induced by binding to substrate, inhibitor, and membranes (Hsu et al., 2013, 2009; Mouchlis et al., 2015; Mouchlis & Dennis, 2016). In general, 40 μg of iPLA$_2$β protein in 25 μL of protein buffer (25 m*M* Tris–HCl, pH 7.5, 50 m*M* NaCl, 10 m*M* urea, 250 m*M* imidazole, 2 m*M* ATP, 30% glycerol) is added to 75 μL of D_2O buffer (50 m*M* MOPS, pH 6.9, 100 m*M* NaCl, 2 m*M* DTT, 2 m*M* ATP; final D_2O concentration = 71%). Enzyme is preincubated with substrates, inhibitors, and membranes before addition of D_2O buffer. For membrane experiments, the ratio of lipid to iPLA$_2$β is typically >60, to ensure that saturation of the enzyme (Cao et al., 2013) and enzyme is preincubated with inhibitor in such experiments to prevent extensive hydrolysis of the membranes (Cao et al., 2013; Hsu et al., 2009). Samples are then incubated at RT (23°C) for 10, 30, 100, 300, 1000, 3000, and 10,000 s.

Phase 2. *Quenching and denaturation.* As exchange of amide protons is negligible at pH 2.5 (Gallagher & Hudgens, 2016), the hydrogen/deuterium exchange reaction is quenched through addition of 100 μL of 0.8% formic acid, 2 *M* guanidine hydrochloride. This solution also denatures the protein, in preparation for protease digestion in the next phase. The quenched reactions are immediately frozen on dry ice and then stored at −80°C.

Phase 3. *Pepsin digestion and collection of peptides.* All subsequent processing steps are performed at 0°C. Given its acid pH optimum, pepsin is among the most commonly used proteases to generate peptides for DXMS analyses (Gallagher & Hudgens, 2016; Percy et al., 2012). Porcine pepsin (Sigma, cat # P6887) is immobilized at 30 mg/mL on an Upchurch Scientific guard column (66 μL bed volume; cat # C.130B) packed with Poros 20 AL-activated affinity medium (ThermoFisher, cat # 1602802). The quenched deuterium exchange reactions are passed over the pepsin affinity column in 0.05% trifluoroacetic acid (TFA) at 100 μL/min for 1 min (resulting digestion time = 13 sc) before resultant peptides are passed onto a 1 × 50 mm reversed-phase C18 column (Vydac columns, Fisher; cat # 501120911). Peptides are eluted with a linear gradient of 0.046% TFA, 6.4% acetonitrile (v/v) to 0.03% TFA, 38.4% acetonitrile at 50 μL/min for 30 min (Hsu et al., 2008). This online digestion approach minimizes back exchange (Cao et al., 2013; Gallagher & Hudgens, 2016; Percy et al., 2012). For reactions containing vesicles (typically at lipid to protein ratios of 60 or greater), it is advisable to pass the samples through a C8 precolumn to remove the lipids (Cao et al., 2013).

Phase 4. *Mass spectrometry analysis of deuterated peptides.* The peptides are analyzed by ESI-MS/MS. Specifically, iPLA$_2$β analyses are performed using a Finnigan LCQ mass spectrometer, capillary temperature 200°C (Burke et al., 2008; Hamuro et al., 2004; Hsu et al., 2008).

Data analysis. The first step in data analysis is to identify and curate the deuterated peptides, using the SEQUEST software from Thermo Finnigan. Once overlapping peptides have been identified/aligned, a variety of software packages can be used to assess rate of hydrogen/deuterium exchange for each peptide (Gallagher & Hudgens, 2016). Solvent exchange rates are higher in solvent-exposed peptides than in peptides buried in the interior of the protein. Changes in the rate of exchange upon protein association with substrate, inhibitor, ligand, membranes, or oligomeric partners can provide mechanistic information about changes in conformation induced by these interactions (Cao et al., 2013; Gallagher & Hudgens, 2016; Percy et al., 2012). As noted earlier, there are alternative explanations for changes in the exchange rate of amide protons that can confound interpretation of DXMS data. As such, the iPLA$_2$β DXMS experiments should be interpreted in the context of data generated in MD simulations, using the homology models described earlier (Hsu et al., 2013, 2009; Mouchlis et al., 2015; Mouchlis & Dennis, 2016). It is also important to include two control experiments to account for any back exchange that occurs during processing. In the first ("back exchange" control), a fully deuterated protein is generated (D_2O labeling of denatured protein) and then carried through the processing step to assess deuterium atoms lost during processing. The second control ("on exchange" control) is generated by adding D_2O to protein in quenching buffer to assess exchange induced during the processing steps (Cao et al., 2013). Back exchange can also be minimized by optimizing experimental conditions (Gallagher & Hudgens, 2016; Percy et al., 2012).

2.3.2 Outcomes of DXMS Analyses of iPLA$_2$β

The protease digestion and processing phases have been optimized to yield 88% coverage of the iPLA$_2$β primary sequence (Hsu et al., 2009, 2013). Several regions of the ankyrin repeat domain (205–211, 390–404, 439–450, 458–470) and two regions in the catalytic domain (617–629; 708–730) show rapid hydrogen/deuterium exchange on the order of >70% at 10 s. These observations suggest that these regions are highly solvent exposed, a supposition that is borne out in the homology model and in additional DXMS experiments which indicate that the amphipathic alpha helix formed by residues 708–730 becomes buried upon exposure to membrane vesicles (Hsu et al., 2009). Binding of a fluoroketone inhibitor (1,1,1,3-tetrafluoro-7-phenylheptan-2-one) induces decreased hydrogen/deuterium exchange in five regions of iPLA$_2$β (residues 483–493, 516–525, 544–549, 631–655, 773–778) that are localized near the active-site serine (S519) and aspartic acid (D652) (Hsu et al., 2013). Comparable changes are observed in several regions of the catalytic domain when iPLA$_2$β is exposed to small unilamellar vesicles composed of 1-palmitoyl-2-arachidonoyl-*sn*-phosphatidylcholine, a membrane mimetic (Hsu et al., 2009). Upon binding to membrane, the catalytic domain of iPLA$_2$β exhibits reduced hydrogen/deuterium exchange in regions composed of residues 631–655, 658–664, 708–730, and 773–778. As noted earlier, the alpha helix in regions 708–730 exhibits the largest reduction (13.2 deuterons, 70% decrease), suggesting a dramatic shift in conformation/localization of this helix when iPLA$_2$β binds to membranes. When taken together and considered in light of molecular dynamics simulations, these observations lead to a model wherein the function of the 708–730 helix region is analogous to that of the "cap" region in group IVA PLA$_2$ (cPLA$_2$) (Hsu et al., 2009; Mouchlis et al., 2015). Upon membrane binding, this region is proposed to convert iPLA$_2$β from a "closed" to an "open" state, by penetrating the membrane and orienting such that its hydrophilic face (R710, K719) is positioned to interact with phospholipid head groups and its hydrophobic face (P711; P714; W715; L717; L721; W722) is positioned to interact with the fatty acid chains (Bucher, Hsu, Mouchlis, Dennis, & McCammon, 2013; Mouchlis et al., 2015). Based on these observations, an intriguing hypothesis has been proposed that the membrane is an allosteric regulator of iPLA$_2$β activity (Mouchlis et al., 2015).

In short, DXMS has provided novel insights into the molecular mechanisms governing iPLA$_2$β interaction with both substrate phospholipids and the membranes in which these molecules reside. A notable corollary to these studies is the potential to apply DXMS (together with other biophysical, structural, and biochemical approaches) to facilitate rational design of selective inhibitors of iPLA$_2$β (Mouchlis, Limnios, et al., 2016).

3. SUMMARY

iPLA$_2$β-derived lipids have critical roles in various biological processes and altered accumulation of these bioactive lipids, due to dysregulation of iPLA$_2$β expression or activation, can have profound consequences. Continued studies of this enzyme and its biology will therefore lead to a greater understanding of the roles that bioactive lipids play in the onset and progression of various disease states. They will also facilitate identification of novel pathways that can be targeted for drug therapy. The approaches described herein will facilitate measurement of iPLA$_2$β activity, purification of iPLA$_2$β from biological samples, and further delineation of conformational changes in iPLA$_2$β subsequent to membrane/substrate/inhibitor binding. Also provided are means to assess membrane dynamics of iPLA$_2$β. For a more comprehensive review of the iPLA$_2$s and related citations, the reader is kindly directed to the recent publication by Ramanadham et al. (2015).

ACKNOWLEDGMENTS

Methods described here were developed in association with projects funded by the American Diabetes Association and NIH/NIDDK (R01-69455).

REFERENCES

Ackermann, E. J., Kempner, E. S., & Dennis, E. A. (1994). Ca^{2+}-independent cytosolic phospholipase A_2 from macrophage-like P388D1 cells. Isolation and characterization. *The Journal of Biological Chemistry*, *269*, 9227–9233.

Ali, T., Kokotos, G., Magrioti, V., Bone, R. N., Mobley, J. A., Hancock, W., et al. (2013). Characterization of FKGK18 as inhibitor of group VIA Ca^{2+}-independent phospholipase A_2 (iPLA$_2$β): Candidate drug for preventing beta-cell apoptosis and diabetes. *PLoS One*, *8*, e71748.

Balboa, M. A., Balsinde, J., Jones, S. S., & Dennis, E. A. (1997). Identity between the Ca^{2+}-independent phospholipase A_2 enzymes from P388D1 macrophages and Chinese hamster ovary cells. *The Journal of Biological Chemistry*, *272*, 8576–8580.

Batchu, K. C., Hokynar, K., Jeltsch, M., Mattonet, K., & Somerharju, P. (2015). Substrate efflux propensity is the key determinant of iPLA$_2$β-mediated glycerophospholipid hydrolysis. *Journal of Biological Chemistry*, *290*(16), 10093–10103.

Bennett-Lovsey, R. M., Herbert, A. D., Sternberg, M. J., & Kelley, L. A. (2008). Exploring the extremes of sequence/structure space with ensemble fold recognition in the program Phyre. *Proteins*, *70*, 611–625.

Bucher, D., Hsu, Y. H., Mouchlis, V. D., Dennis, E. A., & McCammon, J. A. (2013). Insertion of the Ca^{2+}-independent phospholipase A_2 into a phospholipid bilayer via coarse-grained and atomistic molecular dynamics simulations. *PLoS Computational Biology*, *9*, e1003156.

Burke, J. E., Babakhani, A., Gorfe, A. A., Kokotos, G., Li, S., Woods, V. L., Jr., et al. (2009). Location of inhibitors bound to group IVA phospholipase A_2 determined by molecular dynamics and deuterium exchange mass spectrometry. *Journal of the American Chemical Society*, *131*, 8083–8091.

Burke, J. E., Karbarz, M. J., Deems, R. A., Li, S., Woods, V. L., Jr., & Dennis, E. A. (2008). Interaction of group IA phospholipase A_2 with metal ions and phospholipid vesicles probed with deuterium exchange mass spectrometry. *Biochemistry*, *47*, 6451–6459.

Cao, J., Burke, J. E., & Dennis, E. A. (2013). Using hydrogen/deuterium exchange mass spectrometry to define the specific interactions of the phospholipase A_2 superfamily with lipid substrates, inhibitors, and membranes. *The Journal of Biological Chemistry*, *288*, 1806–1813.

Carper, M. J., Zhang, S., Turk, J., & Ramanadham, S. (2008). Skeletal muscle group VIA phospholipase A_2 (iPLA$_2$beta): Expression and role in fatty acid oxidation. *Biochemistry, 47*, 12241–12249.

Dennis, E. A., Cao, J., Hsu, Y. H., Magrioti, V., & Kokotos, G. (2011). Phospholipase A_2 enzymes: Physical structure, biological function, disease implication, chemical inhibition, and therapeutic intervention. *Chemical Reviews, 111*, 6130–6185.

Gallagher, E. S., & Hudgens, J. W. (2016). Mapping protein–ligand interactions with proteolytic fragmentation, hydrogen/deuterium exchange-mass spectrometry. In K. Zvi (Ed.), *Methods in enzymology* (pp. 357–404). Cambridge, MA: Academic Press. [chapter 14].

Ghosh, M., Tucker, D. E., Burchett, S. A., & Leslie, C. C. (2006). Properties of the Group IV phospholipase A_2 family. *Progress in Lipid Research, 45*, 487–510.

Gijon, M. A., & Leslie, C. C. (1997). Phospholipases A_2. *Seminars in Cell & Developmental Biology, 8*, 297–303.

Gross, R. W., Ramanadham, S., Kruszka, K. K., Han, X., & Turk, J. (1993). Rat and human pancreatic islet cells contain a calcium ion independent phospholipase A_2 activity selective for hydrolysis of arachidonate which is stimulated by adenosine triphosphate and is specifically localized to islet beta-cells. *Biochemistry, 32*, 327–336.

Hamuro, Y., Anand, G. S., Kim, J. S., Juliano, C., Stranz, D. D., Taylor, S. S., et al. (2004). Mapping intersubunit interactions of the regulatory subunit (RIalpha) in the type I holoenzyme of protein kinase A by amide hydrogen/deuterium exchange mass spectrometry (DXMS). *Journal of Molecular Biology, 340*, 1185–1196.

Hazen, S., Zupan, L., Weiss, R., Getman, D., & Gross, R. (1991). Suicide inhibition of canine myocardial cytosolic calcium-independent phospholipase A_2. Mechanism-based discrimination between calcium-dependent and -independent phospholipases A_2. *The Journal of Biological Chemistry, 266*, 7227–7232.

Hsu, Y.-H., Bucher, D., Cao, J., Li, S., Yang, S.-W., Kokotos, G., et al. (2013). Fluoroketone inhibition of Ca^{2+}-independent phospholipase A_2 through binding pocket association defined by hydrogen/deuterium exchange and molecular dynamics. *Journal of the American Chemical Society, 135*, 1330–1337.

Hsu, Y. H., Burke, J. E., Li, S., Woods, V. L., Jr., & Dennis, E. A. (2009). Localizing the membrane binding region of Group VIA Ca^{2+}-independent phospholipase A_2 using peptide amide hydrogen/deuterium exchange mass spectrometry. *The Journal of Biological Chemistry, 284*, 23652–23661.

Hsu, Y. H., Burke, J. E., Stephens, D. L., Deems, R. A., Li, S., Asmus, K. M., et al. (2008). Calcium binding rigidifies the C2 domain and the intradomain interaction of GIVA phospholipase A_2 as revealed by hydrogen/deuterium exchange mass spectrometry. *The Journal of Biological Chemistry, 283*, 9820–9827.

Jenkins, C. M., Han, X., Mancuso, D. J., & Gross, R. W. (2002). Identification of calcium-independent phospholipase A_2 (iPLA$_2$)beta, and not iPLA$_2$gamma, as the mediator of arginine vasopressin-induced arachidonic acid release in A-10 smooth muscle cells. Enantioselective mechanism-based discrimination of mammalian iPLA$_2$s. *The Journal of Biological Chemistry, 277*, 32807–32814.

Jenkins, C. M., Mancuso, D. J., Yan, W., Sims, H. F., Gibson, B., & Gross, R. W. (2004). Identification, cloning, expression, and purification of three novel human calcium-independent phospholipase A_2 family members possessing triacylglycerol lipase and acylglycerol transacylase activities. *The Journal of Biological Chemistry, 279*, 48968–48975.

Jenkins, C. M., Yan, W., Mancuso, D. J., & Gross, R. W. (2006). Highly selective hydrolysis of fatty acyl-CoAs by calcium-independent phospholipase A_2beta. Enzyme autoacylation and acyl-CoA-mediated reversal of calmodulin inhibition of phospholipase A_2 activity. *The Journal of Biological Chemistry, 281*, 15615–15624.

Jones, S. S., Tang, J., Kriz, R., Shaffer, M., Knopf, J., & Seehra, J. (1996). Isolation, molecular cloning and expression of a novel calcium-independent phospholipase A_2. *The FASEB Journal, 10*, A977.

Kalyvas, A., Baskakis, C., Magrioti, V., Constantinou-Kokotou, V., Stephens, D., Lopez-Vales, R., et al. (2009). Differing roles for members of the phospholipase A_2 superfamily in experimental autoimmune encephalomyelitis. *Brain: A Journal of Neurology, 132*, 1221–1235.

Kokotos, G., Hsu, Y. H., Burke, J. E., Baskakis, C., Kokotos, C. G., Magrioti, V., et al. (2010). Potent and selective fluoroketone inhibitors of group VIA calcium-independent phospholipase A_2. *Journal of Medicinal Chemistry, 53*, 3602–3610.

Larsson, P. K. A., Claesson, H.-E., & Kennedy, B. P. (1998). Multiple splice variants of the human calcium-independent phospholipase A_2 and their effect on enzyme activity. *The Journal of Biological Chemistry, 273*, 207–214.

Li, H., Zhao, Z., Antalis, C., Zhao, Z., Emerson, R., Wei, G., et al. (2011). Combination therapy of an inhibitor of group VIA phospholipase A_2 with paclitaxel is highly effective in blocking ovarian cancer development. *The American Journal of Pathology, 179*, 452–461.

Lio, Y. C., & Dennis, E. A. (1998). Interfacial activation, lysophospholipase and transacylase activity of group VI Ca^{2+}-independent phospholipase A_2. *Biochimica et Biophysica Acta, 1392*, 320–332.

Lopez-Vales, R., Ghasemlou, N., Redensek, A., Kerr, B. J., Barbayianni, E., Antonopoulou, G., et al. (2011). Phospholipase A_2 superfamily members play divergent roles after spinal cord injury. *The FASEB Journal, 25*, 4240–4252.

Lopez-Vales, R., Navarro, X., Shimizu, T., Baskakis, C., Kokotos, G., Constantinou-Kokotou, V., et al. (2008). Intracellular phospholipase A_2 group IVA and group VIA play important roles in Wallerian degeneration and axon regeneration after peripheral nerve injury. *Brain: A Journal of Neurology, 131*, 2620–2631.

Ma, Z., Ramanadham, S., Hu, Z., & Turk, J. (1998). Cloning and expression of a group IV cytosolic Ca^{2+}-dependent phospholipase A_2 from rat pancreatic islets. Comparison of the expressed activity with that of an islet group VI cytosolic Ca^{2+}-independent phospholipase A_2. *Biochimica et Biophysica Acta, 1391*, 384–400.

Ma, Z., Ramanadham, S., Kempe, K., Chi, X. S., Ladenson, J., & Turk, J. (1997). Pancreatic islets express a Ca^{2+}-independent phospholipase A_2 enzyme that contains a repeated structural motif homologous to the integral membrane protein binding domain of ankyrin. *The Journal of Biological Chemistry, 272*, 11118–11127.

Mancuso, D. J., Jenkins, C. M., & Gross, R. W. (2000). The genomic organization, complete mRNA sequence, cloning, and expression of a novel human intracellular membrane-associated calcium-independent phospholipase A_2. *The Journal of Biological Chemistry, 275*, 9937–9945.

Mouchlis, V. D., Bucher, D., McCammon, J. A., & Dennis, E. A. (2015). Membranes serve as allosteric activators of phospholipase A_2, enabling it to extract, bind, and hydrolyze phospholipid substrates. *Proceedings of the National Academy of Sciences of the United States of America, 112*, E516–E525.

Mouchlis, V. D., & Dennis, E. A. (2016). Membrane and inhibitor interactions of intracellular phospholipases A_2. *Advances in Biological Regulation, 61*, 17–24.

Mouchlis, V. D., Limnios, D., Kokotou, M. G., Barbayianni, E., Kokotos, G., McCammon, J. A., et al. (2016). Development of potent and selective inhibitors for group VIA calcium-independent phospholipase A_2 guided by molecular dynamics and structure-activity relationships. *Journal of Medicinal Chemistry, 59*, 4403–4414.

Mouchlis, V. D., Morisseau, C., Hammock, B. D., Li, S., McCammon, J. A., & Dennis, E. A. (2016). Computer-aided drug design guided by hydrogen/deuterium exchange mass spectrometry: A powerful combination for the development of potent and selective inhibitors of Group VIA calcium-independent phospholipase A_2. *Bioorganic & Medicinal Chemistry*. http://dx.doi.org/10.1016/j.bmc.2016.05.009. pii: S0968-0896(16)30333-9. [Epub ahead of print].

Ong, W. Y., Farooqui, T., Kokotos, G., & Farooqui, A. A. (2015). Synthetic and natural inhibitors of phospholipases A_2: Their importance for understanding and treatment of neurological disorders. *ACS Chemical Neuroscience*, *6*, 814–831.

Percy, A. J., Rey, M., Burns, K. M., & Schriemer, D. C. (2012). Probing protein interactions with hydrogen/deuterium exchange and mass spectrometry—A review. *Analytica Chimica Acta*, *721*, 7–21.

Ramanadham, S., Ali, T., Ashley, J. W., Bone, R. N., Hancock, W. D., & Lei, X. (2015). Calcium-independent phospholipases A_2 and their roles in biological processes and diseases. *Journal of Lipid Research*, *56*, 1643–1668.

Ramanadham, S., Wolf, M. J., Jett, P. A., Gross, R. W., & Turk, J. (1994). Characterization of an ATP-stimulatable Ca^{2+}-independent phospholipase A_2 from clonal insulin-secreting HIT cells and rat pancreatic islets: A possible molecular component of the beta-cell fuel sensor. *Biochemistry*, *33*, 7442–7452.

Rost, B., Yachdav, G., & Liu, J. (2004). The PredictProtein server. *Nucleic Acids Research*, *32*, W321–W326.

Tang, J., Kriz, R. W., Wolfman, N., Shaffer, M., Seehra, J., & Jones, S. S. (1997). A novel cytosolic calcium-independent phospholipase A_2 contains eight ankyrin motifs. *The Journal of Biological Chemistry*, *272*, 8567–8575.

Wolf, M. J., & Gross, R. W. (1996). Expression, purification, and kinetic characterization of a recombinant 80-kDa intracellular calcium-independent phospholipase A_2. *The Journal of Biological Chemistry*, *271*, 30879–30885.

CHAPTER SEVEN

Using Hydrogen–Deuterium Exchange Mass Spectrometry to Examine Protein–Membrane Interactions

O. Vadas*, M.L. Jenkins†, G.L. Dornan†, J.E. Burke†,1
*Pharmaceutical Sciences Section, University of Geneva, Geneva, Switzerland
†University of Victoria, Victoria BC, Canada
[1]Corresponding author: e-mail address: jeburke@uvic.ca

Contents

Abstract

Many fundamental cellular processes are controlled via assembly of a network of proteins at membrane surfaces. The proper recruitment of proteins to membranes can be controlled by a wide variety of mechanisms, including protein lipidation, protein–protein interactions, posttranslational modifications, and binding to specific lipid species present in membranes. There are, however, only a limited number of analytical techniques that can study the assembly of protein–membrane complexes at the

Methods in Enzymology, Volume 583
ISSN 0076-6879
http://dx.doi.org/10.1016/bs.mie.2016.09.008

molecular level. A relatively new addition to the set of techniques available to study these protein–membrane systems is the use of hydrogen–deuterium exchange mass spectrometry (HDX-MS). HDX-MS experiments measure protein conformational dynamics in their native state, based on the rate of exchange of amide hydrogens with solvent. This review discusses the use of HDX-MS as a tool to identify the interfaces of proteins with membranes and membrane-associated proteins, as well as define conformational changes elicited by membrane recruitment. Specific examples will focus on the use of HDX-MS to examine how large macromolecular protein complexes are recruited and activated on membranes, and how both posttranslational modifications and cancer-linked oncogenic mutations affect these processes.

1. INTRODUCTION

Hydrogen–deuterium exchange mass spectrometry (HDX-MS) is an analytical technique that measures the exchange rates of amide hydrogens with solvent. The exchange of amide hydrogens is exquisitely tuned to act as a conformational readout of protein dynamics, as secondary structure is one of the main factors affecting their exchange with solvent. Amide hydrogens form hydrogen bonds in both alpha helices and beta sheets, and current theoretical models of amide exchange suggest that this is the predominant factor in slowing H/D exchange rates of amides in folded proteins (Skinner, Lim, Bédard, Black, & Englander, 2012a, 2012b). HDX-MS has been frequently applied as a technique to examine protein folding, as well as define protein–protein, protein–ligand, and protein–membrane interactions (Gallagher & Hudgens, 2016; Mayne, 2016; Pirrone, Iacob, & Engen, 2014).

Steady improvements in the infrastructure (mass spectrometers, fluidics, liquid handling robotics, and data analysis software) and automation have greatly expanded the throughput and ease of use of HDX-MS to examine protein dynamics (Iacob & Engen, 2012). The basis of the technique is that a protein of interest is incubated in deuterated buffer for a number of time points, and a low temperature low pH "quench" buffer is then added to dramatically slow the exchange rate of amide hydrogens (Fig. 1A). All following steps must be performed as fast as possible at the same low pH and low temperature conditions to minimize spurious exchange of amide hydrogens in a nonphysiological condition. The spatial localization of deuterium exchange can be determined by the use of acid functional proteases that cleave the protein into discrete fragments. These fragments are then separated by reverse phase chromatography, and the masses are measured using mass

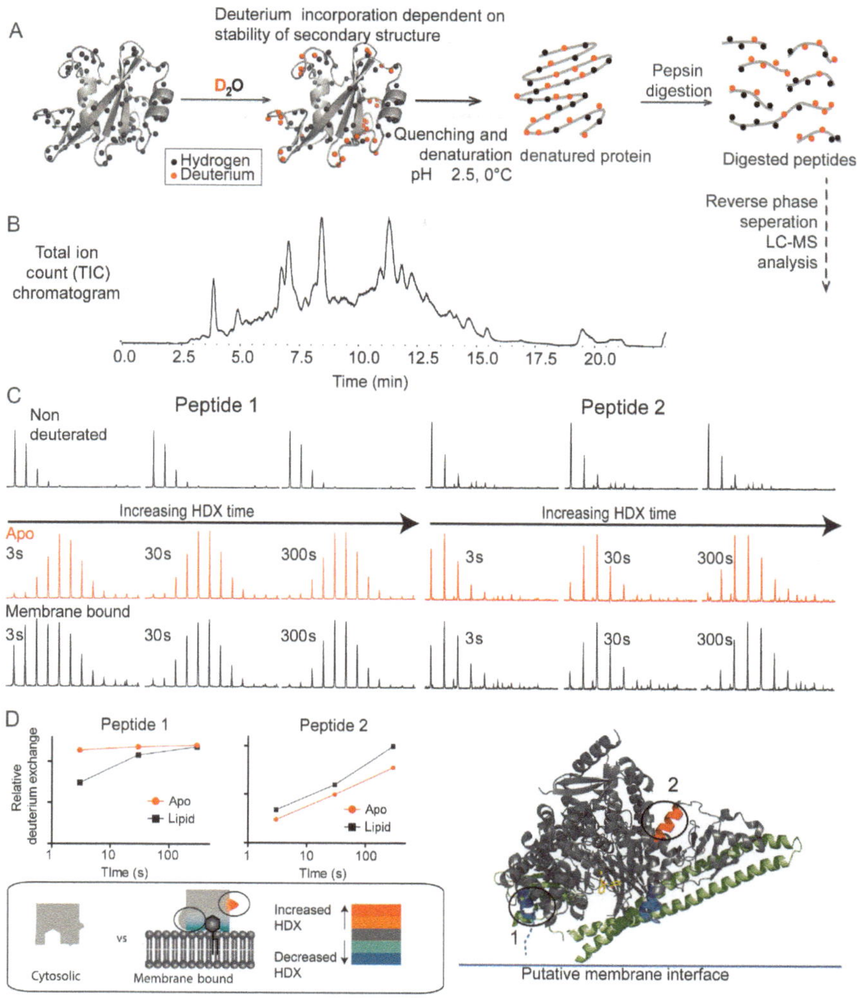

Fig. 1 Overview of hydrogen–deuterium exchange mass spectrometry (HDX-MS) and its application to protein–membrane interactions. (A) Graphical representation of the process of HDX-MS including protein deuteration, digestion, and separation by LC-MS. (B) A representative trace of a UPLC peptide elution profile of a protein–membrane complex by LC-MS. (**C**) HDX-MS data showing the difference in deuterium uptake rates of two representative peptides, in the presence or absence of lipid. The peptide on the left is a mainly unstructured region in the absence of membrane, and upon membrane insertion becomes protected from exchange. The peptide on the right is present in a stable region of secondary structure that is quite distant from the proposed membrane interface. Upon interaction with membranes, there is an increase in deuterium exposure indicating that there is some allosteric change caused by membrane binding. (D) Deuterium incorporation curves for the two representative peptides shown in the presence and absence of lipid. These two peptides show that it is

(Continued)

spectrometry (Zhang & Smith, 1993) (Fig. 1B–D). Fragmentation within the mass spectrometer can also be used to localize deuterium incorporation at single amino acid resolution, using both electron capture dissociation (ECD) (Pan, Han, Borchers, & Konermann, 2009) and electron transfer dissociation (ETD) (Rand, Zehl, Jensen, & Jørgensen, 2009). This approach requires careful optimization of fragmentation parameters to prevent scrambling of the amide hydrogens (Rand & Jørgensen, 2007). These experiments can be performed for the same protein in a variety of different conditions (presence of ligands, protein-binding partners, mutants, etc.).

An area of particular growth in HDX-MS experiments has been in the characterization of biotherapeutics (Wei et al., 2014). In contrast, a still underexploited application of HDX-MS is for the generation of stable truncated protein constructs more suitable for high-resolution structural approaches, including X-ray crystallography(Burke et al., 2014; Fowler et al., 2016; Pantazatos et al., 2004; Spraggon et al., 2004) and nuclear magnetic resonance (NMR) (Khan et al., 2014; Sharma et al., 2009). HDX-MS is also an excellent auxiliary technique for cryo-electron microscopy in the analysis of large protein complexes (Noble et al., 2013; Politis & Borysik, 2015; Shukla et al., 2014).

A steadily growing number of experiments have focused on the use of HDX-MS to characterize both integral membrane proteins (Akgun et al., 2013; Barclay et al., 2015; Busenlehner, Brändén, Namslauer, Brzezinski, & Armstrong, 2008; Busenlehner, Salomonsson, Brzezinski, & Armstrong, 2006; Busenlehner et al., 2004, 2007; Chung et al., 2011; Duc et al., 2015; Hebling et al., 2010; Kim et al., 2011; Koshy, Eyles, Weis, & Thompson, 2013; Orban & Tsybovsky, 2015; Parker et al., 2014; Pirrone, Emert-Sedlak, et al., 2015; Rey, Forest, & Pelosi, 2012; Rey, Man, et al., 2010; Shukla et al., 2014; Vahidi, Bi, Dunn, & Konermann, 2016; Zhang, Chien, et al., 2010), as well as the interaction of peripheral membrane proteins with membranes (Burke, Hsu, et al., 2008; Burke, Karbarz, et al., 2008; Burke, Perisic, Masson, Vadas, &

Fig. 1—Cont'd important to query a number of deuterium exchange time points, as depending on the stability of the amide hydrogen that is perturbed between the two conditions, differences may only be seen at some of the time points of interest. This is particularly true of intrinsically disordered regions within proteins, as standard automated approaches that are only able to sample deuterium incorporation time >10 s may not detect a difference in exchange rates for highly dynamic amides. The two peptides change in deuterium exchange is *colored* on the structure of a lipid-modifying enzyme (in this case PI3Kα).

Williams, 2012; Burke & Williams, 2013; Burke et al., 2011; Cao, Burke, & Dennis, 2013; Cao, Hsu, Li, Woods, & Dennis, 2011, 2013; Dbouk et al., 2012; Hsu, Burke, Li, Woods, & Dennis, 2009; Li, Rodnin, Ladokhin, & Gross, 2014; Man et al., 2007; Masson, Perisic, Burke, & Williams, 2016; Nasr et al., 2013; Rostislavleva et al., 2015; Srinivasan, Dharmarajan, Reed, Griffin, & Schmid, 2016; Vadas & Burke, 2015; Vadas et al., 2013) Defining how peripheral membrane proteins interact with membranes has remained challenging, and this review focuses on the application of HDX-MS to study the molecular basis for how membrane proteins dynamically interact with membrane surfaces. One of the most useful pieces of information generated from HDX-MS, in addition to mapping binding interfaces, is that it allows for the identification of allosteric conformational changes elicited by membrane interaction.

2. MECHANISM OF HYDROGEN EXCHANGE

Numerous reviews have thoroughly reviewed the derivation of the equations that govern the exchange rates of amide hydrogens within proteins (Konermann, Pan, & Liu, 2011; Wales & Engen, 2006). For this reason, this chapter will give only a brief overview of these studies. The use of hydrogen exchange to study proteins was initially used in combination with both the radioactive isotope tritium and gel filtration, followed later by the use of NMR: these studies made major breakthroughs in defining the mechanism of protein folding. The historical basis for hydrogen exchange has been reviewed and provides tremendous detail on the development and application of hydrogen exchange over the last ~60 years (Englander, 2006). The use of hydrogen exchange was expanded to include mass spectrometry in the early 1990s, and this review will only discuss the application of hydrogen exchange coupled with mass spectrometry (Katta & Chait, 1991; Zhang & Smith, 1993).

There are a number of variables that mediate the exchange of amide hydrogens in a protein, and these include temperature, pH, inductive/steric effects from surrounding residues, involvement in secondary structure, and solvent accessibility. For a completely unstructured peptide with no secondary structure the exchange rate is only dependent on temperature, pH, and neighboring residues. Amide exchange is an acid and base catalyzed process, and the global minimum of exchange occurs at a pH around 2.5. The rate of exchange for any given amide hydrogen in a completely unfolded peptide can be defined as the intrinsic rate of exchange, and this is commonly referred to as k_i or as the chemical rate constant k_{ch}. The dependence of the intrinsic exchange rates based on the surrounding residues has been extensively measured using NMR, and for any given peptide with a known sequence this rate can be calculated (Bai, Milne, Mayne, & Englander, 1994; Connelly, Bai, Jeng, & Englander, 1993). In a completely unfolded peptide at standard conditions of pH 7.0 and 25°C, the exchange rate of most amide hydrogens will be on the order of milliseconds to seconds.

The amide exchange rates within proteins are intimately linked to the presence of secondary structure elements, as well as changes in solvent accessibility. Experimental evidence has shown that the exchange rates of amide hydrogens in proteins are primarily influenced by their involvement in hydrogen bonds, specifically their involvement in secondary structure (Skinner et al., 2012a). This slowing of the amide exchange rate in proteins (k_{ex}) can be defined as the protection factor (P) between the experimental amide exchange rate in a folded vs unfolded protein, represented by $P = k_i/k_{ex}$. The intrinsic exchange rate of amide hydrogens within proteins can be decreased by up to a factor of 10^8 (Coales et al., 2010; Milne, Mayne, Roder, Wand, & Englander, 1998). However, even very stable secondary structure elements are able to exchange with solvent due to the innate structural fluctuations of proteins, which can briefly disrupt the hydrogen bond network. The equation governing this phenomenon can be represented as follows:

$$\mathrm{N-H_{cl}} \underset{k_{cl}}{\overset{k_{op}}{\rightleftharpoons}} \mathrm{N-H_{op}} \underset{\mathrm{D_2O}}{\overset{k_{ch}}{\rightarrow}} \mathrm{N-D_{cl}} \tag{1}$$

where k_{op} and k_{cl} (frequently represented as k_1 and k_{-1}) are the rate constants for the opening and closing of the amide hydrogen bond, respectively. N–H_{cl} represents an amide hydrogen that is involved in a hydrogen bond, while N–H_{op} represents an amide hydrogen not involved in hydrogen bonding. The rate constant k_{ch} is the intrinsic rate of exchange for this hydrogen in the absence of secondary structure.

The exchange rates of amide hydrogens in folded proteins can essentially be broken into two possible regimes: EX2 and EX1. In an EX2 exchange regime the intrinsic exchange rate of the amide hydrogen is much slower than the opening and closing rates, and thus numerous opening and closing reactions must occur before the amide hydrogen can exchange. Most amide hydrogens in folded proteins will exchange with solvent within an EX2 kinetics regime. The exchange rate of amide hydrogens by this regime provides an excellent readout of the dynamic stability of secondary structure elements. Alternatively, in an EX1 regime the intrinsic exchange rate is fast compared to the opening and closing rates. This means that once a secondary structure element undergoes an opening event, it will exchange rapidly with solvent. Very few proteins or protein regions exchange with EX1 regimes, although it can be induced using both chemical and thermal denaturants. EX1 kinetics are uncommon in most stable folded proteins, however, many important regulatory regions, such as lids and hinges within proteins, do exchange by this regime. The characteristic experimental readout for an EX1 exchange is the presence of a bimodal isotopic pattern compared to the binomial isotope exchange pattern of EX2 exchange (Weis, Wales, Engen, Hotchko, & Ten Eyck, 2006). This allows for the experimental determination of regions within proteins undergoing different types of hydrogen exchange. However, a major note of caution must be applied when stating a region in a protein exchanges by an EX1 mechanism, as carryover between samples can produce an indistinguishable mass spectrum from actual EX1 kinetics (Fang, Rand, Beuning, & Engen, 2011).

Rates of amide exchange within a folded protein may vary dramatically; therefore, in order to capture the full set of exchange rates in a given protein a large number of exchange times must be tested. A particularly useful approach is in the use of different temperatures and pH values (Coales et al., 2010), as this allows for the modification of amide exchange rates, which when combined with a variety of deuterium exchange times allows for sampling exchange rates that span over eight orders of magnitude. However, caution must be used, as it must be verified that these changes in pH and temperature do not cause changes in protein structure.

3. HYDROGEN–DEUTERIUM EXCHANGE TO STUDY PROTEIN–MEMBRANE INTERACTIONS

3.1 Study of Integral Membrane Proteins

HDX-MS has a long history of studying the interactions of peptides and proteins with membranes. The first experiments examining the interaction of polypeptides with membranes were carried out with model transmembrane peptides and their insertion into membranes (Demmers, Haverkamp, Heck, Koeppe, & Killian, 2000; Demmers et al., 2001; Hansen, Broadhurst, Skelton, & Arkin, 2002). A few representative examples of projects using HDX-MS to study membrane proteins will be highlighted in this review.

Technically, one of the major difficulties of studying membrane proteins by HDX-MS is dealing with highly hydrophobic and abundant lipids and detergents. The approaches that may be used to address this problem will be discussed extensively in the experimental section.

The first study of an integral membrane protein with HDX-MS was carried out on the microsomal glutathione transferase 1 (MGS1) integral membrane protein. In order to study the protein in its native membrane bound state, it was embedded within phosphatidylcholine vesicles (Busenlehner et al., 2007). These studies revealed conformational changes upon binding of a glutathione ligand and demonstrated the possibility of probing integral membrane proteins in their native membrane environment. Studies have also been carried out on Cytochrome C oxidase in the presence of detergents and revealed the dynamic conformational changes that can occur within different catalytic intermediates of Cytochrome C oxidase (Busenlehner et al., 2006).

There was a delay in the wide scale use of HDX-MS to probe integral membrane proteins, primarily due to the complicated removal of either lipid or detergent contaminants. However, in recent years, a number of groups carried out innovative studies on the conformational changes within membrane-embedded protein complexes. These included the mitochondrial adenine nucleotide carrier, and the different conformations it adopts to carry out nucleotide transport (Rey, Man, et al., 2010). This protein plays key roles in the transport of nucleotides across mitochondrial membranes, and during transport it switches between two conformational states. Using toxins to selectively lock the transporter in each of these states, differences in H/D exchange were determined between the two conditions. These experiments were carried out in the presence of detergent, and key to their success was the separation of detergent before injection onto the fluidics system using a combination of chlorinated solvents and reversed phase particles (Rey, Mrazek, et al., 2010). Another approach to studying protein–membrane assemblies used protein–lipid nanodiscs as a methodological platform. Nanodiscs are composed of membrane lipids encircled by a membrane scaffolding protein (Denisov, Grinkova, Lazarides, & Sligar, 2004). As nanodiscs are soluble, stable, and monodisperse, they are useful as a platform for HDX experiments (Hebling et al., 2010). They have been used to study the conformational changes of gamma-glutamyl carboxylase upon ligand binding (Parker et al., 2014). Complications of the nanodisc approach include the potential interference from the membrane scaffolding protein itself, although they can serve as a positive control of proper nanodisc assembly as their conformational transitions upon membrane binding have been studied by HDX (Morgan et al., 2011).

One of the most extensively studied class of integral membrane proteins by HDX has been the G protein-coupled receptors (GPCRs) and their complexes with heterotrimeric G proteins and arrestins. GPCRs are among the most important signaling molecules within cells, and are able to transmit signals through agonist binding, which induces a conformational change in both the GPCR and the G alpha subunit of the heterotrimeric G protein, leading to exchange of GDP for GTP. Upon GTP binding, the heterotrimeric G protein dissociates into G alpha and G beta gamma subunits, leading to the activation or inhibition of a number of downstream signaling pathways (Katritch, Cherezov, & Stevens, 2013). HDX was used to examine conformational changes upon GPCRs binding to both agonists and antagonists and revealed a number of conformationally distinct states depending on the ligand bound (West et al., 2011). HDX experiments were also carried out in tandem with the X-ray crystallographic analysis of the intact beta adrenergic receptor in complex with heterotrimeric G proteins (Chung et al., 2011). This study revealed an opening of the helical domain of the G alpha subunit upon interaction with the agonist bound activated form of the GPCR. This provided further experimental evidence for the opening of this domain as suggested by electron microscopy. Both of these studies were conducted in detergent micelles, however, recent reports have shown increased digestion efficiency and sequence coverage when GPCRs are presented in lipid bicelles (Duc et al., 2015). HDX experiments, in combination with chemical cross-linking and electron microscopy on a GPCR-beta arrestin complex, defined the orientation and dynamics of this membrane complex (Shukla et al., 2014). These studies highlight the powerful application of HDX-MS in tandem with other structural methodologies to probe structure and dynamics of large protein–membrane complexes.

3.2 Study of Peripheral Membrane–Protein Interactions

Peripheral membrane proteins cycle between a soluble state and a membrane bound state. In many cases their recruitment to the membrane is under tight regulation, and their coordinated recruitment and assembly play fundamental roles in key cellular processes. This is particularly true for enzymes that catalyze reactions on membranes, as they can also be selectively activated on membranes. In that sense, these proteins represent an ideal system to study protein–membrane interactions. As misregulation of the activity of these enzymes is closely linked to a number of human diseases, including cancer, inflammation, immune diseases, and developmental diseases, understanding the molecular basis for how these enzymes are recruited is essential (Wymann & Schneiter, 2008). This section will specifically focus on the use of HDX-MS to probe the molecular basis for how proteins interact with membranes, as well as the parameters that modify how activation on a membrane surface is mediated, including protein–protein interactions, posttranslational modifications, and disease-linked mutations. Studying the molecular details of the interaction of lipid-modifying enzymes with potent small molecule inhibitors using HDX-MS is also a powerful approach (Burke et al., 2009; Clarke et al., 2015; Karageorgos et al., 2013; Nasr et al., 2013).

The first experiments examining the interaction of a cytosolic protein with membranes were carried out using apomyoglobin (Man et al., 2007). These experiments determined the pH-dependent interaction of apomyoglobin with membranes, with pH-dependent unfolding leading to exposure of protein regions that are able to bind and interact with lipid membranes. The phospholipase A_2 (PLA_2) family of enzymes was the first family of enzymes that act on membrane substrates to be studied at a large-scale using HDX. Specifically, Group IA PLA_2 (Burke, Karbarz, et al., 2008), Group IVA PLA_2 (Burke, Hsu, et al., 2008), Group VIA PLA_2 (Hsu et al., 2009), and the lipoprotein-associated Group VII PLA_2 (Cao, Hsu, et al., 2013; Cao et al., 2011) have been studied. These studies have been reviewed elsewhere (Cao, Burke, et al., 2013), and will not be extensively discussed. Other lipases have also been examined, including the enzyme monoacylglycerol lipase, and its association with protein–membrane nanodiscs (Nasr et al., 2013).

Some of the largest and most dynamic protein–membrane complexes to be examined using HDX-MS are the class IA, class IB, and class III phosphoinositide 3-kinases (PI3Ks). PI3Ks are cytosolic lipid kinases that phosphorylate phosphoinositide lipids, promoting the recruitment and activation of effectors that contain binding domains specific to these modified phospholipids. These events trigger a signaling cascade that regulates essential cellular functions including cell growth, proliferation, and migration (Burke & Williams, 2015; Vanhaesebroeck, Guillermet-Guibert, Graupera, & Bilanges, 2010). The PI3K classes are defined by the differences in their protein regulatory partners, and their lipid substrate preference. There are several lipid-anchored proteins that can activate PI3Ks, ranging from activated receptor tyrosine kinases (RTKs), small GTPases, and G protein beta gamma heterodimers ($G\beta\gamma$) released by GPCRs (Vadas, Burke, Zhang,

Berndt, & Williams, 2011). The activation of class IA PI3Ks by somatic mutation are key drivers of oncogenesis (Samuels et al., 2004), and the antagonist of PI3K signaling, the lipid phosphatase PTEN, is often down-regulated in cancer (Song, Salmena, & Pandolfi, 2012). The activity of PI3Ks and PTEN must be tightly regulated, and the disruption of this equilibrium is a frequent driver of human disease.

Major goals of HDX studies on enzymes that act on membrane surfaces are to determine the molecular mechanisms that mediate membrane recruitment. The lipid phosphatase PTEN is composed of two globular domains, with a disordered tail that can be phosphorylated, and plays a key role in mediating phosphatase activity. Phosphorylation of the tail prevents interaction of PTEN with membranes. HDX-MS experiments examining PTEN in the presence and absence of tail phosphorylation revealed a large increase in exchange in the putative membrane-binding interface upon dephosphorylation (Fig. 2A) (Masson et al., 2016). HDX-MS experiments examining the membrane association of PTEN showed decreases in exchange at both the C2 and phosphatase domains in many of the same regions that were exposed upon dephosphorylation. This indicates that the phosphorylated tail acts as an autoinhibitory binding partner to these membrane-binding regions, and these interactions are released upon dephosphorylation. PTEN also acts as a useful example of a membrane-binding protein that only shows decreases in exchange along the proposed membrane-binding surface and no changes in deuterium exchange at potential allosteric regions distant from the active site or membrane-binding interface. This HDX-MS exchange behavior appears to be common in small (<50 kDa) and stable peripheral membrane proteins, including GIA, GVIA, and GVII PLA_2, and monoacylglycerol lipase (Burke, Karbarz, et al., 2008; Cao et al., 2011; Hsu et al., 2009; Nasr et al., 2013).

The interaction of all class I and class III PI3Ks with membranes has been examined using HDX-MS. The class IA PI3Ks can all be activated downstream of phosphopeptides derived from RTKs, and HDX-MS revealed that this is mediated by breaking the inhibitory protein interface between the SH2 domains of the p85 PI3K regulatory subunit with the catalytic subunit (Burke & Williams, 2013) (Fig. 2C). Many of the regions exposed upon removal of the inhibitory SH2 domains are the same regions that mediate interaction with membrane substrate (Fig. 2D). Intriguingly, all of the class I and class III PI3Ks studied by HDX-MS undergo a large increase in exchange at regions distant from both the membrane-binding site and active site (see Figs. 1D and 2D for examples). This includes

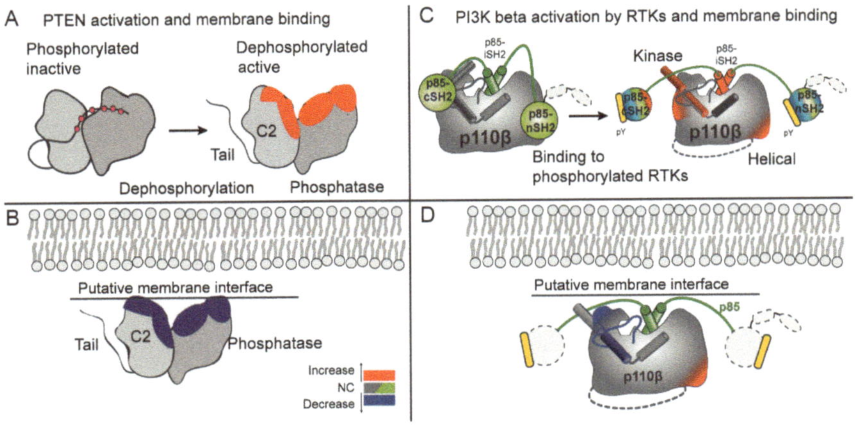

Fig. 2 Examples of using HDX-MS to probe the activation of the lipid-modifying enzymes PTEN (Masson et al., 2016) and PI3Kβ (Dbouk et al., 2012), and their interaction with membranes. (A) Cartoon depictions of the use of HDX-MS to probe the changes in H/D exchange upon dephosphorylation and activation of the lipid phosphatase PTEN. Dephosphorylation leads to increased deuterium exposure along the putative membrane interfacial surface on both the C2 and phosphatase domains of PTEN. (B) Depiction of the use of HDX-MS to probe the interaction of PTEN with membrane vesicles. Many of the same regions exposed upon release of the c-terminal phosphorylated tail of PTEN are the same regions that are protected upon membrane interaction. PTEN also acts as an example of a peripheral membrane protein that does not have large-scale allosteric conformational changes stimulated upon membrane binding. (C) Depiction of the use of HDX-MS to probe the conformational changes induced upon class IA PI3Kβ binding to phosphorylated peptides derived form RTKs (pY). Increased exposure was seen along the interface of catalytic subunit of PI3K that interact with the SH2 domains of the regulatory subunit that interact with pY. (D) Many of the same regions that are exposed upon pY binding are the same regions that are protected from exchange upon membrane binding. PI3Kβ also has a number of allosteric conformational changes showing indicated by an increased level of deuterium exchange at regions distinct from both the active site and membrane-binding interface.

experiments on class I PI3Kα (Burke et al., 2012), PI3Kβ (Dbouk et al., 2012), PI3Kδ (Burke et al., 2011), PI3Kγ (Vadas et al., 2013), and complex II of class III PI3K (Rostislavleva et al., 2015). The relevance of these distant allosteric conformational changes is exemplified by the fact that the majority of these regions in PI3Kα are at the location of some of the most frequent oncogenic linked somatic mutations, and that these mutations appear to facilitate the transition to an activated state on membrane vesicles (Burke et al., 2012). Intriguingly in class I PI3Ks that are able to be activated downstream of GPCRs, the allosteric conformational changes upon membrane binding appear to mediate conformational changes that facilitate its interaction with activating G beta gamma subunits derived from GPCRs (Dbouk et al., 2012; Vadas et al., 2013).

The fact that changes in deuterium exchange upon membrane interaction can be strongly dependent on both direct and indirect allosteric conformational changes requires that all HDX-MS results are carefully interpreted, and that many possible structural and dynamic possibilities for changes in deuterium exchange are interrogated.

4. EXPERIMENTAL CONSIDERATIONS FOR HDX-MS MEMBRANE EXPERIMENTS

4.1 General Considerations

HDX-MS experiments using continuous labeling conditions are carried out by exposing a protein of interest to deuterated buffer, and multiple time points of exposure are used to track the rate of deuterium incorporation over time. Experiments can also be carried out measuring the rate of deuterium off exchange, where a fully deuterated protein can be exposed to non-deuterated buffer, and the loss of deuterium can be monitored. Another type of HDX experiment can be conducted through pulse labeling, where instead of monitoring deuterium incorporation over time, the protein is treated with short pulses of deuterium so that only fully exposed amides are exchanged. Pulse-labeling experiments have been particularly useful in understanding the mechanisms of protein folding (Englander & Mayne, 2014).

Deuterium incorporation can be localized either through digestion of the protein of interest into peptides with an acid functional protease (bottom up) or fragmentation within the mass spectrometer using ECD or ETD fragmentation (top down). In a bottom up approach, a protein of interest is incubated with deuterium for a specific amount of time, and immediately quenched in a low pH (~2.5) low temperature (~0°C) buffer to minimize back exchange. The protein is proteolyzed by an acid functional protease either in solution or on an immobilized column. Once digested, the peptides are desalted, separated on a reverse-phase column, and analyzed using a mass spectrometer. This approach is repeated for all sets of conditions and time points (protein in apo state, protein bound to small molecule, protein bound to membranes).

One of the most important considerations for any deuterium exchange experiment is the extreme importance of both pH and temperature in the experimental flow path. This is especially critical in the setup of differential

HDX samples, where the dynamics of the same protein in distinct conditions is compared. Minimal changes in buffer pH will have major effects on the deuteration level of amide hydrogens, as the rate of exchange changes around 10-fold with a single pH unit difference. Temperature should remain constant while performing HDX experiments as well, as the rate of exchange is strongly dependent on temperature. Therefore, experiments looking at many different protein states require stringent control of the temperature, pH, protein concentration, and buffer composition. All efforts should be directed into controlling these factors as closely as possible to maximize reproducibility and ensure the ability to detect conformational differences between states.

The following sections will focus on key methods within the HDX flow path that should be optimized for any protein studied, as well as specific considerations that need to be addressed for the examination of peripheral membrane proteins and their association with membranes. The majority of experiments looking at protein–membrane interactions have been carried out with continuous labeling conditions combined with a bottom up flow path for deuterium incorporation, and for this reason this section will focus on this specific application of HDX-MS.

4.2 Optimization of Quench, Protease Digestion, and Peptide Identification

The first step when working on any new protein is the generation of a high confidence set of peptides, as this is the determining factor in both the sequence coverage and resolution to localize deuterium exchange information. One of the most important considerations is the amount of protein that should be used per sample. In our hands the amount of protein required to get good signal and decent peptide coverage (>90% of the sequence) is between 5 and 150 pmol of protein. This is highly dependent on the sensitivity of the mass spectrometer to be used, as well as the digestion efficiency of the protease on the specific protein of interest. On highly sensitive mass spectrometers, less material can be used dependent on the stability of the protein upon a freeze/thaw cycle. For initial experiments only dedicated to generating peptide coverage maps, a nondeuterated buffer of the same pH and salt composition as the deuterated buffer is used. After the protein is exposed to the mock D_2O buffer, the sample must be "quenched" with a low pH, low temperature buffer. Although it is unnecessary to prevent back exchange when using the mock D_2O buffer, the quench buffer should be optimized at this stage for digestion efficiency before analyzing any deuterated samples. It is essential that the final pH (after addition of quench to the exchange buffer) should be as close as possible to pH 2.5. This can be tested by combining the quench solution with the protein buffer (in the absence of protein) and the mock D_2O buffer and measuring the final pH value. The final concentration of acid is dependent on the buffering capability of the exchange buffer, but it is frequently around 0.3–0.8% formic acid or 0.1% TFA.

The quench solution will often contain a denaturant such as guanidine, and a reducing agent (if disulfides are present) that helps initiate unfolding of the protein before exposure to the acid functional protease. A number of different concentrations of denaturant can be used in the quench buffer to test for optimal digestion efficiency. For heavily disulfide bound proteins, Tris(2-carboxyethyl)phosphine (TCEP) is the best reductant option as it provides efficient disulfide breakage at the low pH and low temperature required (Burke, Karbarz, et al., 2008; Zhang, McLoughlin, et al., 2010).

For protease treatment, the protein can be directly incubated with a soluble acid functional protease, or passed through an immobilized protease column. An advantage of using immobilized protease columns is that they can be incorporated directly into the fluidics flow path. In this setup, the protease column is inserted before a reverse phase trapping column, allowing for online digestion and desalting. This greatly decreases the generation of autolysis peptide fragments derived from the protease itself, and experiments have shown that the increased back pressure during immobilized digestion can greatly increase digestion efficiency (Ahn, Jung, Wyndham, Yu, & Engen, 2012; Jones, Zhang, Vidavsky, & Gross, 2010). The majority of HDX-MS experiments have used pepsin as the acid functional protease. However, there are a variety of other acid functional proteases that have been used including fungal XIII and XVI, plasmepsin, and nepenthesin (Kadek et al., 2014; Marcoux et al., 2010; Mayne et al., 2011; Rey et al., 2013).

Although convenient, every column incorporated on the fluidics system increases the chance of carryover between samples, and the incorporation of an immobilized protease cartridge into the fluidics flow path can lead to dramatic increases in the amount of carryover between samples. Carryover causes major problems in downstream data analysis, and therefore washing protocols have been developed for the reverse phase trap and LC columns (Fang et al., 2011), as well as the immobilized protease column (Majumdar et al., 2012). Different proteins will have different levels of carryover, and thus initial tests should be carried out to ensure a limited level of carryover between samples. Carryover can be reduced through the incorporation of blank samples between runs, as well as injection of different buffers/solvents to clean both the protease and trap columns. It is extremely important that any cleaning buffer/solvent should be compatible with the immobilized protease column. For any protein of interest, the washing methods should be optimized so that no peptide has greater than 5% level of carryover between runs.

Once digestion and desalting is complete, the trapping column is switched in line with an analytical column and peptides are eluted using a water–acetonitrile gradient. Optimal HDX experiments are carried out on ultra performance liquid chromatography (UPLC) at pressures up to 15,000 psi. Peptide separations carried out at UPLC pressures have greatly increased separation efficiency (Wales, Fadgen, Gerhardt, & Engen, 2008). This leads to a lower frequency of overlapped fragments, and this decreased complexity allows for the study of larger protein complexes using HDX. Ion mobility MS also provides a useful approach to decrease sample complexity in short LC gradients (Iacob, Murphy, & Engen, 2008), as overlapping peptides can be separated by their drift time in an ion mobility cell. It is critically important that quench conditions (0°C and pH 2.5) are maintained throughout the liquid chromatography stage, in order to prevent back exchange of deuterated amides with nondeuterated solvents. This is achieved by including FA or TFA in all buffers, and having all columns and valves immersed in ice or stored in a refrigerated cold box. The design of multivalve ice box/refrigerated systems is a relatively straightforward and affordable approach for performing semiautomated HDX-MS (Keppel, Jacques, Young, Ratzlaff, & Weis, 2011), and commercial fully automated solutions are also currently available from both LEAP technologies and Waters. Experiments examining the different chromatography and MS parameters that contribute to back exchange in online digestion and separation showed a surprisingly low correlation between LC gradient time and back exchange rates, with a threefold decrease in LC elution time only decreasing back exchange ~2% (Walters, Ricciuti, Mayne, & Englander, 2012). We have found that using longer gradient elutions (a gradient of 5–45% over 20 min, with Buffer A being 0.1% formic acid and Buffer B being 100% acetonitrile) is useful for the study of large proteins >250 kDa, and that shorter gradients (~10 min) are more suitable for smaller proteins with less peptide fragments. Ideally gradients should be optimized to maintain a constant level of peptides eluting over the entire time course, and this can be accomplished through

Table 1 Shaped UPLC Gradient for Rapid Elution of Peptides During LC Elution

Time (m)	Buffer B (%)
0	3
1.5	10
11	35
12	45
13	80

Buffer A is composed of 0.1% formic acid and buffer B is composed of 100% Acetonitrile.

the use of specifically shaped LC gradients. Our standard gradient for rapid HDX separations is shown in Table 1.

Initial peptide identification of proteolyzed fragments is carried out using tandem MS/MS experiments (most commonly collision-induced dissociation or higher-energy collisional dissociation, but ETD/ECD can also be used). The tandem MS/MS datasets can then be searched across reference databases using standard peptide identification software (MASCOT, SEQUEST, PEAKS, etc.). Reference databases should contain common contaminants including the protease used in the analysis, and common chaperones from the organism that the proteins were purified from. For large protein complexes it is exceptionally important to verify that peptides are correctly identified, especially due to the relatively nonspecific cleavage of many commonly used proteases. Database searches should be carried out with tight mass tolerances matching the accuracy of the mass spectrometer used. For all tandem MS/MS experiments performed in our laboratories, we externally calibrate the mass spectrometer immediately before the run. If possible, the use of an internal calibration lock mass measurement during the run also helps in the accuracy of peptide identification.

Ideally peptide fragments will be identified that span the entire sequence of the protein of interest, however, this can be challenging especially for large multiprotein assemblies. We aim to have at least 90% sequence coverage, and optimization of coverage can be achieved by adjusting the quench composition, incubation time with quench before flash freezing, pepsin digestion time, and protein concentration. The use of multiple proteases, along with carefully shaped acetonitrile gradients can lead to the identification of numerous overlapping peptide fragments that can be used to resolve deuterium exchange information at close to single amide resolution (Kan, Walters, Mayne, & Englander, 2013; Walters et al., 2012).

4.3 Optimization of Deuteration and Quench Conditions

Once peptide fragmentation maps that adequately cover the protein of interest have been generated, experiments examining the rate of deuterium exchange can be initiated. The hydrogen–deuterium exchange reaction is initiated by diluting the protein of interest with D_2O buffer. The dilution fold can vary from 2 to 20, depending on both protein and ligand concentration. D_2O buffer must be made fresh before every experiment and kept sealed to avoid exchange with moisture in the air. As the rate amide hydrogens exchange with solvent can vary by many orders of magnitude, a variety of time points of deuterium exposure should be sampled. We frequently perform experiments with deuterium exposure times varying from 3 to 3000 s. When preparing samples, the timing of the deuterium incubation should be tightly regulated. Once the incubation is complete, the optimized low pH, low temperature quench buffer is added. The sample can then be flash frozen in liquid nitrogen and stored for up to a month at −80°C without affecting amide HDX levels. In automated robotic systems the deuterium exchange experiments can be made, quenched and immediately injected, digested, separated, and analyzed on an automated fluidics MS system. Although this approach provides highly reproducible conditions for several sample conditions tested as well as noninterrupted sample analysis over a number of hours, a major limitation is inherent to this approach: that proteins studied have to be stable for the entire course of the experiment in the same tube. Even at low temperatures, proteins may undergo aggregation or large-scale conformational changes that will radically shift the hydrogen exchange rates. Thus, the use of this automated reproducible and flexible approach will need to be screened for suitability for every protein analyzed.

To minimize variability, all sets of conditions that will be compared should be deuterated simultaneously and under the same conditions. It is vital to make sure that temperature is stable, and if experiments prepared at different times are to be directly compared, it is essential that carefully chosen control samples are run. It is also very important to test that the protein of interest is stable throughout the time course of the reaction. A simple way to check this is by allowing the protein to sit at the exchange temperature for the longest time point to be examined in a nondeuterated buffer, and at the end of this time the protein can be treated with a short pulse of deuteration (10 s). Ideally this experiment should give the same deuterium exchange results as a protein that has been allowed to equilibrate to temperature and then immediately treated with a short pulse of deuteration (10 s). If these results are different it suggests that the protein is not remaining stable under the experimental conditions, and different conditions (buffers, temperature) should be evaluated to promote protein stability.

Ideally all deuterium exchange reactions should be carried out in triplicate. The generation of controls is also useful for the analysis of both spurious on exchange and back exchange during sample processing. On exchange can be controlled by adding quench buffer to the deuterated buffer before addition of protein. This mimics any deuterium on exchange that may occur upon quenching, or injection onto the fluidics LC–MS system. This step is particularly important if the protein of interest requires a longer than normal amount of time with the denaturant/reductant in the quench condition to obtain sufficient digestion. Back exchange can be controlled through the generation of a fully deuterated sample. The analysis of a fully deuterated sample allows for the determination of the level of back exchange that occurs during sample handling, digestion, separation, and mass spectrometry analysis. The generation of a fully deuterated sample protein sample is nontrivial, and this is particularly true for large stable protein complexes. Proteins can be treated under denaturing conditions, followed by addition of deuterated buffer for an amount of time, specific to each protein that leads to full incorporation of deuterium followed by standard quench, digestion, separation, and data analysis. Denaturing conditions and deuterium exposure time need to be empirically determined for each protein to be studied. Experiments can also be performed where the protein of interest is proteolyzed and the peptide fragments themselves are deuterated, and then undergo the standard flow path of HDX-MS experiments to probe back exchange. Factors that need to be considered in the generation of fully deuterated samples is that fully denatured proteins may undergo differential digestion, and for particularly stable proteins, the back exchange rates may be different than the peptide fragments. More than one strategy should be used in the determination of full deuteration incorporation, and ideally if performed properly these should all give similar rates of back exchange.

4.4 Optimization of Protein–Membrane Interactions

When studying the interaction of proteins with membranes, one of the most important steps in the HDX-MS experiment is the determination of optimal membrane conditions to promote both protein stability and protein–membrane interactions. This requires that initial biophysical studies be conducted to identify the composition, and concentration of membranes that will provide sufficient binding to the protein of interest to generate a reproducible HDX signal. The composition of membranes and their presentation in vesicles, bicelles, nanotubes, or nanodiscs are important parameters that need to be optimized for every protein studied using HDX-MS. For experiments using vesicles, the size of the lipid vesicles must be considered before starting. We have found that an excellent tool to begin examination of any protein–membrane system is the use of protein–lipid fluorescence resonance energy transfer (FRET) assays to probe the affinity of proteins to membranes of different sizes and compositions. This assay relies on the FRET signal between tryptophan residues in the protein acting as donors and dansyl-containing lipids within the vesicles as acceptors (Nalefski, Slazas, & Falke, 1997). Alternatively, a number of other protein–lipid affinity measurement techniques including lipid sedimentation, lipid flotation, or Surface Plasmon Resonance can be used to optimize the affinity of protein to membranes of different composition.

Deuteration of samples is generally achieved by diluting the concentrated sample with a D_2O-containing buffer and thus, interaction of protein with lipid must be strong enough to be maintained in the diluted conditions. However, one of the frequent experimental observations of our laboratories has been the instability of membrane-interacting proteins when incubated in the presence of membranes at high concentrations (For studies on PI3Ks, protein concentrations $>10\ \mu M$, with lipid concentrations $>500\ \mu M$ caused large-scale precipitation/aggregation of protein out of solution). This must be experimentally examined for each protein under study.

One of the approaches that can be used to avoid this problem is including membranes in the deuterated buffer, in which case both deuteration and membrane interaction will be initiated simultaneously (Burke et al., 2011, 2012; Dbouk et al., 2012; Vadas et al., 2013). Alternatively, if no binding events are observed, protein and lipids can be preincubated at high concentrations prior to deuteration initiation (Rostislavleva et al., 2015). Altogether, every HDX-MS experiment looking at protein–lipid interaction requires optimization of the conditions to maximize affinity between protein and lipids, while at the same time verifying stability of the protein of interest with membrane substrate.

4.5 Optimization of Membrane/Detergent Management for LC and MS Analysis

One of the major considerations for experiments examining protein–membrane assemblies is how to treat the excess of either lipid or detergent (or both) that is required for the deuterium exchange step before injection onto the LC-MS. Direct injection of detergent and lipids onto the LC will lead to large amounts of the hydrophobic material sticking to the reverse phase material in the trapping column, with the potential risk of clogging columns after a few sample injections. This clogging will be dependent on lipid concentration but also on column wash procedures. Lipid and detergent on the MS eluting onto the MS system can also lead to ion suppression and loss of signal for peptides of interest. A number of approaches have been attempted to allow for protein–membrane assemblies to be studied using semiautomated fluidics HDX-MS approaches. For experiments using membrane vesicles, the simplest approach is to carry out all experiments containing membrane vesicles back to back, with a minimal reversed phase column wash protocol. In our experience up to six membrane samples (10–50 μg of lipid) can be run in tandem without any detrimental effects on peptide separation, column pressure, or ion suppression. After these samples are run, the reverse phase system is unplugged from the mass spectrometer and is extensively washed until all contaminating lipids are removed. Careful monitoring of LC performance (pressures, retention times) is critical for this approach.

An alternative approach to directly injecting lipids onto the LC system is to selectively remove any detergent or lipid before injection onto the LC system. A number of different approaches have been used to extract membranes. Chlorinated solvents have been used to specifically elute detergents from a reverse phase column (Rey, Mrazek, et al., 2010); however, caution must be applied for any extremely hydrophobic peptides. ZrO_2-coated silica beads have also been used to selectively remove phospholipid contaminants in experiments using protein–membrane nanodiscs as a platform (Hebling et al., 2010). Recent experiments have also shown the feasibility of Langmuir monolayers as a membrane suitable for HDX-MS applications, and this technology appears particularly well suited for studying the insertion of lipidated proteins into membrane bilayers (Pirrone, Emert-Sedlak, et al., 2015; Pirrone, Vernon, Kent, & Engen, 2015). Experiments carried out with the Langmuir monolayer approach also used an offline cleaning of the reversed phase trapping column after a certain threshold of lipid is introduced into the system.

4.6 Data Analysis

One of the most important aspects of any HDX-MS experiment is the analysis of deuterium incorporation. An excellent review has recently been written on all aspects of HDX data analysis (Wales, Eggertson, & Engen, 2013).

The simplest analysis of HDX data analysis is that deuterium exchange can be represented as the % hydrogen exchange between the nondeuterated profile, and a fully deuterated isotopic profile. For experiments where a fully deuterated control cannot be generated relative levels of deuterium incorporation can also be used, particularly if the main focus is differential HDX rates between two different conditions. Most HDX-MS studies represent deuterium incorporation through the centroid of mass of the isotopic distribution of the deuterated peptide. However, this approach loses any information present in the shape of the isotopic distribution of the peptide. The shape of the isotopic distribution contains information on the mechanism of deuterium incorporation, specifically for regions undergoing deuterium exchange under EX1 exchange conditions (Weis et al., 2006), and can also be used as a tool for the determination of residue-specific deuterium occupancy within a peptide (Kan et al., 2013). A wide variety of software packages, both freely accessible and commercial, are currently available and have been recently reviewed (Gallagher & Hudgens, 2016).

5. CASE STUDY OF A PROTEIN–MEMBRANE HDX-MS SETUP EXPERIMENT: PI3Kδ

5.1 Preparation of Protein Samples for HDX-MS Analysis

Protein samples of PI3Kδ (heterodimer of p110δ/p85α protein subunits, >95% pure) are prepared in gel filtration buffer (20 m*M* Hepes pH 7.5, 150 m*M* NaCl, 1 m*M* TCEP). Once purified, the protein is snap frozen in liquid nitrogen at ~1 mg/mL (~5 μ*M*) and stored at −80°C until HDX sample preparation. Depending on the sensitivity of the mass spectrometer used for the analysis different concentrations of protein may need to be explored. On both a Bruker Impact HD qTOF MS or an Orbitrap XL MS, we have found the lower limit of protein giving a decent signal to be ~6 pmol, or 2 μL of a 3 μ*M* solution. Before setup of HDX samples the protein is thawed on ice, and then aggregates that may clog the fluidics system are removed by using either a 0.22 μm filter, or spun at 21,000 × *g*. After thawing, protein is diluted in identical gel filtration buffer to a final concentration of 3 μ*M* in a fresh tube.

5.2 Preparation of Membrane Vesicles for HDX-MS Analysis

Vesicles of the following composition (5% Brain PIP_2, 30% Brain PS, 50% Brain PE, 15% Egg PC, Avanti polar lipids, % mole composition) are prepared at 1 mg/mL by adding the respective lipid populations suspended in chloroform together in a glass vial. The organic solvent is evaporated by a gentle stream of N_2 gas, followed by 30–60 min in a dessicator. Once dried, the lipid film is rehydrated in lipid buffer (20 m*M* HEPES pH 7.5 (RT), 100 m*M* KCl, 0.5 m*M* EDTA) for 10 min, followed by 5 min of bath sonication. Lipid solutions were then subjected to 10 freeze–thaw cycles by alternating between liquid nitrogen and a warm water bath. Lipid solutions are then extruded 11 × through a 100-nm polycarbonate membrane using a mini extruder (Avanti). Vesicles were aliquoted and then snap frozen in liquid nitrogen before storing at −80°C. Before setup of HDX samples aliquots were thawed and allowed to equilibrate to room temperature.

5.3 Preparation of HDX-MS Samples

To fully examine amide exchange rates for amides that have a range of protection factors, sampling a variety of different time points of deuterium incorporation is required. We routinely use five different combinations of time and temperature for deuterium incorporation experiments (3, 30, 300, 3000 s at 23°C and 3 s at 0°C) to fully probe amide exchange rates.

Tight binding of PI3Kδ to membranes is dependent on its interaction with phosphorylated residues present on RTKs, and this can be mimicked through the use of a bisphosphorylated peptide derived from the PDGFR receptor (pY peptide, PDGFR residues 735–767, with pY740 and pY751) (Burke et al., 2012; Hon, Berndt, & Williams, 2012). For this reason all experiments examining membrane binding were carried out with a saturating concentration of this pY peptide (5 μ*M* final in the exchange reaction).

The buffer used for deuterium exchange experiments is 10 m*M* Hepes pH 7.5, 50 m*M* NaCl, and 1 m*M* TCEP in 98% D_2O. To prepare each deuterium exchange experiment 2 μL of the 3 μ*M* PI3Kδ protein stock with 25 μ*M* pY peptide is added into a clean 0.6-mL Eppendorf tube. To this solution 5 μL of the membrane solution (or corresponding lipid buffer blank) is added. This solution is allowed to equilibrate at 23°C for 2 min. To initiate deuterium exchange 48 μL of the deuterated buffer pre-equilibrated at 23°C is added, and deuteration is allowed to proceed for the desired amount of time. At the end of this time course, the deuteration is ended by rapid mixing with 20 μL of ice cold quench buffer (2.4% Formic acid, 2 *M* Guanidinium chloride), followed by freezing in liquid nitrogen, and storage for up to a week at −80°C. For setup of the 3 s samples two pipettes are used, one pipette containing prealiquoted ice cold quench buffer, and the other pipette containing deuterated buffer. This allows for addition of deuterium and quench to occur within 3 s. For experiments carried out at 0°C all samples were preequilibrated on ice, and setup was also performed in a 4°C cold room.

6. CONCLUSIONS

Hydrogen–deuterium exchange methods have proven to be a powerful addition to other biophysical techniques available to study protein structure and dynamics. Structural and dynamic studies of both peripheral and integral membrane proteins and their association with membranes and membrane bound proteins have lagged behind the study of their soluble counterparts. The improvements that have occurred in HDX-MS in every facet of sample preparation, fluidics, mass spectrometers, and data analysis has greatly expanded the size and complexity of systems that can be studied, as well as allowing for the study of protein–membrane assemblies. We can expect that an increasing number of researchers working on complicated protein–membrane systems will add HDX-MS to their toolkit of biophysical methods to address structural and dynamic questions of biological and therapeutic relevance.

ACKNOWLEDGMENTS

J.E.B. is supported by a new investigator grant from CIHR, a discovery research grant from the Natural Sciences and Engineering Research Council of Canada (NSERC-2014-05218), as well as a small project health research grant from the British Columbia Proteomics Network (BCPN). O.V. is supported by a Swiss National Science Foundation fellowship (Ambizione PZ00P3_148269).

REFERENCES

Ahn, J., Jung, M. C., Wyndham, K., Yu, Y. Q., & Engen, J. R. (2012). Pepsin immobilized on high-strength hybrid particles for continuous flow online digestion at 10,000 psi. *Analytical Chemistry*, *84*(16), 7256–7262.

Akgun, B., Satija, S., Nanda, H., Pirrone, G. F., Shi, X., Engen, J. R., et al. (2013). Conformational transition of membrane-associated terminally acylated HIV-1 Nef. *Structure (London, England: 1993)*, *21*(10), 1822–1833.

Bai, Y., Milne, J. S., Mayne, L., & Englander, S. W. (1994). Protein stability parameters measured by hydrogen exchange. *Proteins*, *20*(1), 4–14.

Barclay, L. A., Wales, T. E., Garner, T. P., Wachter, F., Lee, S., Guerra, R. M., et al. (2015). Inhibition of pro-apoptotic BAX by a noncanonical interaction mechanism. *Molecular Cell*, *57*(5), 873–886.

Burke, J. E., Babakhani, A., Gorfe, A. A., Kokotos, G., Li, S., Woods, V. L., et al. (2009). Location of inhibitors bound to group IVA phospholipase A2 determined by molecular dynamics and deuterium exchange mass spectrometry. *Journal of the American Chemical Society, 131*(23), 8083–8091.

Burke, J. E., Hsu, Y.-H., Deems, R. A., Li, S., Woods, V. L., & Dennis, E. A. (2008). A phospholipid substrate molecule residing in the membrane surface mediates opening of the lid region in group IVA cytosolic phospholipase A2. *The Journal of Biological Chemistry, 283*(45), 31227–31236.

Burke, J. E., Inglis, A. J., Perisic, O., Masson, G. R., McLaughlin, S. H., Rutaganira, F., et al. (2014). Structures of PI4KIIIβ complexes show simultaneous recruitment of Rab11 and its effectors. *Science (New York, N.Y.), 344*(6187), 1035–1038.

Burke, J. E., Karbarz, M. J., Deems, R. A., Li, S., Woods, V. L., & Dennis, E. A. (2008). Interaction of group IA phospholipase A2 with metal ions and phospholipid vesicles probed with deuterium exchange mass spectrometry. *Biochemistry, 47*(24), 6451–6459.

Burke, J. E., Perisic, O., Masson, G. R., Vadas, O., & Williams, R. L. (2012). Oncogenic mutations mimic and enhance dynamic events in the natural activation of phosphoinositide 3-kinase p110α (PIK3CA). *Proceedings of the National Academy of Sciences of the United States of America, 109*(38), 15259–15264.

Burke, J. E., Vadas, O., Berndt, A., Finegan, T., Perisic, O., & Williams, R. L. (2011). Dynamics of the phosphoinositide 3-kinase p110d interaction with p85a and membranes reveals aspects of regulation distinct from p110a. *Structure (London, England: 1993), 19*(8), 1127–1137.

Burke, J. E., & Williams, R. L. (2013). Dynamic steps in receptor tyrosine kinase mediated activation of class IA phosphoinositide 3-kinases (PI3K) captured by H/D exchange (HDX-MS). *Advances in Biological Regulation, 53*(1), 97–110.

Burke, J. E., & Williams, R. L. (2015). Synergy in activating class I PI3Ks. *Trends in Biochemical Sciences, 40*(2), 88–100.

Busenlehner, L. S., Alander, J., Jegerscöhld, C., Holm, P. J., Bhakat, P., Hebert, H., et al. (2007). Location of substrate binding sites within the integral membrane protein microsomal glutathione transferase-1. *Biochemistry, 46*(10), 2812–2822.

Busenlehner, L. S., Brändén, G., Namslauer, I., Brzezinski, P., & Armstrong, R. N. (2008). Structural elements involved in proton translocation by cytochrome c oxidase as revealed by backbone amide hydrogen-deuterium exchange of the E286H mutant. *Biochemistry, 47*(1), 73–83.

Busenlehner, L. S., Codreanu, S. G., Holm, P. J., Bhakat, P., Hebert, H., Morgenstern, R., et al. (2004). Stress sensor triggers conformational response of the integral membrane protein microsomal glutathione transferase 1. *Biochemistry, 43*(35), 11145–11152.

Busenlehner, L. S., Salomonsson, L., Brzezinski, P., & Armstrong, R. N. (2006). Mapping protein dynamics in catalytic intermediates of the redox-driven proton pump cytochrome c oxidase. *Proceedings of the National Academy of Sciences of the United States of America, 103*(42), 15398–15403.

Cao, J., Burke, J. E., & Dennis, E. A. (2013). Using hydrogen/deuterium exchange mass spectrometry to define the specific interactions of the phospholipase A2 superfamily with lipid substrates, inhibitors, and membranes. *The Journal of Biological Chemistry, 288*(3), 1806–1813.

Cao, J., Hsu, Y.-H., Li, S., Woods, V. L., & Dennis, E. A. (2011). Lipoprotein-associated phospholipase A(2) interacts with phospholipid vesicles via a surface-disposed hydrophobic α-helix. *Biochemistry, 50*(23), 5314–5321.

Cao, J., Hsu, Y.-H., Li, S., Woods, V. L., & Dennis, E. A. (2013). Structural basis of specific interactions of Lp-PLA2 with HDL revealed by hydrogen deuterium exchange mass spectrometry. *Journal of Lipid Research, 54*(1), 127–133.

Chung, K. Y., Rasmussen, S. G. F., Liu, T., Li, S., DeVree, B. T., Chae, P. S., et al. (2011). Conformational changes in the G protein Gs induced by the β2 adrenergic receptor. *Nature*, *477*(7366), 611–615.

Clarke, J. H., Giudici, M. L., Burke, J. E., Williams, R. L., Maloney, D. J., Marugan, J., et al. (2015). The function of phosphatidylinositol 5-phosphate 4-kinase γ (PI5P4Kγ) explored using a specific inhibitor that targets the PI5P-binding site. *The Biochemical Journal*, *466*(2), 359–367.

Coales, S. J., E, S. Y., Lee, J. E., Ma, A., Morrow, J. A., & Hamuro, Y. (2010). Expansion of time window for mass spectrometric measurement of amide hydrogen/deuterium exchange reactions. *Rapid Communications in Mass Spectrometry*, *24*(24), 3585–3592.

Connelly, G. P., Bai, Y., Jeng, M. F., & Englander, S. W. (1993). Isotope effects in peptide group hydrogen exchange. *Proteins*, *17*(1), 87–92.

Dbouk, H. A., Vadas, O., Shymanets, A., Burke, J. E., Salamon, R. S., Khalil, B. D., et al. (2012). G protein-coupled receptor-mediated activation of p110β by Gβγ is required for cellular transformation and invasiveness. *Science Signaling*, *5*(253), ra89.

Demmers, J. A., Haverkamp, J., Heck, A. J., Koeppe, R. E., & Killian, J. A. (2000). Electrospray ionization mass spectrometry as a tool to analyze hydrogen/deuterium exchange kinetics of transmembrane peptides in lipid bilayers. *Proceedings of the National Academy of Sciences of the United States of America*, *97*(7), 3189–3194.

Demmers, J. A., van Duijn, E., Haverkamp, J., Greathouse, D. V., Koeppe, R. E., Heck, A. J., et al. (2001). Interfacial positioning and stability of transmembrane peptides in lipid bilayers studied by combining hydrogen/deuterium exchange and mass spectrometry. *Journal of Biochemistry*, *276*(37), 34501–34508.

Denisov, I. G., Grinkova, Y. V., Lazarides, A. A., & Sligar, S. G. (2004). Directed self-assembly of monodisperse phospholipid bilayer nanodiscs with controlled size. *Journal of the American Chemical Society*, *126*(11), 3477–3487.

Duc, N. M., Du, Y., Thorsen, T. S., Lee, S. Y., Zhang, C., Kato, H., et al. (2015). Effective application of bicelles for conformational analysis of G protein-coupled receptors by hydrogen/deuterium exchange mass spectrometry. *Journal of the American Society for Mass Spectrometry*, *26*, 808–817.

Englander, S. W. (2006). Hydrogen exchange and mass spectrometry: A historical perspective. *Journal of the American Society for Mass Spectrometry*, *17*(11), 1481–1489.

Englander, S. W., & Mayne, L. (2014). The nature of protein folding pathways. *Proceedings of the National Academy of Sciences of the United States of America*, *111*(45), 15873–15880.

Fang, J., Rand, K. D., Beuning, P. J., & Engen, J. R. (2011). False EX1 signatures caused by sample carryover during HX MS analyses. *International Journal of Mass Spectrometry*, *302*(1–3), 19–25.

Fowler, M. L., McPhail, J. A., Jenkins, M. L., Masson, G. R., Rutaganira, F. U., Shokat, K. M., et al. (2016). Using hydrogen deuterium exchange mass spectrometry to engineer optimized constructs for crystallization of protein complexes: Case study of PI4KIIIβ with Rab11. *Protein Science*, *25*(4), 826–839.

Gallagher, E. S., & Hudgens, J. W. (2016). Mapping protein-ligand interactions with proteolytic fragmentation, hydrogen/deuterium exchange-mass spectrometry. *Methods in Enzymology*, *566*, 357–404.

Hansen, R. K., Broadhurst, R. W., Skelton, P. C., & Arkin, I. T. (2002). Hydrogen/deuterium exchange of hydrophobic peptides in model membranes by electrospray ionization mass spectrometry. *Journal of the American Society for Mass Spectrometry*, *13*(12), 1376–1387.

Hebling, C. M., Morgan, C. R., Stafford, D. W., Jorgenson, J. W., Rand, K. D., & Engen, J. R. (2010). Conformational analysis of membrane proteins in phospholipid bilayer nanodiscs by hydrogen exchange mass spectrometry. *Analytical Chemistry*, *82*(13), 5415–5419.

Hon, W. C., Berndt, A., & Williams, R. L. (2012). Regulation of lipid binding underlies the activation mechanism of class IA PI3-kinases. *Oncogene*, *31*(32), 3655–3666.

Hsu, Y.-H., Burke, J. E., Li, S., Woods, V. L., & Dennis, E. A. (2009). Localizing the membrane binding region of Group VIA Ca2+-independent phospholipase A2 using peptide amide hydrogen/deuterium exchange mass spectrometry. *The Journal of Biological Chemistry*, *284*(35), 23652–23661.

Iacob, R. E., & Engen, J. R. (2012). Hydrogen exchange mass spectrometry: Are we out of the quicksand? *Journal of the American Society for Mass Spectrometry*, *23*(6), 1003–1010.

Iacob, R. E., Murphy, J. P., & Engen, J. R. (2008). Ion mobility adds an additional dimension to mass spectrometric analysis of solution-phase hydrogen/deuterium exchange. *Rapid Communications in Mass Spectrometry*, *22*(18), 2898–2904.

Jones, L. M., Zhang, H., Vidavsky, I., & Gross, M. L. (2010). Online, high-pressure digestion system for protein characterization by hydrogen/deuterium exchange and mass spectrometry. *Analytical Chemistry*, *82*(4), 1171–1174.

Kadek, A., Mrazek, H., Halada, P., Rey, M., Schriemer, D. C., & Man, P. (2014). Aspartic protease nepenthesin-1 as a tool for digestion in hydrogen/deuterium exchange mass spectrometry. *Analytical Chemistry*, *86*(9), 4287–4294.

Kan, Z.-Y., Walters, B. T., Mayne, L., & Englander, S. W. (2013). Protein hydrogen exchange at residue resolution by proteolytic fragmentation mass spectrometry analysis. *Proceedings of the National Academy of Sciences of the United States of America*, *110*(41), 16438–16443.

Karageorgos, I., Wales, T. E., Janero, D. R., Zvonok, N., Vemuri, V. K., Engen, J. R., et al. (2013). Active-site inhibitors modulate the dynamic properties of human monoacylglycerol lipase: A hydrogen exchange mass spectrometry study. *Biochemistry*, *52*(29), 5016–5026.

Katritch, V., Cherezov, V., & Stevens, R. C. (2013). Structure-function of the G protein-coupled receptor superfamily. *Annual Review of Pharmacology and Toxicology*, *53*, 531–556.

Katta, V., & Chait, B. T. (1991). Conformational changes in proteins probed by hydrogen-exchange electrospray-ionization mass spectrometry. *Rapid Communications in Mass Spectrometry*, *5*(4), 214–217.

Keppel, T. R., Jacques, M. E., Young, R. W., Ratzlaff, K. L., & Weis, D. D. (2011). An efficient and inexpensive refrigerated LC system for H/D exchange mass spectrometry. *Journal of the American Society for Mass Spectrometry*, *22*(8), 1472–1476.

Khan, A. G., Whidby, J., Miller, M. T., Scarborough, H., Zatorski, A. V., Cygan, A., et al. (2014). Structure of the core ectodomain of the hepatitis C virus envelope glycoprotein 2. *Nature*, *509*(7500), 381–384.

Kim, M., Sun, Z.-Y. J., Rand, K. D., Shi, X., Song, L., Cheng, Y., et al. (2011). Antibody mechanics on a membrane-bound HIV segment essential for GP41-targeted viral neutralization. *Nature Structural & Molecular Biology*, *18*(11), 1235–1243.

Konermann, L., Pan, J., & Liu, Y.-H. (2011). Hydrogen exchange mass spectrometry for studying protein structure and dynamics. *Chemical Society Reviews*, *40*(3), 1224–1234.

Koshy, S. S., Eyles, S. J., Weis, R. M., & Thompson, L. K. (2013). Hydrogen exchange mass spectrometry of functional membrane-bound chemotaxis receptor complexes. *Biochemistry*, *52*(49), 8833–8842.

Li, J., Rodnin, M. V., Ladokhin, A. S., & Gross, M. L. (2014). Hydrogen-deuterium exchange and mass spectrometry reveal the pH-dependent conformational changes of diphtheria toxin T domain. *Biochemistry*, *53*(43), 6849–6856.

Majumdar, R., Manikwar, P., Hickey, J. M., Arora, J., Middaugh, C. R., Volkin, D. B., et al. (2012). Minimizing carry-over in an online pepsin digestion system used for the H/D exchange mass spectrometric analysis of an IgG1 monoclonal antibody. *Journal of the American Society for Mass Spectrometry*, *23*(12), 2140–2148.

Man, P., Montagner, C., Vernier, G., Dublet, B., Chenal, A., Forest, E., et al. (2007). Defining the interacting regions between apomyoglobin and lipid membrane by hydrogen/deuterium exchange coupled to mass spectrometry. *Journal of Molecular Biology*, *368*(2), 464–472.

Marcoux, J., Thierry, E., Vivès, C., Signor, L., Fieschi, F., & Forest, E. (2010). Investigating alternative acidic proteases for H/D exchange coupled to mass spectrometry: Plasmepsin 2 but not plasmepsin 4 is active under quenching conditions. *Journal of the American Society for Mass Spectrometry*, *21*(1), 76–79.

Masson, G. R., Perisic, O., Burke, J. E., & Williams, R. L. (2016). The intrinsically disordered tails of PTEN and PTEN-L have distinct roles in regulating substrate specificity and membrane activity. *The Biochemical Journal*, *473*(2), 135–144.

Mayne, L. (2016). Hydrogen exchange mass spectrometry. *Methods in Enzymology*, *566*, 335–356.

Mayne, L., Kan, Z.-Y., Chetty, P. S., Ricciuti, A., Walters, B. T., & Englander, S. W. (2011). Many overlapping peptides for protein hydrogen exchange experiments by the fragment separation-mass spectrometry method. *Journal of the American Society for Mass Spectrometry*, *22*(11), 1898–1905.

Milne, J. S., Mayne, L., Roder, H., Wand, A. J., & Englander, S. W. (1998). Determinants of protein hydrogen exchange studied in equine cytochrome c. *Protein Science*, 7(3), 739–745.

Morgan, C. R., Hebling, C. M., Rand, K. D., Stafford, D. W., Jorgenson, J. W., & Engen, J. R. (2011). Conformational transitions in the membrane scaffold protein of phospholipid bilayer nanodiscs. *Molecular & Cellular Proteomics*, *10*(9). M111.010876-(1–11).

Nalefski, E. A., Slazas, M. M., & Falke, J. J. (1997). Ca2+-signaling cycle of a membrane-docking C2 domain. *Biochemistry*, *36*(40), 12011–12018.

Nasr, M. L., Shi, X., Bowman, A. L., Johnson, M., Zvonok, N., Janero, D. R., et al. (2013). Membrane phospholipid bilayer as a determinant of monoacylglycerol lipase kinetic profile and conformational repertoire. *Protein Science*, *22*(6), 774–787.

Noble, A. J., Zhang, Q., O'Donnell, J., Hariri, H., Bhattacharya, N., Marshall, A. G., et al. (2013). A pseudoatomic model of the COPII cage obtained from cryo-electron microscopy and mass spectrometry. *Nature Structural & Molecular Biology*, *20*(2), 167–173.

Orban, T., & Tsybovsky, Y. (2015). Probing conformational changes in rhodopsin using hydrogen-deuterium exchange coupled to mass spectrometry. *Methods in Molecular Biology (Clifton, NJ)*, *1271*, 113–121.

Pan, J., Han, J., Borchers, C. H., & Konermann, L. (2009). Hydrogen/deuterium exchange mass spectrometry with top-down electron capture dissociation for characterizing structural transitions of a 17 kDa protein. *Journal of the American Chemical Society*, *131*(35), 12801–12808.

Pantazatos, D., Kim, J. S., Klock, H. E., Stevens, R. C., Wilson, I. A., Lesley, S. A., et al. (2004). Rapid refinement of crystallographic protein construct definition employing enhanced hydrogen/deuterium exchange MS. *Proceedings of the National Academy of Sciences of the United States of America*, *101*(3), 751–756.

Parker, C. H., Morgan, C. R., Rand, K. D., Engen, J. R., Jorgenson, J. W., & Stafford, D. W. (2014). A conformational investigation of propeptide binding to the integral membrane protein γ-glutamyl carboxylase using nanodisc hydrogen exchange mass spectrometry. *Biochemistry*, *53*(9), 1511–1520.

Pirrone, G. F., Emert-Sedlak, L. A., Wales, T. E., Smithgall, T. E., Kent, M. S., & Engen, J. R. (2015). Membrane-associated conformation of HIV-1 Nef investigated with hydrogen exchange mass spectrometry at a Langmuir monolayer. *Analytical Chemistry*, *87*(14), 7030–7035.

Pirrone, G. F., Iacob, R. E., & Engen, J. R. (2014). Applications of hydrogen/deuterium exchange MS from 2012 to 2014. *Analytical Chemistry*, *87*, 99–118.

Pirrone, G. F., Vernon, B. C., Kent, M. S., & Engen, J. R. (2015). Hydrogen exchange mass spectrometry of proteins at Langmuir monolayers. *Analytical Chemistry*, *87*(14), 7022–7029.

Politis, A., & Borysik, A. J. (2015). Assembling the pieces of macromolecular complexes: Hybrid structural biology approaches. *Proteomics*, *15*, 2792–2803.

Rand, K. D., & Jørgensen, T. J. D. (2007). Development of a peptide probe for the occurrence of hydrogen (1H/2H) scrambling upon gas-phase fragmentation. *Analytical Chemistry*, *79*(22), 8686–8693.

Rand, K. D., Zehl, M., Jensen, O. N., & Jørgensen, T. J. D. (2009). Protein hydrogen exchange measured at single-residue resolution by electron transfer dissociation mass spectrometry. *Analytical Chemistry*, *81*(14), 5577–5584.

Rey, M., Forest, E., & Pelosi, L. (2012). Exploring the conformational dynamics of the bovine ADP/ATP carrier in mitochondria. *Biochemistry*, *51*(48), 9727–9735.

Rey, M., Man, P., Clémençon, B., Trézéguet, V., Brandolin, G., Forest, E., et al. (2010). Conformational dynamics of the bovine mitochondrial ADP/ATP carrier isoform 1 revealed by hydrogen/deuterium exchange coupled to mass spectrometry. *The Journal of Biological Chemistry*, *285*(45), 34981–34990.

Rey, M., Mrazek, H., Pompach, P., Novák, P., Pelosi, L., Brandolin, G., et al. (2010). Effective removal of nonionic detergents in protein mass spectrometry, hydrogen/deuterium exchange, and proteomics. *Analytical Chemistry*, *82*(12), 5107–5116.

Rey, M., Yang, M., Burns, K. M., Yu, Y., Lees-Miller, S. P., & Schriemer, D. C. (2013). Nepenthesin from monkey cups for hydrogen/deuterium exchange mass spectrometry. *Molecular & Cellular Proteomics*, *12*(2), 464–472.

Rostislavleva, K., Soler, N., Ohashi, Y., Zhang, L., Pardon, E., Burke, J. E., et al. (2015). Structure and flexibility of the endosomal Vps34 complex reveals the basis of its function on membranes. *Science (New York, NY)*, *350*(6257), aac7365.

Samuels, Y., Wang, Z., Bardelli, A., Silliman, N., Ptak, J., Szabo, S., et al. (2004). High frequency of mutations of the PIK3CA gene in human cancers. *Science (New York, N.Y.)*, *304*(5670), 554.

Sharma, S., Zheng, H., Huang, Y. J., Ertekin, A., Hamuro, Y., Rossi, P., et al. (2009). Construct optimization for protein NMR structure analysis using amide hydrogen/deuterium exchange mass spectrometry. *Proteins*, *76*(4), 882–894.

Shukla, A. K., Westfield, G. H., Xiao, K., Reis, R. I., Huang, L.-Y., Tripathi-Shukla, P., et al. (2014). Visualization of arrestin recruitment by a G-protein-coupled receptor. *Nature*, *512*, 218–222.

Skinner, J. J., Lim, W. K., Bédard, S., Black, B. E., & Englander, S. W. (2012a). Protein dynamics viewed by hydrogen exchange. *Protein Science*, *21*(7), 996–1005.

Skinner, J. J., Lim, W. K., Bédard, S., Black, B. E., & Englander, S. W. (2012b). Protein hydrogen exchange: Testing current models. *Protein Science*, *21*(7), 987–995.

Song, M. S., Salmena, L., & Pandolfi, P. P. (2012). The functions and regulation of the PTEN tumour suppressor. *Nature Reviews. Molecular Cell Biology*, *13*(5), 283–296.

Spraggon, G., Pantazatos, D., Klock, H. E., Wilson, I. A., Woods, V. L., & Lesley, S. A. (2004). On the use of DXMS to produce more crystallizable proteins: Structures of the T. maritima proteins TM0160 and TM1171. *Protein Science*, *13*(12), 3187–3199.

Srinivasan, S., Dharmarajan, V., Reed, D. K., Griffin, P. R., & Schmid, S. L. (2016). Identification and function of conformational dynamics in the multidomain GTPase dynamin. *The EMBO Journal*, *35*(4), 443–457.

Vadas, O., & Burke, J. E. (2015). Probing the dynamic regulation of peripheral membrane proteins using hydrogen deuterium exchange-MS (HDX-MS). *Biochemical Society Transactions*, *43*(5), 773–786.

Vadas, O., Burke, J. E., Zhang, X., Berndt, A., & Williams, R. L. (2011). Structural basis for activation and inhibition of class I phosphoinositide 3-kinases. *Science Signaling*, *4*(195), 1–13.

Vadas, O., Dbouk, H. A., Shymanets, A., Perisic, O., Burke, J. E., Abi Saab, W. F., et al. (2013). Molecular determinants of PI3Kγ-mediated activation downstream of G-protein-coupled receptors (GPCRs). *Proceedings of the National Academy of Sciences of the United States of America*, *110*(47), 18862–18867.

Vahidi, S., Bi, Y., Dunn, S. D., & Konermann, L. (2016). Load-dependent destabilization of the γ-rotor shaft in FOF1 ATP synthase revealed by hydrogen/deuterium-exchange mass spectrometry. *Proceedings of the National Academy of Sciences of the United States of America*, *113*(9), 2412–2417.

Vanhaesebroeck, B., Guillermet-Guibert, J., Graupera, M., & Bilanges, B. (2010). The emerging mechanisms of isoform-specific PI3K signalling. *Nature Reviews. Molecular Cell Biology*, *11*(5), 329–341.

Wales, T. E., Eggertson, M. J., & Engen, J. R. (2013). Considerations in the analysis of hydrogen exchange mass spectrometry data. *Methods in Molecular Biology (Clifton, N.J.)*, *1007*, 263–288.

Wales, T. E., & Engen, J. R. (2006). Hydrogen exchange mass spectrometry for the analysis of protein dynamics. *Mass Spectrometry Reviews*, *25*(1), 158–170.

Wales, T. E., Fadgen, K. E., Gerhardt, G. C., & Engen, J. R. (2008). High-speed and high-resolution UPLC separation at zero degrees Celsius. *Analytical Chemistry*, *80*(17), 6815–6820.

Walters, B. T., Ricciuti, A., Mayne, L., & Englander, S. W. (2012). Minimizing back exchange in the hydrogen exchange-mass spectrometry experiment. *Journal of the American Society for Mass Spectrometry*, *23*(12), 2132–2139.

Wei, H., Mo, J., Tao, L., Russell, R. J., Tymiak, A. A., Chen, G., et al. (2014). Hydrogen/deuterium exchange mass spectrometry for probing higher order structure of protein therapeutics: Methodology and applications. *Drug Discovery Today*, *19*(1), 95–102.

Weis, D. D., Wales, T. E., Engen, J. R., Hotchko, M., & Ten Eyck, L. F. (2006). Identification and characterization of EX1 kinetics in H/D exchange mass spectrometry by peak width analysis. *Journal of the American Society for Mass Spectrometry*, *17*(11), 1498–1509.

West, G. M., Chien, E. Y. T., Katritch, V., Gatchalian, J., Chalmers, M. J., Stevens, R. C., et al. (2011). Ligand-dependent perturbation of the conformational ensemble for the GPCR β2 adrenergic receptor revealed by HDX. *Structure (London, England: 1993)*, *19*(10), 1424–1432.

Wymann, M., & Schneiter, R. (2008). Lipid signalling in disease. *Nature Reviews. Molecular Cell Biology*, *9*(2), 162–176.

Zhang, X., Chien, E. Y. T., Chalmers, M. J., Pascal, B. D., Gatchalian, J., Stevens, R. C., et al. (2010). Dynamics of the beta2-adrenergic G-protein coupled receptor revealed by hydrogen-deuterium exchange. *Analytical Chemistry*, *82*(3), 1100–1108.

Zhang, H.-M., McLoughlin, S. M., Frausto, S. D., Tang, H., Emmett, M. R., & Marshall, A. G. (2010). Simultaneous reduction and digestion of proteins with disulfide bonds for hydrogen/deuterium exchange monitored by mass spectrometry. *Analytical Chemistry*, *82*(4), 1450–1454.

Zhang, Z., & Smith, D. L. (1993). Determination of amide hydrogen exchange by mass spectrometry: A new tool for protein structure elucidation. *Protein Science*, *2*(4), 522–531.

CHAPTER EIGHT

Reversible Chemical Dimerization by rCD1

M. Schifferer*, S. Feng*,†, F. Stein*, C. Schultz*,[1]
*European Molecular Biology Laboratory (EMBL), Heidelberg, Germany
†NCCR Chemical Biology, University of Geneva, Geneva, Switzerland
[1]Corresponding author: e-mail address: schultz@embl.de

Contents

Abstract

An optimal tool to unravel the role of a specific player within a cellular network or process requires its spatiotemporally resolved perturbation. Chemically induced dimerization (CID) by the rapamycin system has proven useful to induce protein dimerization or translocation with high spatiotemporal precision. Recently, we and others have added reversibility of the dimerization event as a novel feature to CID approaches. Among those, our reversible chemical dimerizer (rCD1) shows the fastest release kinetics observed, comparable to optogenetic methods. Induction and termination of enzyme activities, including phosphatidylinositol 3-kinase (PI3K) and 5-phosphatase (5Ptase), therefore allowed us to monitor the relaxation of the downstream effectors within living

Methods in Enzymology, Volume 583
ISSN 0076-6879
http://dx.doi.org/10.1016/bs.mie.2016.10.035

cells by imaging and traditional biochemical methods. Because switching off the rCD1-induced enzyme activity is sufficiently fast, it is possible to estimate kinetic parameters for enzyme activity and metabolism. Fast reversible CIDs are therefore unique tools for performing semiquantitative biochemistry in intact cells. In this chapter, we discuss advantages and constraints for the design of reversible CID applications. We provide detailed protocols for rCD1 synthesis, CID component expression in and delivery to mammalian cells and the determination of enzyme kinetics inside intact cells by a specially designed image acquisition and data analysis method.

1. INTRODUCTION

1.1 Chemically Induced Dimerization as a Switching Tool

Biological processes are tightly regulated to restore cellular homeostasis and to respond to environmental cues. In recent years, the validity of experimental results from bulk methods that strongly alter the levels of a certain cellular component and ignore cell-to-cell variability is questioned. Even inducible systems to up- or downregulate proteins suffer from limitations in precision of timing (min h) or the fine-tuning of dosage. The spatiotemporal resolution of the process of interest and the methods applied are strongly interdependent. Only the development of more sophisticated tools that allow acute and precise perturbation at the single-cell level will give rise to a deeper understanding of highly dynamic systems. In essence, tools are needed to rapidly and noninvasively switch enzyme activities on and off.

In order to meet these requirements, chemically inducible dimerizer (CID) systems have been developed as an alternative to the modulation of protein quantities (Inoue, Heo, Grimley, Wandless, & Meyer, 2005; Liang, Ho, & Crabtree, 2011; Putyrski & Schultz, 2012). With the addition of a bioorthogonal (meaning nontoxic) dimerizing molecule to the cellular system under investigation, two proteins are brought into close proximity allowing them to interact (Fig. 1). Using this principle, the natural or designed anchorage of one protein to a certain cellular site allows the induced relocalization of another molecule of interest to an ectopic site. The protein of interest faces a different microenvironment regarding pH and the presence of substrates or inhibitors. This change in location can be exploited to study protein functionality (Clackson, 2008).

Traditionally and most commonly, rapamycin has been used to induce the dimerization of FKBP- and FRB-fusion proteins (Putyrski & Schultz, 2012; Rivera et al., 1996). With a translocation speed in the seconds range,

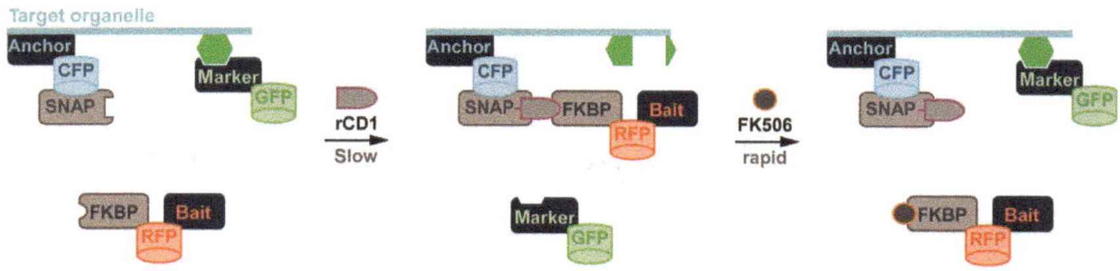

Fig. 1 The reversible chemical dimerizer approach based on rCD1 uses FK506 to outcompete the chemical dimerizer and switch-off the enzyme function. In the case shown, the bait is an enzyme destroying a membrane-bound molecule such as a lipid. After adding FK506, this molecule is rapidly resynthesized.

rapamycin has been employed to control enzyme activities within a signaling cascade or posttranslational modifications of interest. In many applications, the protein is recruited to the plasma membrane or to another organelle such as the nucleus where it meets its substrate or activator. Recent engineering of switchable kinases and phosphatases by insertion of FKBP allows the direct manipulation of enzyme activity upon dimerization (Chu et al., 2014). As rapamycin causes off-target effects by binding to and inhibiting endogenous TORC1 (Zoncu, Efeyan, & Sabatini, 2011), rapamycin analogs (rapalogs) have been developed which exhibit a conserved affinity to a mutated FRB without affecting mTOR signaling (Bayle et al., 2006).

However, the high-affinity binding of rapamycin or a rapalog still restricts the method as quasi-irreversible. This is due to the subnanomolar affinity of the ternary FRB–rapamycin–FKBP complex. Similarly, the use of annexin A4 for translocating baits to the plasma membrane, by elevating the intracellular calcium concentration, is usually irreversible due to annexin polymerization (Skrahina, Piljic, & Schultz, 2008). As most cellular processes are reversible by nature (e.g., protein phosphorylation), their complexity is better dissected by bifunctional tools allowing to rapidly switch between the on- and the off-state. To meet this requirement, two kinds of reversible chemical dimerizer systems, based on photoinduction or competition, have been developed (Voß, Klewer, & Wu, 2015). As an alternative, optogenetic methods have also been used to switch enzyme activity (Tischer & Weiner, 2014).

1.2 The Reversible Chemical Dimerizer System rCD1: Mode of Action

Recently, we developed a reversible chemical dimerizer system based on chemical competition for the FKBP-binding site (Feng et al., 2014).

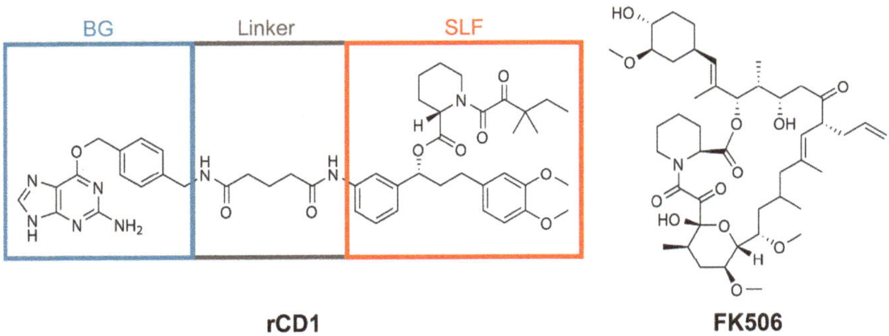

Fig. 2 Chemical structures of rCD1 and FK506.

The approach is based on rCD1 (reversible chemical dimerizer 1), a bifunctional dimerizer comprising a BG (benzylguanine) and an SLF (synthetic ligand of FKBP) moiety with a linker inbetween (Fig. 2). The BG moiety covalently and irreversibly binds to the SNAP-tag. The SLF component is designed to bind to FKBP with submicromolar affinity. It is outcompeted by addition of FK506 (Fig. 2) due to its 50-fold lower dissociation constant (0.4 n*M* vs 20 n*M*) (Briesewitz, Ray, Wandless, & Crabtree, 1999). The protein determining the cellular target localization, called anchor protein, is fused to a SNAP-tag and a fluorescent protein (FP). The second bait construct combines the FKBP tag, a bait protein (for example, an enzyme) and another FP color variant. Upon addition of rCD1 to cells that transiently express both constructs, the bait is targeted to the anchor within 10–30 min via bound SLF–FKBP interaction (Figs. 1 and 3). Fast release of the bait is initiated by FK506 supplementation that replaces SLF at the FKBP-binding site (Fig. 3). Consequently, rCD1 remains attached to the SNAP-tag.

SNAP-tag is an engineered mutant of the DNA repair protein O^6-alkylguanine-DNA alkyltransferase (AGT) that forms a covalent thioether bond with BG (Juillerat et al., 2003). It has been extensively used for protein labeling (Keppler et al., 2003). One of the benefits from SNAP-tagging originates from its small size (20 kDa), leading to minimal interference with the functionality of the fused protein. It has been shown that BG modifications show little alterations in the reaction rates (Bosch et al., 2014). BG derivatives are fairly bioorthogonal in a sense that they do not unspecifically bind to endogenous cellular components although some coupling to endogenous AGT cannot be fully excluded.

As the FKBP-FRB system is well established for protein dimerization and has been widely applied, we took advantage of many available bait

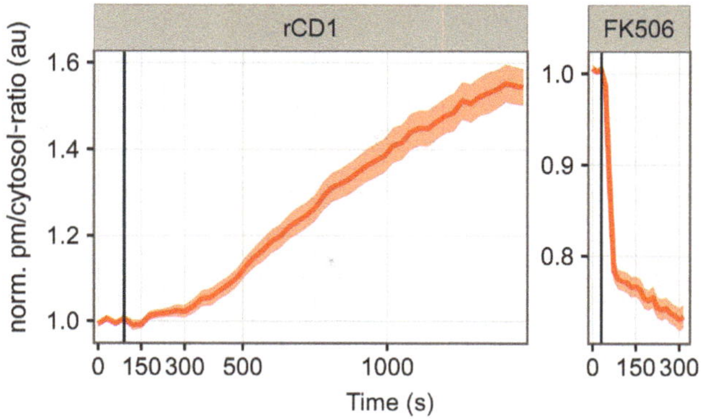

Fig. 3 Translocation kinetics of iSH2-FKBP to the plasma membrane upon rCD1 treatment and release upon FK506 treatment. This construct stimulates PI3K activity when translocated to the plasma membrane (Suh, Inoue, Meyer, & Hille, 2006). Traces show the mean of plasma membrane (pm) to cytosol ratio of HeLa Kyoto cells which was normalized to the average of data points prior to stimulation with rCD1 or FK506 as indicated with a vertical *black line*. *Error bars* indicate standard error of the mean.

constructs ready for usage with our rCD1 system. Additionally, beside FK506, there are several FKBP-binding small molecules, such as rapamycin and rapalogs, that can be used to fine-tune the competition with the SLF moiety of rCD1.

1.3 Measuring the Effects of Reversible Chemical Dimerization

The reversible CID experiment is designed for imaging applications. Direct or indirect effects of bait dimerization with the anchor protein are observed by suitable reporters. When translocating enzymes, the substrate or product levels are monitored. Usually, marker proteins are fused to FPs. It has to be considered that the marker itself can compete with the bait for substrate or product binding thereby inhibiting its function.

As a major characteristic, chemical dimerizers can only be used in cell batch experiments as the addition will reach all cells in a dish. The dimerizer molecule enters all cells in a dish simultaneously at a defined time point. This allows for the screening of downstream effects by biochemical methods. For studying the effects of protein kinase or phosphatase activity on a signaling network, the enzyme of interest is reversibly translocated and protein levels or activities are investigated by Western blotting, for example. For single-cell applications, a photoactivatable chemical dimerizer is used (Ballister, Aonbangkhen, Mayo, Lampson, & Chenoweth, 2014).

1.4 Kinetic Parameters of Translocation and Release

The rCD1 translocation speed of about 20–30 min is orders of magnitude slower than the irreversible rapamycin-inducible system, which is completed after several seconds. So far, only modest efforts were spent to understand the relatively slow labeling process. As increasing rCD1 concentrations do not alter translocation kinetics we hypothesize that cell entry is not the rate-limiting step. Direct delivery, by microinjecting rCD1 into the cytoplasm, for example, would allow direct comparison of dimerizer-binding kinetics, independent of their cellular uptake characteristic. In our hands, this did not increase the translocation speed (Feng et al., 2014). As rCD1 can be regarded as a modified BG substrate for AGT, its binding kinetics could possibly be improved by systematic modification of rCD1 linker length (Erhart et al., 2013) or the molecular evolution of the SNAP-tag (Gronemeyer, Chidley, Juillerat, Heinis, & Johnsson, 2006; Juillerat et al., 2003). It is further possible that the orientation or distance of the SNAP-tag from the membrane is unfavorable for fast labeling.

On the other hand, outcompeting SLF by FK506 initiates the fastest reversion of a dimerization process reported so far. We took advantage of this extraordinary speed for relaxation experiments enabling the analysis of fast processes like lipid metabolism. As a prerequisite, FK506 is a small hydrophobic molecule that quickly enters cells. We assume that fast kinetics originate from a 50-fold higher binding affinity of FK506 to FKBP ($K_d = 0.4$ nM) compared to SLF in rCD1 ($K_d = 20$ nM) (Briesewitz et al., 1999). This was confirmed by slower release rates of FKBP-fusion proteins by rapalogs with intermediate affinity (unpublished results).

Table 1 and Fig. 4 directly compare the translocation and release of either iSH2 or 5Ptase to and from the plasma membrane. As an anchor we employed the myristoylated and palmitoylated N-terminal tail of lymphocyte-specific protein tyrosine kinase Lck (Lck anchor). Half times were estimated by fitting monoexponential decay curves to single-cell traces (see Section 2.6 for details). The translocation of iSH2 is slightly slower than translocation of the 5Ptase to the plasma membrane, although their release kinetics are comparable. The production of PIP_3 upon iSH2 translocation or depletion of PIP_2 upon 5Ptase translocation is faster than the bait translocation indicating that the fully translocated bait is not necessary to induce enzyme activity. After addition of FK506, the release kinetics between 5Ptase and resynthesis of PIP_2 are comparable (~20 s). However, the dephosphorylation of PIP_3 by PTEN after the release of iSH2 shows the

Table 1 Translocation and Release Half-Life Times for Translocation of iSH2 and 5-Phosphatase to the Plasma Membrane
Plasma Membrane Anchor Lck-eCFP-SNAP

Bait	iSH2	5Ptase
Molecular weight	64 kDa	109 kDa
rCD1 addition	PIP_3 generation started	$PI(4,5)P_2$ depletion started
Translocation bait ($t_{1/2}$)	20 ± 14 min	6 ± 5 min
Translocation marker ($t_{1/2}$)	5 ± 3 min	3 ± 2
FK506 addition	PIP_3 generation stopped	$PI(4,5)P_2$ depletion stopped
Monitored process	PIP_3 metabolism (PTEN)	$PI(4,5)P_2$ recovery (PI5K)
Marker	PH_{AKT}	$PH_{PLC\delta}$
Release bait ($t_{1/2}$)	20 ± 7 s	19 ± 7 s
Release marker ($t_{1/2}$)	52 ± 12 s	18 ± 9 s

slowest decay of all FK506 induced processes. Nevertheless, that K_{cat}/K_m is still estimated to be about 10,000 $mol^{-1}\ s^{-1}$ (Feng et al., 2014).

2. METHODS

2.1 rCD1 Applications to Monitor Lipid Metabolism in Living Cells

Studying lipid metabolism and transport has been hindered by the lack of methods for rapid intervention. The rCD1 system has been applied to modulate phosphoinositide (PI) levels in the plasma membrane. The Lck anchor has been expressed as neutral anchor fusion to eCFP and SNAP for plasma membrane translocation (Zlatkine, Mehul, & Magee, 1997). We took advantage of the fast release kinetics of rCD1 to reversibly recruit PI kinases and phosphatases as bait constructs to the plasma membrane and study downstream effects (Feng et al., 2014; Schifferer, Feng, Stein, Tischer, & Schultz, 2015).

In a first study, we locally and transiently generated PIP_3 from $PI(4,5)P_2$ by reversibly activating PI3K activity (Feng et al., 2014) (Fig. 5A). This was achieved by translocating the iSH2 (inter-Src homology 2, bait) domain to

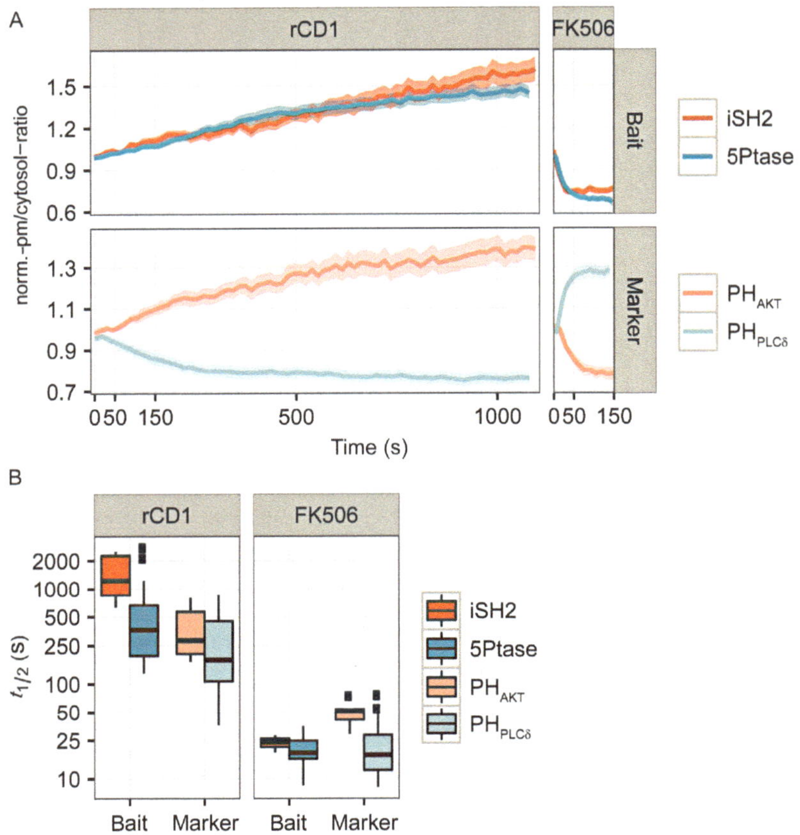

Fig. 4 Analysis of lipid metabolism by reversible dimerization at the plasma membrane using iSH2 and 5Ptase as baits. (A) Averaged plasma membrane/cytosol-ratio traces for the translocation of the bait to the plasma membrane upon rCD1 treatment and their releases upon FK506 treatment. The marker is PH_{AKT} for iSH2 and $PH_{PLC\delta}$ for the 5Ptase. (B) Corresponding half times extracted from single-cell traces of (A). Please note that the absolute values of two enzyme systems are not directly comparable because the number of substrate and enzyme molecules is very different.

the plasma membrane, where it in turn switched on the PI3K activity of the p110α domain that converts $PI(4,5)P_2$ into PIP_3 (Suh et al., 2006). Similarly, translocation of another lipid metabolizing enzyme, the phosphatidylinositol 5-phosphatase (5Ptase), was applied to generate PI(4)P from $PI(4,5)P_2$ (Schifferer et al., 2015) (Fig. 5B).

In both experiments, lipid turnover was observed by the pleckstrin homology (PH) domains PH_{Akt} and $PH_{PLC\delta}$ (phospholipase C δ) that specifically recognize PIP_3 or $PI(4,5)P_2$, respectively (Fig. 4). Beyond direct

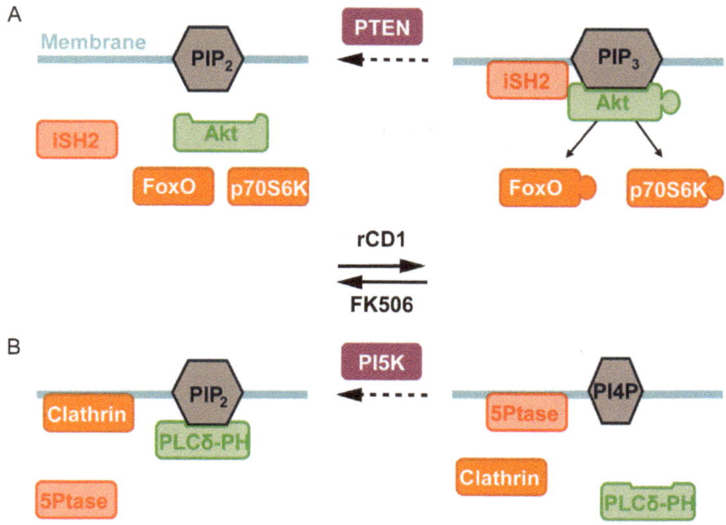

Fig. 5 Analysis of PIP_3 metabolism (A) and PI(4)P phosphorylation (B) by relaxation experiments induced by reversible dimerization at the plasma membrane. Upon removing the enzyme from the plasma membrane, PH domains report on phosphoinositide turnover in real-time, permitting the determination of kinetic parameters for the metabolizing enzymes in intact cells.

imaging of PI levels, the fast enzyme removal from the plasma membrane within 15–25 s permits relaxation experiments. As soon as the enzyme was removed, we exclusively monitored the metabolism of the lipid via its specific reporter domain. In the case of PIP_3, we observed the dephosphorylation by the cysteine phosphatase PTEN; in the case of PI 5-phosphatase translocation, we monitored the rapid resynthesis of PIP_2 from PI(4)P. In regard to PIP_3 metabolism, it should be noted that addition of H_2O_2, a reversible PTEN inhibitor, fully blocked PIP_3 turnover, suggesting that only PTEN and not the PI 5-phosphatase is dominating the PIP_3 metabolism (Feng et al., 2014). Therefore, we were able to estimate the kinetic K_{cat}/K_m parameters for PTEN and found PIP_3 turnover rates of roughly 10,000 $mol^{-1}\ s^{-1}$ (Feng et al., 2014) (for details, see Section 2.6). This value is in strong contrast to the much more sluggish turnover rates reported in the BRENDA database. Studies to determine the kinetic values for PI 5-kinase in intact cells are under way.

In addition to monitoring PIP_3 turnover, iSH2 translocation was used to dissect downstream effects of PIP_3 synthesis, and breakdown in the PI3K/Akt pathway. It is known that PIP_3 can modulate the activity of Akt and downstream effectors including FoxO and p70S6K. This was assessed *in vitro* by Western blotting. The results demonstrated how the hierarchy within the phosphorylation cascade downstream of Akt is organized and what delay mechanisms are involved in transducing a signaling event.

Among many other processes, PI(4,5)P_2 plays a major role in endocytosis, as the recruitment of several endocytic proteins including adaptor protein 2 (AP2) and subsequently clathrin is PI(4,5)P_2 dependent. By using the translocatable PI 5-phosphatase system, we reduced PIP_2 levels and released AP2 and consequently clathrin from the plasma membrane. After removing the 5-phosphatase from the plasma membrane, the immediate resynthesis of PIP_2 levels permitted the reformation of clathrin-coated pits. By using clathrin-GFP fusions we monitored this event in real-time for the first time (Schifferer et al., 2015).

2.2 Cell Culture Preparation for Reversible Translocation

Reversible dimerization has been successfully tested in different cell lines, each requiring specifications in the translocation protocol. Here we describe a reversible dimerization procedure in HeLa cells with general model constructs involving an anchor, bait, and marker construct. Cell confluency, the amount of DNA, and the transfection method need to be adapted to the cell type and the constructs of interest. We discuss crucial aspects and give examples for variations in a separate section.

Protocol

1. Under laminar flow hood prepare DMEM containing 4.5 g/L glucose (Gibco, cat. no. 41965-062), supplemented with 10% FBS (Gibco, cat. no. 10270-106),1% Penicillin-streptomycin (Gibco). Sterile-filter, aliquot, and store at 4°C. Prewarm to 37°C (water bath) before usage.
2. Fill the 8-well LabTek chamber slides (Thermo Scientific, Waltham, cat. no. 155411) with 300 μL of prewarmed full growth medium. Add appropriate amount of the HeLa cell suspension to the dish in a way that the cells reach about 80% confluency on the next day. Place the dishes in the incubator and let the cells attach overnight.
3. Prepare two tubes containing 25 μL Opti-MEM (without serum and any other additives) each for every dish of cells to be transfected. In one of the two tubes, add Lipofectamine 2000 (Invitrogen, Waltham, cat. no. 11668-019), in the other one add appropriate amounts of the DNA constructs (Table 2). It is crucial to adjust the ratio of DNA to anchor:bait: marker to 8:2:1. Use 1 μL of Lipofectamine 2000 per LabTek well. Mix the solution gently and let stand under the hood for 5 min. Combine

Table 2 Transfection and Imaging of Translocation Components

DNA Construct	DNA Amount for Transfection (μg)	Excitation Wavelength (nm)	Emission Wavelength (nm)
Lck-eCFP-anchor	1	405	415–500
eGFP-marker	0.3	488	510–600
mRFP-FKBP-(bait)	0.2	561	610–700

the contents of both tubes (the one containing DNA and the one with the transfection reagent), mix gently, and let stand under the hood for another 20 min. During the incubation time gently exchange the full growth medium in the LabTek wells with 250 μL of prewarmed Opti-MEM (Thermo Fisher Scientific). Do not wash the cells. Add the DNA–Lipofectamine complexes to the dishes. Mix gently and place the dishes in the incubator. Exchange the transfection solution for the growth medium after 3–12 h.

4. Allow the cells to express the constructs for at least 15 h. Remove the medium from the dish. Add 100 μL imaging buffer (20 m*M* HEPES (Biomol, Hamburg, cat. no. 05288), 115 m*M* NaCl (Fluka, St. Louis, cat. no. 71381), 1.2 m*M* $CaCl_2$ (Merck, Darmstadt, cat. no. 1.02382.1000), 1.2 m*M* $MgCl_2$ (Merck, cat. no. 1.05833.1000), 1.2 m*M* K_2HPO_4 (Merck, cat. no. 1.05099.1000), pH 7.4. sterile-filter, and aliquot under laminar flow hood), add 100 μL rCD1 stock solution to the cells, and place the dish in the 37°C incubator for 30 min.

2.3 Live Cell Imaging of Translocation and Release

For these experiments we used a LSM 780 confocal microscope (Zeiss, Jena) equipped with an environmental chamber to maintain 37°C and stable humidity and a 63 × 1.40 oil objective.

Protocol

1. Prepare the imaging buffer, sterile-filter, and aliquot under laminar flow hood. Before usage, supplement the medium with glucose up to 2 g/L (from sterile 20% stock solution (Merck, cat. no. 1.04074.1000)) and prewarm to 37°C (water bath). Remove the dish with cells from the incubator. Aspirate the medium and gently add 150 μL of prewarmed imaging buffer.

2. Stock solutions for dimerizing reagents (2 m*M* rCD1 (SiChem GmbH, Bremen, cat. no.SC-7000), 10 m*M* FK506 (Tocris Bioscience, Bristol)) are prepared in DMSO (Sigma, cat. no. D2438), aliquotted, and stored at −20°C. Repetitive freezing–thawing cycles have to be avoided. Please make sure to prepare rCD1 and FK506 working solutions freshly for usage at the same day. rCD1 (5 μ*M*) and FK506 (5 μ*M*) working solutions are prepared in the imaging buffer.
3. Quickly transfer the dish to the microscope with preheated (37°C) and humidified environmental chamber. Make sure to minimize the incubation time in the microscope chamber. If no time-lapse of the translocation experiments before the release is needed cells have to be stored in the regular cell culture incubator for that time.
4. Set the image acquisition parameters: scanning frequency 400 Hz, bidirectional scanning, 8-bit mode, pinhole opened to 1.0 airy. Choose proper laser lines for fluorophore excitation and suitable spectral windows for collection of the emitted light (Table 1). Set the laser power and adjust gain and offset for every channel. Use sequential scanning option to minimize the bleed-through of the present fluorophores into other channels. It is strongly recommended to use a software-based sample drift correction or an active drift stabilization if applicable.
5. Start the live imaging experiment by choosing a proper ROI. Translocation is most probable to occur in cells that show high anchor expression compared to bait. This criterion has to be in compromise with a sufficient amount of bait compared to marker to destroy its binding capacity. The lower limit of marker expression level is the signal-to-noise taking into account the bleaching of the fluorophore.
6. Start imaging at 2 frames per minute and add 50 μL warm rCD1 working solution after 5–10 frames. Prewarming and gently pipetting by shortening of a pipette tip to generate a larger tip diameter are crucial to avoid focus shifts. The time-lapse is imaged for 30–40 min until translocation is complete.
7. Release is most probable to be monitored in cells with a maximum colocalization between the bait and the anchor which indicates complete translocation. If there is no need for release data from the same cells imaged during translocation, it is recommended to choose another ROI accordingly.
8. Translocation reversibility is optimal if the release is started immediately after completion of translocation. Start imaging in all three channels with a frame rate of one per 1–5 s for initially 1–2 min of base line acquisition.

Stimulate cells with 1–5 μ*M* FK506 and run time-lapse for another 5–15 min depending on the timeline of the observed process.

2.4 Image Processing

Image processing was performed using ImageJ or Fiji and our "Translocation-Quantification Macro" which is available from https://github.com/fstein. A detailed R markdown file explaining the downstream data analysis using the R programming language is also available for download. The macro combines the algorithms for analyzing protein translocation that have been described (Feng et al., 2014; Nadler et al., 2015). The basic principle relies on a pixel classification into two different classes (e.g., plasma membrane and cytosol), calculation of the class averages separately for each cell and finally computation of a ratio (e.g., plasma membrane/cytosol-ratio). Different algorithms are available to classify the pixels. The user can either manually define the two pixel classes using ImageJ's "Pencil Tool," use the outer rim of each cell as the plasma membrane or employ a "marker" channel (an additional channel showing the localization of a compartment, e.g., nucleus by DAPI staining or plasma membrane by Lck-ECFP expression) to classify pixels.

Protocol

1. Load images into ImageJ or Fiji.
2. Subtract background (Process>Subtract Background...).
3. Apply median filter if necessary (Process>Filters>Median...).
4. Convert image to 32-bit (Image>Type>32-bit).
5. Threshold image to indicate low intensity pixels (Image>Adjust>Threshold...).
6. Remove low intensity pixels (Process>Math>NaN Background).
7. Load "Translocation-Quantification Macro" into Fiji (Plugins>Macros>Install...)
8. Start Macro (Plugins>Macros>Translocation-Quantification Macro), choose options and hit "OK" button to perform analysis (see Fig. 6).
9. The macro guides the user interactively through the analysis and gives multiple additional imaging channels, graphical plots, and data tables as an output to the screen.

Depending on the image quality, calculation of plasma membrane–cytosol ratios might be difficult, for example, if the translocation results in only subtle changes to the pixel intensity distribution. Defining the plasma membrane maybe difficult, particularly if cells are moving over time (in this

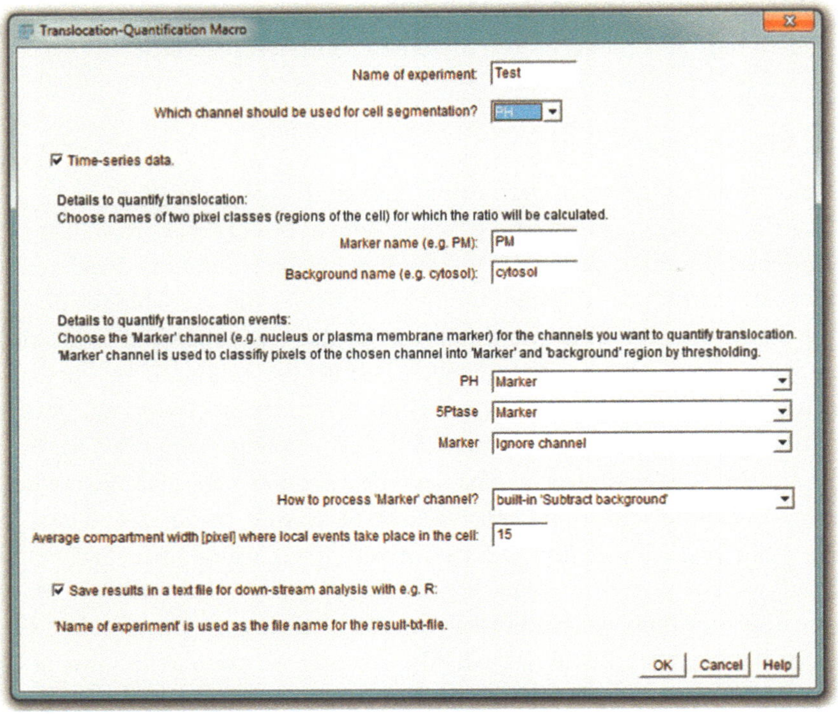

Fig. 6 Screenshot of the graphical user interface of the "Translocation-Quantification-Macro."

case a marker channel such as the Lck-ECFP is needed to accurately define the plasma membrane in each time frame). TIRF microscopy (TIRFM) is an alternative.

2.5 Imaging Plasma Membrane Translocation by TIRF Microscopy

A large number of cellular processes occur at the plasma membrane. Many signaling pathways are controlled by enzymes that are active only if they are located there, for example, they require activation, or because the substrate location is restricted to the plasma membrane. Reversible translocation experiments to plasma membrane anchors can be monitored either by confocal or evanescent field microscopy (TIRFM). Beyond eliminating out-of-focus fluorescence and increasing the signal-to-noise ratio, TIRF microscopy enables simplification of image analysis. An increase of the fluorescence at the plasma membrane can be simply analyzed as an increase in the

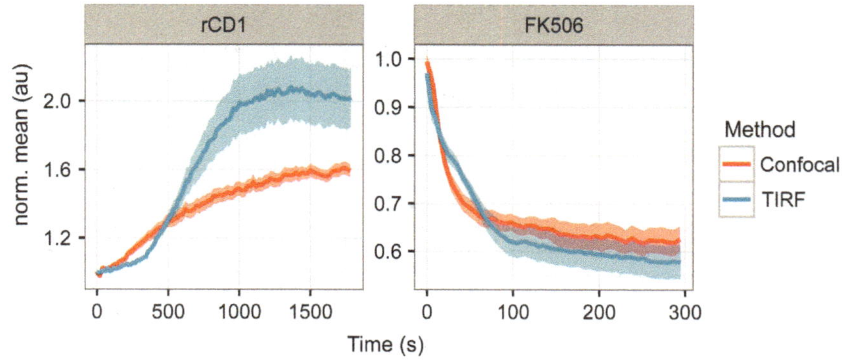

Fig. 7 5Ptase translocation and release to and from the plasma membrane and comparison between normalized plasma membrane/cytosol-ratio traces derived from confocal images and normalized fluorescent intensities derived from TIRF microscopy images.

average pixel intensity since only the plasma membrane fluorescence is monitored. Therefore, quantification is not dependent on the error-prone classification of pixels into plasma membrane and cytosol. This obviates the need for a FP tag in the anchor construct that simplifies the assay or could potentially be used for another marker protein of interest. We used Lck-ECFP-SNAP and ECFP-SNAP-CAAX anchor constructs with N-terminal or C-terminal plasma membrane anchorage, respectively. Fig. 7 shows a comparison between the two complementary analysis methods for the 5Ptase translocation. The release from the plasma membrane upon FK506 treatment gives almost identical release kinetics using both methods. No differences in translocation efficiency of bait constructs have been observed. However, the spatial separation of the SNAP-tag from the plasma membrane is crucial for translocation. We found that baits could not be translocated to Lck-SNAP constructs without the ß-barrel of a FP. For that reason, the anchoring function was rescued by inserting a dark mutant of ECFP between Lck and SNAP (Lck-$ECFP_{dark}$-SNAP) (Putyrski & Schultz, 2011). The performance of other dimerizing systems was limited by steric hindrance or accessibility to the substrate (Buckley, Petersen, Roe, Douce, & Christie, 2015; Pathak, Vrana, & Tucker, 2013). Similarly, the functionality of the bait protein could potentially be affected by its distance to the plasma membrane.

2.6 Estimation of Enzyme Kinetics in Living Cells

An easy way to describe the rate of a reaction is the time required until 50% of the reporter translocated or released from the plasma membrane (half time). This was done by fitting exponential curves to the single-cell plasma membrane/cytosol-ratio traces. For traces showing a decrease over time we used the following function to describe the kinetic:

$$\text{norm.ratio}(t) = ae^{-bt} + c$$

for traces showing an increase over time we used the following function:

$$\text{norm.ratio}(t) = a\left(1 - e^{-bt}\right) + c$$

The half time was then determined as following:

$$t_{1/2} = \frac{\log(0.5)}{b} - 1$$

Reversible iSH2 translocation has been applied to modulate PIP_3 levels at the plasma membrane (Feng et al., 2014). Fast release kinetics allowed the observation of PIP_3 breakdown at the site where it had been ectopically generated. We plotted the iSH2 release (kinase activity) and product decrease (PH_{Akt} monitoring PIP_3) at different FK506 concentrations. Both traces were used to determine half times ($t_{1/2}$) for PIP_3 and iSH2 for the FK506 titration. We found that PIP_3 levels were independent of PI3K activity at a FK506 concentration higher than 500 n*M* (see Fig. 8). For the 5Ptase we did not observe this effect. Here the release speed of the 5Ptase correlated with the synthesis speed of PIP_2. In contrast to iSH2, an overexpressed and active enzyme was released from the plasma membrane. In case of iSH2,

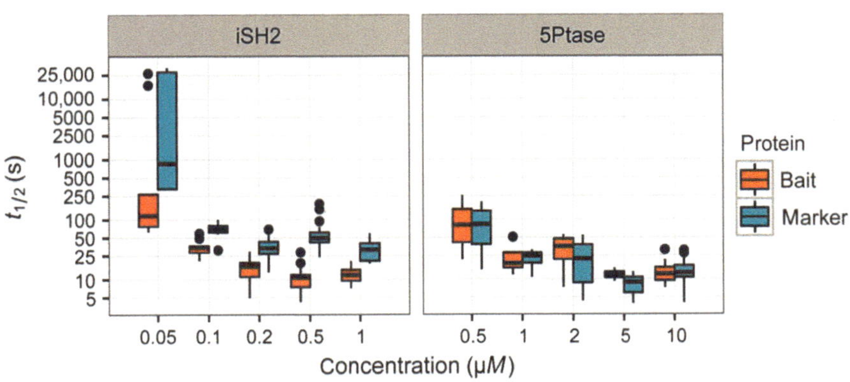

Fig. 8 Half times $t_{1/2}$ for translocation of bait (iSH2 or 5Ptase) and marker (PH_{AKT} for iSH2 and $PH_{PLC\delta}$ for 5Ptase) to and from the plasma membrane upon release with different FK506 concentrations.

endogenous PI3K was not further activated determining the limit of the catalytic activity.

Because PIP_3 dephosphorylation was mainly catalyzed by endogenous PTEN (Feng et al., 2014) its catalytic activity could be calculated according to the Schnell-Mendoza equation (Goudar, Sonnad, & Duggleby, 1999):

$$[S] = K_M W\left(\frac{[S]_0}{K_M} e^{\frac{[S]_0}{K_M} - \frac{v_{max}}{K_M} t}\right)$$

$$v_{max} = k_{cat}[E]_{tot}$$

W = Lambert-W function; [S] = substrate concentration; $[S]_0$ = initial substrate concentration; v_{max} = maximal rate; k_{cat} = turnover number; K_M = Michaelis constant; $[E]_{tot}$ = total enzyme concentration measured by Western blotting.

First, the plasma membrane to cytosol ratio for each single cell was normalized by defining the average of the last 10 data points (the plateau) as 0 μM and the start (maximum value) as the initial PIP_3 concentration ($[S]_0 = 0.5$ or 1 μM). By fitting the Schnell-Mendoza equation to these normalized traces the Michaelis Menten constant K_M and the maximal turnover rate v_{max} of PTEN were estimated. Together with the total enzyme concentration of PTEN, which was determined to be 0.08 μM by quantitative Western blotting, the turnover number of PTEN molecules was estimated from quantitative Western blots by calculating the k_{cat}/K_M ratios. We found that k_{cat}/K_M values for PTEN that we measured in living cells were significantly higher compared to *in vitro* measurements extracted from the BRENDA database. This observation holds true for two different PIP_3 concentrations used as the initial start values $[S]_0$. This demonstrates that PI3K and phosphatase activities are both high and likely neutralize each other to a large extended in order to maintain lipid homeostasis and, consequently, downstream effectors of PIP_3.

3. LIMITATIONS

3.1 Critical Parameters for Setting up a Novel Translocation System

In order to validate a new bait construct, it is useful to test different validated SNAP-anchor constructs including Lck, CAAX, or KDEL for their ability to translocate. As a control experiment it is recommended to use an FRB anchor to perform rapid rapamycin-mediated translocation. The functional anchor–bait system can subsequently be combined with the appropriate expression levels of the marker protein of interest. Such constructs are available from the Schultz lab upon request.

Most bait proteins include enzymes that are only functional when localized to a certain subcellular site. At this location, these enzymes are either activated by a local factor or they meet their substrate molecules. If bait expression is toxic for the time between transfection and translocation, the amount of cDNA must be reduced. Another way to circumvent this is to place the bait under the control of an inducible or weak promoter. It has been shown that trapping a bait at another organelle, e.g., in the nucleus by NLS fusion, successfully quenches the detrimental effect before the actual translocation (Goodchild et al., 2015). Timing of the relaxation experiment also depends on the toxicity of the bait at the anchorage site.

A properly validated translocation experiment requires an optimal reporter that is sensitive and specific to the bait protein substrate or product. Many marker proteins bind to their target with high affinity thereby limiting the accessibility of the bait to its substrate as it has been shown for the PH domain of PLCδ (Rosivatz & Woscholski, 2011).

We found that several other parameters must be adapted for an optimal functioning of the reversible translocation experiment in different cell types. These include cell confluency, the total, and relative amounts of DNA transfected and the transfection method itself. For example, we used one dose of 0.5 μg DNA for a genome edited SK-MEL2 cell line that expresses a marker protein (clathrin) fusion to RFP. In that case, the bait was controlled by a CMV promoter and was available in excess compared to the endogenous promoter control of the marker protein.

3.2 Comparison of rCD1 With Other Reversible CID Systems

The combination of two binding modules has been exploited for the development of other reversible dimerization systems. Similar to rCD1, an SLF′-TMP compound has been successfully applied for dimerization experiments (Liu et al., 2014). Trimethoprim (TMP) selectively binds *Escherichia coli* dihydrofolate reductase (eDHFR) thereby ensuring bioorthogonality in mammalian cells (Chen, Jing, Gallagher, Sheetz, & Cornish, 2012). Consequently, TMP shows minimal off-target effects comparable to SNAP-tags. In SLF′-TMP-mediated release, the SLF′ serves as the permanent binding interface, while TMP is used as a competitor similar to FK506 in rCD1 release. SLF′-TMP-mediated translocation is about 10–20 times faster compared to rCD1; however, out competition by TMP is inherently slow with a release occurring within minutes. In contrast to rCD1, SLF′-TMP allows several rounds of translocation but requires washing steps (Czlapinski et al., 2008). The two techniques should be orthogonal and have the potential to be used simultaneously.

Besides diffusive chemical dimerization, light-induced dimerizer uncaging has been exploited in order to achieve fast kinetics (Karginov et al., 2011; Umeda, Ueno, Pohlmeyer, Nagano, & Inoue, 2011; Wright, Guo, & Liang, 2015; Zeng, Zhang, Xuan, Wang, & Liang, 2015). Equipped with a photoprotecting group, the caged dimerizer is only allowed to trigger protein dimerization upon UV illumination. However, this strategy lacks spatial specificity due to the diffusive motion of liberated dimerizers. More importantly, the dimerization process is not reversible. Recently, this was overcome by the development of a probe called MeNV-HaXS. Here, a photocleavable linker is placed between the Halo- and SNAP-tags that covalently bind to their respective ligands (Suh et al., 2006). Similar to the rCD1 system, the MeNV-HaXS method is still limited to one round of reversion after photolytic cleavage. This was overcome by optogenetic switching strategies based on phytochromes or LOV domains that allow repetitive switching (Ballister et al., 2014). However, while the release efficiency can be improved by excess chemical competitor, the optimization of optical induction is limited by phototoxic effects. A major difference of optogenetic tools compared to chemical dimerizers is the locally generated effect induced by light that cannot be mimicked by batch experiments.

In order to combine the advantages of chemical competition with spatial precision of photoinduction, Ballister et al. designed the photocleavable reversible chemical dimerizer cTMP-Htag (Ballister et al., 2014). The TMP domain is caged (cTMP) and attached to a Halotag ligand (Htag). The reversion by TMP addition takes up to 10 min (Ballister et al., 2014).So far the rCD1 approach shows the fastest relaxation kinetics of any chemical-induced dimerization methods allowing the monitoring of fast processes (Feng et al., 2014).

Repetitive switching has been achieved by optogenetic methods based on phytochromes or LOV domains (Pathak et al., 2013). Compared to CID approaches, optogenetic methods have the advantage of generating local effects with precise spatiotemporal resolution (Fig. 9). However, the bulky size of at least one bait reduces the transfection efficiency, and stable cell lines are often required for successful experiments. Furthermore,

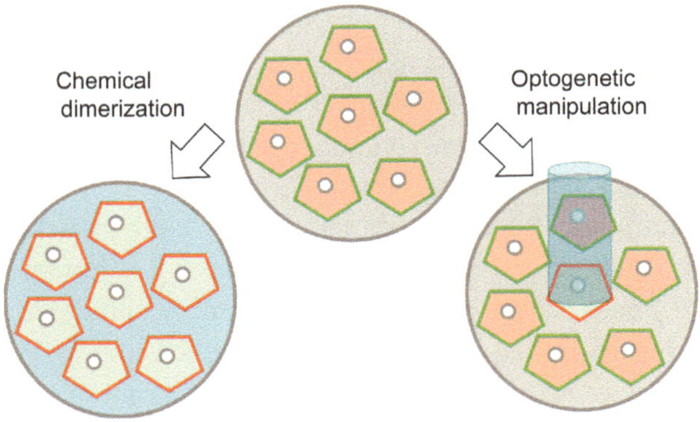

Fig. 9 Monitoring downstream effects of reversible dimerization and optogenetic manipulation. Chemical dimerizers usually work on an entire dish while optogenetic tools are switchable with spatial precision.

compared with the rCD1 system, it is more challenging to apply these methods for biochemical experiments.

In conclusion, it is crucial to carefully select the proper dimerization method suitable for the investigation of a biological process of interest.

4. PERSPECTIVES

4.1 Improvements of rCD1

No toxicity of rCD1 or FK506 up to a concentration of 10 μ*M* each was observed. However, the application of FK506 may cause off-target effects in some cell types. FK506 is able to interact with calcineurin and inhibits T-lymphocytes signal transduction and IL-2 transcription (Kissinger et al., 1995). This can be prevented by using rapalogs for the release of FKBP-bait proteins if moderate release kinetics are sufficient. Bioorthogonality is further limited by the use of FKBP-fusion proteins that potentially compete with endogenous FKBP proteins thereby interfering with processes like intracellular calcium release (Wehrens, Lehnart, & Marks, 2005). Additionally, both FK506 and the SLF part of rCD1 could be sequestered by those FKBP domains.

In order to further increase translocation speed, optimization of rCD1 performance regarding cell permeability and solubility is required. Different linker lengths and types have proven successful in the improvement of other CIDs (Erhart et al., 2013). Further fine-tuning involves the anchor construct design regarding the accessibility of the SNAP-tag or the distance to the anchor site. This is strongly dependent on the relevant bait and anchor construct of choice and needs to be optimized by targeted mutagenesis or molecular evolution.

In order to standardize CID experiments in a given cell type, it is convenient to establish a stable cell line or a viral vector. Given an optimized ratio of anchor to bait construct levels, it is potentially advantageous to drive their expression by different promoters or an internal ribosome entry site.

4.2 Future Applications

So far the rCD1 system has mostly been exploited for translocation to the plasma membrane. Similar to other CIDs, it can be adopted not only for translocation to different subcellular anchor sites such as the nucleus or the mitochondria surface but also for dimerization of two protein domains or the reversible activation of catalytic domains by conformation modulation (Camacho-Soto, Castillo-Montoya, Tye, & Ghosh, 2014; Chu et al., 2014).

It will be especially interesting to apply this reversible CID to the study of fast processes in intact cells that were inaccessible by previously used methods. Among those, the *in vivo* determination of catalytic activities by rCD1 requires many approximations. Genome editing of labeled enzymes using techniques such as CRISPR-Cas9 (Doudna & Charpentier, 2014) could facilitate the access to variables needed for the estimation of catalytic parameters.

Optimal reversibility is achieved immediately after translocation. However, it can be relevant to vary the translocation period in order to study short- and long-term effects of products being present at a certain subcellular site.

In order to directly compare the dynamics of two different processes in parallel, the rCD1 approach should be combined with other orthogonal methods in the future.

ACKNOWLEDGMENTS

The authors acknowledge funding by the EMBL and Transregio 83, funded by the DFG. M.S. is a fellow of the EMBL Interdisciplinary Postdoc Programme, partly funded by the EU Marie-Curie Programme. We thank Nicole Heath and Madeleine Schultz for critically reading the manuscript.

REFERENCES

Ballister, E. R., Aonbangkhen, C., Mayo, A. M., Lampson, M. A., & Chenoweth, D. M. (2014). Localized light-induced protein dimerization in living cells using a photocaged dimerizer. *Nature Communications*, *5*, 5475.

Bayle, J. H., Grimley, J. S., Stankunas, K., Gestwicki, J. E., Wandless, T. J., & Crabtree, G. R. (2006). Rapamycin analogs with differential binding specificity permit orthogonal control of protein activity. *Chemistry & Biology, 13*, 99–107.

Bosch, P. J., Corrêa, I. R., Sonntag, M. H., Ibach, J., Brunsveld, L., Kanger, J. S., et al. (2014). Evaluation of fluorophores to label SNAP-Tag fused proteins for multicolor single-molecule tracking microscopy in live cells. *Biophysical Journal, 107*, 803–814.

Briesewitz, R., Ray, G. T., Wandless, T. J., & Crabtree, G. R. (1999). Affinity modulation of small-molecule ligands by borrowing endogenous protein surfaces. *Proceedings of the National Academy of Sciences of the United States of America, 96*, 1953–1958.

Buckley, A. M., Petersen, J., Roe, A. J., Douce, G. R., & Christie, J. M. (2015). LOV-based reporters for fluorescence imaging. *Current Opinion in Chemical Biology, 27*, 39–45.

Camacho-Soto, K., Castillo-Montoya, J., Tye, B., & Ghosh, I. (2014). Ligand-gated split-kinases. *Journal of the American Chemical Society, 136*, 3995–4002.

Chen, Z., Jing, C., Gallagher, S. S., Sheetz, M. P., & Cornish, V. W. (2012). Second-generation covalent TMP-tag for live cell imaging. *Journal of the American Chemical Society, 134*, 13692–13699.

Chu, P.-H., Tsygankov, D., Berginski, M. E., Dagliyan, O., Gomez, S. M., Elston, T. C., et al. (2014). Engineered kinase activation reveals unique morphodynamic phenotypes and associated trafficking for Src family isoforms. *Proceedings of the National Academy of Sciences of the United States of America, 111*, 12420–12425.

Clackson, T. (2008). Controlling protein–protein interactions using chemical inducers and disrupters of dimerization. In *Chemical biology* (pp. 227–249). Weinheim: Wiley-VCH Verlag GmbH.

Czlapinski, J. L., Schelle, M. W., Miller, L. W., Laughlin, S. T., Kohler, J. J., Cornish, V. W., et al. (2008). Conditional glycosylation in eukaryotic cells using a biocompatible chemical inducer of dimerization. *Journal of the American Chemical Society, 130*, 13186–13187.

Doudna, J. A., & Charpentier, E. (2014). The new frontier of genome engineering with CRISPR-Cas9. *Science, 346*(6213), 1258096, review.

Erhart, D., Zimmermann, M., Jacques, O., Wittwer, M. B., Ernst, B., Constable, E., et al. (2013). Chemical development of intracellular protein heterodimerizers. *Chemistry & Biology, 20*, 549–557.

Feng, S., Laketa, V., Stein, F., Rutkowska, A., MacNamara, A., Depner, S., et al. (2014). A rapidly reversible chemical dimerizer system to study lipid signaling in living cells. *Angewandte Chemie, International Edition, 53*, 6720–6723.

Goodchild, R. E., Buchwalter, A. L., Naismith, T. V., Holbrook, K., Billion, K., Dauer, W. T., et al. (2015). Access of torsinA to the inner nuclear membrane is activity dependent and regulated in the endoplasmic reticulum. *Journal of Cell Science, 128*, 2854–2865.

Goudar, C. T., Sonnad, J. R., & Duggleby, R. G. (1999). Parameter estimation using a direct solution of the integrated Michaelis-Menten equation. *Biochimica et Biophysica Acta, 1429*, 377–383.

Gronemeyer, T., Chidley, C., Juillerat, A., Heinis, C., & Johnsson, K. (2006). Directed evolution of O6-alkylguanine-DNA alkyltransferase for applications in protein labeling. *Protein Engineering Design and Selection, 19*, 309–316.

Inoue, T., Heo, W. D., Grimley, J. S., Wandless, T. J., & Meyer, T. (2005). An inducible translocation strategy to rapidly activate and inhibit small GTPase signaling pathways. *Nature Methods, 2*, 415–418.

Juillerat, A., Gronemeyer, T., Keppler, A., Gendreizig, S., Pick, H., Vogel, H., et al. (2003). Directed evolution of O6-alkylguanine-DNA alkyltransferase for efficient labeling of fusion proteins with small molecules in vivo. *Chemistry & Biology, 10*, 313–317.

Karginov, A. V., Zou, Y., Shirvanyants, D., Kota, P., Dokholyan, N. V., Young, D. D., et al. (2011). Light regulation of protein dimerization and kinase activity in living cells using photocaged rapamycin and engineered FKBP. *Journal of the American Chemical Society, 133*, 420–423.

Keppler, A., Gendreizig, S., Gronemeyer, T., Pick, H., Vogel, H., & Johnsson, K. (2003). A general method for the covalent labeling of fusion proteins with small molecules in vivo. *Nature Biotechnology*, *21*, 86–89.
Kissinger, C. R., Parge, H. E., Knighton, D. R., Lewis, C. T., Pelletier, L. A., Tempczyk, A., et al. (1995). Crystal structures of human calcineurin and the human FKBP12-FK506-calcineurin complex. *Nature*, *378*, 641–644.
Liang, F. S., Ho, W. Q., & Crabtree, G. R. (2011). Engineering the ABA plant stress pathway for regulation of induced proximity. *Science Signaling*, *4*, rs2.
Liu, P., Calderon, A., Konstantinidis, G., Hou, J., Voss, S., Chen, X., et al. (2014). A bioorthogonal small-molecule-switch system for controlling protein function in live cells. *Angewandte Chemie, International Edition*, *53*, 10049–10055.
Nadler, A., Yushchenko, D. A., Muller, R., Stein, F., Feng, S., Mulle, C., et al. (2015). Exclusive photorelease of signalling lipids at the plasma membrane. *Nature Communications*, *6*, 10056.
Pathak, G. P., Vrana, J. D., & Tucker, C. L. (2013). Optogenetic control of cell function using engineered photoreceptors. *Biology of the Cell*, *105*, 59–72.
Putyrski, M., & Schultz, C. (2011). Switching heterotrimeric G protein subunits with a chemical dimerizer. *Chemistry & Biology*, *18*, 1126–1133.
Putyrski, M., & Schultz, C. (2012). Protein translocation as a tool: The current rapamycin story. *FEBS Letters*, *586*, 2097–2105.
Rivera, V. M., Clackson, T., Natesan, S., Pollock, R., Amara, J. F., Keenan, T., et al. (1996). A humanized system for pharmacologic control of gene expression. *Nature Medicine*, *2*, 1028–1032.
Rosivatz, E., & Woscholski, R. (2011). Removal or masking of phosphatidylinositol(4,5) bisphosphate from the outer mitochondrial membrane causes mitochondrial fragmentation. *Cellular Signalling*, *23*, 478–486.
Schifferer, M., Feng, S., Stein, F., Tischer, C., & Schultz, C. (2015). Reversible chemical dimerizer-induced recovery of PIP2 levels moves clathrin to the plasma membrane. *Bioorganic & Medicinal Chemistry*, *23*, 2862–2867.
Skrahina, T., Piljic, A., & Schultz, C. (2008). Heterogeneity and timing of translocation and membrane-mediated assembly of different annexins. *Experimental Cell Research*, *314*, 1039–1047.
Suh, B. C., Inoue, T., Meyer, T., & Hille, B. (2006). Rapid chemically induced changes of PtdIns(4,5)P2 gate KCNQ ion channels. *Science*, *314*, 1454–1457.
Tischer, D., & Weiner, O. D. (2014). Illuminating cell signalling with optogenetic tools. *Nature Reviews Molecular Cell Biology*, *15*, 551–558.
Umeda, N., Ueno, T., Pohlmeyer, C., Nagano, T., & Inoue, T. (2011). A photocleavable rapamycin conjugate for spatiotemporal control of small GTPase activity. *Journal of the American Chemical Society*, *133*, 12–14.
Voß, S., Klewer, L., & Wu, Y.-W. (2015). Chemically induced dimerization: Reversible and spatiotemporal control of protein function in cells. *Current Opinion in Chemical Biology*, *28*, 194–201.
Wehrens, X. H., Lehnart, S. E., & Marks, A. R. (2005). Intracellular calcium release and cardiac disease. *Annual Review of Physiology*, *67*, 69–98.
Wright, C. W., Guo, Z.-F., & Liang, F.-S. (2015). Light control of cellular processes by using photocaged abscisic acid. *ChemBioChem*, *16*, 254–261.
Zeng, G., Zhang, R., Xuan, W., Wang, W., & Liang, F.-S. (2015). Constructing de novo H_2O_2 signaling via induced protein proximity. *ACS Chemical Biology*, *10*, 1404–1410.
Zlatkine, P., Mehul, B., & Magee, A. I. (1997). Retargeting of cytosolic proteins to the plasma membrane by the Lck protein tyrosine kinase dual acylation motif. *Journal of Cell Science*, *110*(Pt. 5), 673–679.
Zoncu, R., Efeyan, A., & Sabatini, D. M. (2011). mTOR: From growth signal integration to cancer, diabetes and ageing. *Nature Reviews. Molecular Cell Biology*, *12*, 21–35.

CHAPTER NINE

Interfacial Enzymes: Membrane Binding, Orientation, Membrane Insertion, and Activity

S.A. Tatulian[1]

College of Sciences, University of Central Florida, Orlando, FL, United States

[1]Corresponding author: e-mail address: statulia@ucf.edu

Contents

Abstract

Most interfacial enzymes undergo activation upon membrane binding. Interfacial activation is determined not only by the binding strength but also by the specific mode of protein–membrane interactions, including the angular orientation and membrane insertion of the enzymes. This chapter describes biophysical techniques to quantitatively evaluate membrane binding, orientation, membrane insertion, and activity of secreted phospholipase A_2 (PLA_2) and lipoxygenase (LO) enzymes. Procedures for recombinant production and purification of human pancreatic PLA_2 and human 5-lipoxygenase (5-LO) are also presented. Several methods for measurements of membrane binding of peripheral proteins are described, i.e., fluorescence resonance energy transfer (FRET) from tryptophan or tyrosine residues of the protein to a fluorescent lipid in vesicles, changes in fluorescence of an environment-sensitive fluorescent lipid upon binding of proteins to membranes, and attenuated total reflection Fourier transform infrared (ATR-FTIR) spectroscopy. These methods produce the apparent binding constant, the protein-to-lipid binding stoichiometry, and the Hill cooperativity coefficient.

Methods in Enzymology, Volume 583
ISSN 0076-6879
http://dx.doi.org/10.1016/bs.mie.2016.09.009

Experimental procedures for segmental isotope labeling of proteins and determination of the orientation of membrane-bound proteins by polarized ATR-FTIR spectroscopy are described. Furthermore, evaluation of membrane insertion of peripheral proteins by a fluorescence quenching technique is outlined. Combination of the orientation and membrane insertion provides a unique configuration of the protein–membrane complex and hence elucidates certain details of the enzyme function, such as the modes of acquisition of a membrane-residing substrate and product release. Finally, assays for determination of the activities of secreted PLA_2, soybean LO, and human 5-LO are described.

1. INTRODUCTION

Interfacial enzymes are water-soluble proteins that bind to cellular membranes to reach their membrane-residing substrates, such as lipids or fatty acids, and thereby gain optimal activity (Damodaran, 2015; Emptage, Tonthat, York, Schumacher, & Zhou, 2014; Gelb, Min, & Jain, 2000; Lambeau & Gelb, 2008; Lomize, Pogozheva, Joo, Mosberg, & Lomize, 2012; Newie et al., 2016; Weiss, 2015). Membrane binding itself is not enough for activity. The enzyme should bind to the membrane with a unique configuration, including the three-dimensional angular orientation and the degree of membrane insertion, which facilitate formation of a productive enzyme–substrate complex. Analysis of the molecular basis of membrane binding and resulting activation of interfacial enzymes, known as interfacial activation, has been the subject of extensive studies for several decades. Interfacial activation of certain enzymes involves conformational changes or displacement of a lid covering the active site slot, while the others become activated via yet unidentified mechanisms. In many cases the target membrane is viewed as an allosteric effector causing subtle or large conformational changes and leading to enzyme activation. Clearly, the interfacial enzymes, and membrane proteins in general, have an additional, functionally important structural level, i.e., the angular orientation and insertion depth, which can be viewed as the quinary structure of membrane proteins. This chapter describes biophysical methods for identification of the geometric mode of membrane binding of secreted phospholipase A_2 (PLA_2) and Lipoxygenase (LO) enzymes and its role in enzyme function.

Selected methods to probe the mode of membrane binding of peripheral proteins have been reviewed by Scott, Musselman, Adu-Gyamfi, Kutateladze, and Stahelin (2012). Gelb and coworkers (Canaan, Nielsen, Ghomashchi, Robinson, & Gelb, 2002; Lin et al., 1998) determined the high-resolution modes of membrane binding of bee venom PLA_2 and human group IIA PLA_2, utilizing spin-label/spin-relaxation electron paramagnetic resonance. Bee venom PLA_2 was bound to the membrane superficially, without any appreciable insertion, using mostly hydrophobic and only one cationic (lysine) residue, with the catalytic orifice turned to the membrane but not in physical contact with it (Lin et al., 1998). The nonpolar nature of the interface-binding surface (IBS) explained the failure of mutations of the cationic surface residues to impair membrane binding of bee venom PLA_2 (Gelb, Cho, & Wilton, 1999; Ghomashchi et al., 1998). Despite the strongly cationic nature of human IIA PLA_2, its IBS was composed mainly of hydrophobic residues as well, involving only two basic residues (Arg^7 and Lys^{10}), providing tight binding to the membrane with a slight tilt of the catalytic slot relative to the membrane plane (Canaan et al., 2002).

Another important headway toward determination of protein–membrane-binding mode was made by Kutateladze and others using NMR methods (Brunecky et al., 2005; Kutateladze, 2012; Kutateladze et al., 2004; Prior, Fulcher, Koppisetti, Jurkevich, & Van Doren, 2015). In these studies, a uniformly ^{15}N-labeled protein is allowed to bind to micelles containing a low mole-fraction of a doxyl spin-labeled lipid. Paramagnetic relaxation enhancement and resulting NMR line broadening occurs for protein residues positioned close to the doxyl group. Using two or more lipids spin-labeled at defined locations along the acyl chain allows determination of the orientation of the micelle-bound protein. Similar methods have been utilized to identify membrane- or micelle-binding of proteins, using water-soluble (rather than lipophilic) paramagnetic probes such as chelated manganese or gadolinium (Cook, Dawson, Tian, & Opella, 2013; Li, Li, Wong, Liew, & Kang, 2015). Here, protein residues exposed to the aqueous phase are identified by reduced resonance intensities relative to those embedded in the micelle core. An electron paramagnetic resonance-based method has been employed where the effects of both membrane-soluble (oxygen) and water-soluble (chelated nickel) paramagnetic relaxants on the spin–lattice relaxation rates of spin-labeled proteins were used to measure the locations of the labeled amino acid residues with respect to the membrane center (Nielsen, Che, Gelb, & Robinson, 2005). A C2 domain of cytosolic PLA_2 was positioned in phospholipid membranes by this method.

Computational methods have provided significant insights into interfacial enzymology. Using molecular dynamics (MD) simulations, Sansom and co-workers determined the membrane-binding mode of porcine pancreatic PLA_2 (Wee, Balali-Mood, Gavaghan, & Sansom, 2008). Several nonpolar residues were found to serve as hydrophobic membrane anchors, whereas the cationic residues of the IBS were involved in either H-bonding or ionic interactions with membrane lipid, consistent with earlier experimental work (Tatulian, Qin, Pande, & He, 2005). Lomize, Lomize, Pogozheva, and Mosberg (2006), Lomize et al. (2012), Lomize, Pogozheva, Lomize, and Mosberg (2006, 2007) developed a database named Orientation of Proteins in Membranes (OPM) that contains over 1200 membrane-bound integral and peripheral membrane proteins. A number of web-based servers have been developed to build a membrane or a micelle of desired lipid or detergent composition and to locate a protein into it. CHARMM-GUI allows construction of a membrane of any shape and size in a realistic aqueous environment with various salts at arbitrary concentrations and pairs it with a membrane protein using an orientation predicted by OPM (Jo, Kim, Iyer, & Im, 2008). Further improvements of the micelle builder or membrane builder functions enable detailed characterization of proteins embedded in or bound to respective assemblies (Cheng, Kim, Kim, Bowie, & Im, 2016; Wu et al., 2014). Furthermore, a Protein and Bilayer Lipid Membrane (ProBLM) web server has been developed to position proteins in membranes with a predicted orientation (Kimmett et al., 2014).

Development and utilization of new, reliable, and accurate methods capable of detailed structural characterization of protein–membrane complexes are needed for a better understanding of the mechanisms of interfacial enzymes. A multiscale approach involving peptide ligation, segmental isotopic labeling, polarized attenuated total reflection Fourier transform infrared (ATR-FTIR), fluorescence quenching, and computation has been developed to study the membrane-binding mode of peripheral proteins. This method was employed to determine the membrane-binding mode of human pancreatic (group IB) PLA_2 (Tatulian et al., 2005) and a chimeric PLA_2, i.e., a hybrid between human pancreatic and group IIA PLA_2s (Qin, Pande, Nemec, He, & Tatulian, 2005). The power of this method is that it provides the structure of the membrane-bound protein, where the *z*-coordinates of protein atoms are the distance from membrane center. Thus, the resulting model allows easy assessment of the location of each amino acid residue relative to the membrane. Section 2 describes recombinant expression and purification of human pancreatic PLA_2 and 5-LO and determination of their membrane-binding modes and activities in more detail.

2. METHODS

2.1 Recombinant Production and Purification of Human Pancreatic PLA_2

The prokaryotic expression vector of human pancreatic PLA_2 has been constructed using a human pancreatic cDNA library (Qin, Pande, Nemec, & Tatulian, 2004). Procedures involved in plasmid construction, overexpression of the protein into inclusion bodies within *E. coli*, followed by purification and refolding, are described below:

- Amplify the gene encoding PLA_2 from a human pancreatic cDNA library (Edge BioSystems, Gaithersburg, MD) by touchdown polymerase chain reaction (PCR) using a proof-reading pfu Turbo DNA polymerase (Stratagene, La Jolla, CA) and appropriate oligonucleotide primers containing *Nde*I and *Bam*HI restriction sites (Sambrook & Russell, 2001).
- Purify the PCR fragments of PLA_2 (410 bp) by a PCR Purification Kit (Qiagen, Valencia, CA).
- Process with *Nde*I and *Bam*HI restriction enzymes (New England Biolabs, Mississauga, ON) and clone into the prokaryotic expression vector pET-21a(+) (Clontech, Mountain View, CA).
- Check gene insertion into the *Nde*I/*Bam*HI site of pET-21a(+) vector by agarose gel electrophoresis.
- Transform the plasmid into *E. coli* DH5α (ThermoFisher Scientific, Grand Island, NY) using a calcium chloride method (Sambrook & Russell, 2001).
- Amplify the plasmid, extract, and purify with Qiagen Miniprep Kit.
- Verify the sequence of the open reading frame of the purified construct.
- Transform the purified plasmid into *E. coli* BL21(DE3) (Novagen, Madison, WI).
- Inoculate into 100 mL of Luria-Bertani (LB) medium, containing 0.1 mg/mL carbenicillin (Sigma-Aldrich, St. Louis, MO), and grow the starter culture overnight at 37°C.
- Transfer the cells into LB medium containing 0.1 mg/mL carbenicillin and grow in a bacterial incubator-shaker.

- At $A_{600}\approx 0.6$, reduce the temperature to 30°C and induce protein expression by 1 m*M* isopropyl β-D-thiogalactopyranoside (IPTG) (Sigma-Aldrich).
- Grow the culture for 8 h at 30°C. The protein will be overexpressed into inclusion bodies.
- Harvest the cells by centrifugation (6000 rpm, 10 min, 4°C) and resuspend in 200 mL of buffer A (50 m*M* NaCl, 1 m*M* EDTA, 1 m*M* NaN_3, 50 m*M* Tris–HCl, pH 8.0) supplemented with 0.4% (w/v) Na-deoxicholate, 0.4% (v/v) Triton X-100, and a protease inhibitor cocktail (Sigma-Aldrich).
- Aliquot the cell suspension in ~30 mL fractions in Falcon plastic tubes (Fisher Scientific, Suwanee, GA) and sonicate on ice for 6 min (three 2-min cycles).
- Centrifuge for 30 min at 12,000 rpm, at 4°C.
- Resuspend the pellets in 190 mL buffer A containing 0.8% Na-deoxicholate, 0.8% Triton X-100, and protease inhibitor cocktail. Sonicate and centrifuge as earlier.
- Wash the pellets of inclusion bodies twice with pure buffer A, resuspend in a denaturing buffer (buffer A containing 6 *M* guanidinum-HCl and 5% β-mercaptoethanol), and stir at 4°C overnight.
- Centrifuge the sample as earlier and dialyze the supernatant, containing the denatured protein, four times against 3 L of refolding buffer (5 m*M* $CaCl_2$, 5 m*M* L-cysteine, 0.9 *M* guanidinum-HCl, 25 m*M* Tris–HCl, pH 8.0) at 4°C, allowing ~8 h per cycle.
- Separate the refolded protein from the precipitate by centrifugation and dialyze against a buffer containing 2.5 m*M* KCl and 20 m*M* Tris–HCl, pH 5.5.
- Purify the protein using a cation-exchange column (e.g., HiTrap SP FF, GE Healthcare, Pittsburgh, PA).
- Elute the protein using a linear gradient between the above buffer and that supplemented with 1.5 *M* KCl.
- Collect the elution fractions containing PLA_2 and dialyze against a buffer containing 150 m*M* KCl, 25 m*M* Tris–HCl, pH 7.4.
- Conduct a second round of purification using a size-exclusion column, e.g., HiLoad 16/60 Superdex 75 column, using an ÄKTA protein purifier (GE Healthcare) or another column chromatography system.
- Analyze the elution fractions by sodium dodecyl sulfate-polyacrylamide gel electrophoresis (SDS-PAGE), using 15% (w/v) polyacrylamide gels.

- Collect pure PLA_2 fractions and dialyze against pure water.
- The yield of pure protein per liter of culture is typically ~2 mg.
- The identity and the enzymatic activity of the protein can be verified by mass-spectroscopy, N-terminal sequencing, Western blot, and by activity assays (see below):
- The pure protein can be used in membrane-binding, activity, or other assays immediately or it can be lyophilized and stored in a freezer for further use. Secreted PLA_2s are notoriously stable and maintain structure and activity for several years when stored under these conditions.

2.2 Expression and Purification of Human 5-Lipoxygenase

Human 5-Lipoxygenase (5-LO) plasmid, pT3-5LO, was provided by Y.-Y. Zhang (Boston University School of Medicine, Boston, MA) (Zhang, Rådmark, & Samuelsson, 1992). The procedures for expression and purification of the protein have been described by Pande et al. (2004) and are outlined below:

- Insert the 5-LO open reading frame into the pET-21 expression vector and amplify by PCR.
- Place the PCR product into a pGem-T vector (Promega, Madison, WI) and confirm the sequence.
- Subclone into pET-21 using the *Nde*I and *Hind*III sites engineered in the primers.
- Digest the insert (5-LO) and pET-21 with *Nde*I and *Hind*III restriction enzymes.
- Purify the digestion products from a 1% agarose gel using a gel extraction kit (Qiagen) and ligate using a T4 ligase (USB Corp., Cleveland, OH).
- Transform the ligated product into *E. coli* DH5α and confirm the sequence.
- Place positive clones in *E. coli* BL21(DE3) for T7-driven expression.
- Grow a 50-mL culture of transformed cells overnight in LB medium, supplemented with 0.1 mg/mL carbenicillin.
- Dilute the starter culture into 1 L M9CA medium (Amresco, Solon, OH) (42 m*M* Na_2HPO_4, 24 m*M* KH_2PO_4, 9 m*M* NaCl, 19 m*M* NH_4Cl, 1 m*M* $MgSO_4$, 0.1 m*M* $CaCl_2$, 0.2% D-glucose, 0.1% casamino acids, 5 μ*M* $FeSO_4$) supplemented with 0.1 mg/mL carbenicillin.
- Grow the culture at 25°C and induce with 0.25 m*M* IPTG at $A_{600} \approx 0.6$.
- Harvest the cells by centrifugation at 5000 × *g* for 10 min at $A_{600} \approx 2$.

- Resuspended the cell pellet in buffer B (10 m*M* β-mercaptoethanol, 0.1 m*M* EDTA, and 50 m*M* triethanolamine, pH 7.3) supplemented with a protease inhibitor cocktail and 0.5 mg/mL lysozyme.
- Stir on ice for 30 min and break the cells by sonication (see earlier).
- Spin the cell lysate at 15,000 × *g* for 15 min.
- Precipitate the proteins in the supernatant with 50% saturated $(NH_4)_2SO_4$ and spin at 16,000 × *g* for 15 min. (Proteins in the salt pellet can be stored at −80°C without considerable loss of 5-LO activity for several months.)
- Resuspend the salt pellet from 1 L culture in 25 mL buffer B, centrifuge at 40,000 × *g* for 15 min, and pass the supernatant through a 0.22-μm polyethersulfone filter (ThermoFisher Scientific).
- All buffers used for 5-LO purification should be extensively degassed to remove dissolved oxygen, which is known to inactivate 5-LO. This can be done easily using a vacuum flask, as described in Thermo Scientific Tech Tip #29.
- Purify the 5-LO sample with an ATP-agarose column (Sigma-Aldrich). Wash the column with 1 column volume of a gradient between buffer B and the same buffer supplemented with 1 *M* NaCl to remove nonspecifically bound proteins.
- Elute 5-LO with buffer B containing 100 m*M* NaCl and 12 m*M* ATP.
- Analyze the elution fractions by SDS-PAGE and purify the fraction containing a major band at ~75 kDa with a HiLoad Superdex-75 or another size exclusion column.
- In our lab, this procedure yielded two major protein peaks, both of which were 5-LO as identified by SDS-PAGE and Western blot analysis using antihuman 5-LO antibody (Cayman Chemical Co., Ann Arbor, MI). The second peak, corresponding to longer retention times, showed significantly higher 5-LO activity.
- Collect the second elution peak, which has been shown to contain iron unlike the first peak (Pande et al., 2004), transfer to the 5-LO activity assay buffer (see Section 2.6.2), and use immediately for activity experiments as 5-LO loses activity over time.
- Alternatively, transfer the protein into pure deionized water by a desalting column, lyophilize, and store at −80°C for structural studies. In our hands, the lyophilized sample retained 60–80% of initial activity for 1 week under these conditions.

2.3 Membrane-Binding Measurements

2.3.1 Fluorescence Resonance Energy Transfer

Fluorescence spectroscopy has been used to quantitatively measure binding of interfacial enzymes to lipid membranes (Nemec et al., 2006; Pande, Qin, & Tatulian, 2005; Qin et al., 2005, 2004). Preparation of unilamellar lipid vesicles and measurements and analysis of protein–membrane interactions are described below:

- To make unilamellar vesicles, prepare lipid stock solutions in chloroform or chloroform/methanol (2:1, v/v). While phosphatidylcholine (PC) dissolves in chloroform without problems, some acidic lipids have limited solubility in pure chloroform and may require methanol as cosolvent.
- Mix lipid solutions in a desired proportion, such as 70 mol% PC +30 mol % phosphatidylserine or phosphatidylglycerol. Cholesterol or other lipids can be added to mimic lipid compositions of certain organelles.
- Most lipids are available from Avanti Polar Lipids (Alabaster, AL). Store lipids in a freezer (−20 to −80°C). Use chloroform and methanol only in a chemical fume hood; both are harmful if inhaled. Methanol is toxic by inhalation or if swallowed. Store in a cabinet for flammables at room temperature. Chloroform can slowly convert to phosgene upon prolonged storage in the presence of oxygen unless stabilized with ethanol by the provider. Handle with gloves.
- Dry the lipids by a gentle stream of nitrogen, followed by desiccation.
- Following desiccation, suspend the dry lipid in an aqueous buffer, such as 100 m*M* NaCl, 20 m*M* Hepes, 0.2 m*M* EGTA or EDTA, pH 7.2, by extensive vortexing for 5 min. The Ca^{2+} chelator is added to prevent lipid hydrolysis by PLA_2 during the experiment. The total lipid concentration is typically between 1.0 and 10 m*M*. Keep the containers of salts, chelators, and buffers tightly closed, in a dry place, at room temperature.
- Extrude the lipid suspension through a 100-nm pore-size polycarbonate filter using a mini-extruder (Avanti Polar Lipids, or Avestin, Ottawa, Canada).
- Prepare identical vesicles containing 2 mol% of 1,2-dioleoyl-*sn*-glycero-3-phosphoethanolamine-*N*-1-pyrenesulfonyl (PyPE, Avanti Polar Lipids) for fluorescence resonance energy transfer (FRET) experiments.

- Polycarbonate filters of smaller (50 nm) or larger (200–400 nm) pore size can be used, but 100 nm vesicles are more stable and homogeneous and generate reproducible data.

Membrane binding of a protein can be measured by FRET from Trp or Tyr to a fluorescent lipid, as described below:

- Prepare protein solution in same buffer as the vesicles, at a concentration of ~10 μ*M* and measure Trp (or Tyr, if the protein is Trp-less) fluorescence using excitation at 292 nm (or 275 nm for Tyr) in a quartz cuvette. In both cases, the emission spectra can be measured between 290 and 380 nm.
- Titrate the protein solution with PyPE-containing vesicles, with gentle stirring, by adding small volumes (e.g., 1–5 μL, not to exceed 0.5–2% of total sample volume) of extruded vesicles from a stock suspension. Total lipid concentration may change in a range from 1.0 μ*M* to ~2.0 m*M*.
- After each addition of vesicles, let the sample equilibrate for 2–3 min and measure a fluorescence emission spectrum, as described earlier.
- Conduct control experiments using plain vesicles, without PyPE.
- In control experiments, protein binding to the vesicles will result in an increase in the fluorescence emission intensity and/or a blue shift. In FRET experiments with PyPE-containing vesicles, the emission intensity of the donor will decrease and that of the acceptor (PyPE) will increase (Fig. 1).
- Correct the donor emission intensities for the dilution effect and for the increase in fluorescence intensity in control experiments.
- Calculate changes in peak emission intensities (likely around 330–340 nm for Trp and 304–307 nm for Tyr) for each lipid concentration relative to the zero lipid level, i.e., $\Delta F = F_0 - F_L$, where F_0 and F_L are fluorescence intensities without and with lipid.
- Analysis of FRET data and determination of the protein-membrane-binding parameters is conducted as follows. The binding process is described with a Langmuir-type binding isotherm:

$$\Delta F = \Delta F_{max} \frac{[L]}{[L]_{1/2} + [L]} \tag{1}$$

where ΔF_{max} is the saturating value of ΔF at high lipid concentration, $[L]$ is the lipid concentration, and $[L]_{1/2}$ is the lipid concentration at half-maximal FRET effect, when $\Delta F = 0.5\Delta F_{max}$. Both $[L]_{1/2}$ and ΔF_{max} can be determined from the slope and the vertical axis intercept of a Scatchard plot, i.e., a plot of ΔF vs $\Delta F/[L]$

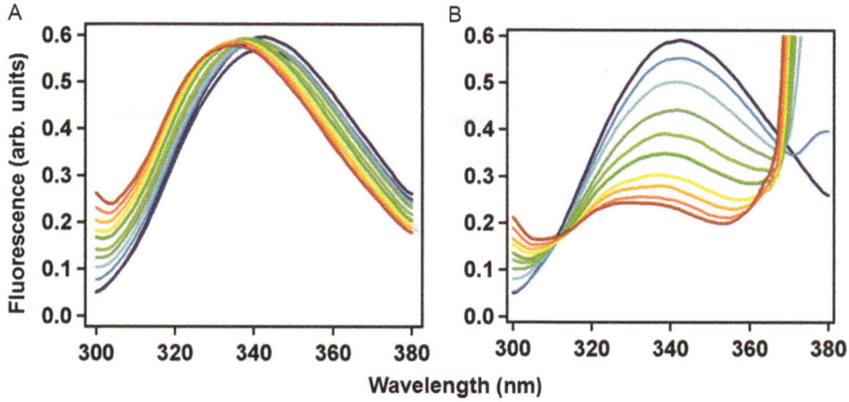

Fig. 1 Fluorescence emission spectra of human pancreatic PLA_2 in the presence of unilamellar vesicles without (A) and with PyPE (B) in 10 m*M* Hepes, 1 m*M* EDTA (pH 7.4). Excitation was at 290 nm. Increasing amounts of vesicles composed of 60 mol% POPC and 40 mol% POPG (A) or 58 mol% POPC, 40 mol% POPG, and 2 mol% PyPE (B) were added to a protein solution with continuous stirring at 25°C. The change in color from *blue* to *red* corresponds to increasing lipid concentration from 0 to 1 m*M*. *Adapted from Qin, S., Pande, A. H., Nemec, K. N., & Tatulian, S. A. (2004). The N-terminal α-helix of pancreatic phospholipase A_2 determines productive-mode orientation of the enzyme at the membrane surface.* Journal of Molecular Biology, 344, *71–89.*

- Eq. (1) yields the following (see Qin et al., 2004 for details):

$$\frac{\Delta F}{\Delta F_{\max}} = \frac{1}{2}\left(1 + \frac{\delta[L]}{N[P]} + \frac{K_D}{[P]}\right) - \sqrt{\frac{1}{4}\left(1 + \frac{\delta[L]}{N[P]} + \frac{K_D}{[P]}\right)^2 - \frac{\delta[L]}{N[P]}} \quad (2)$$

  where δ is the fraction of the lipid in the external leaflet of the vesicles ($\delta \approx 0.52$ for 100 nm vesicles), N is the number of lipids per protein-binding site, K_D is the dissociation constant, and $[P]$ is the total protein concentration.
- The following expression for the dissociation constant (K_D) can be obtained (Qin et al., 2004):

$$K_D = \frac{\delta}{N}[L]_{1/2} - \frac{[P]}{2} \quad (3)$$

A reasonable starting value for N can be selected, e.g., 40 lipids per PLA_2-binding site (Jain, DeHaas, Marecek, & Ramirez, 1986), and used to evaluate a respective K_D. Then, theoretical-binding curves ($\Delta F/\Delta F_{\max}$ vs $[L]$) are constructed through Eq. (2), and the final, best fit values of K_D and N are determined through a series of iterations.

How to measure membrane binding of a protein or a peptide that has no fluorophores? This has been done using membranes containing 2 mol% of a fluorescent lipid, *N*-(fluorescein-5-thiocarbamoyl)-1,2-dihexadecanoyl-*sn*-glycero-3-phosphoethanolamine (FPE, Invitrogen/ThermoFisher Scientific) (Qin et al., 2005). Fluorescence spectra of the vesicles are measured in the absence of protein, between 500 and 540 nm with excitation at 490 nm, then the vesicle suspension is titrated with protein. The fluorescence intensity of FPE is sensitive to the microenvironment and increases with increasing deprotonation of the carboxyl group of fluorescein. Binding of a cationic protein to an anionic membrane disturbs the local pH at the membrane surface, resulting in deprotonation of fluorescein and an increase in the fluorescence intensity. This effect was used to quantitatively characterize membrane binding of PLA_2s and their N-terminal peptides (Nemec et al., 2006; Qin et al., 2005).

2.3.2 Infrared Spectroscopy

Binding of proteins to lipid membranes can be measured in a quantitative manner by ATR-FTIR spectroscopy (Tatulian, 2001, 2003). An infrared spectrometer equipped with ATR accessories and a polarizer will be required. These can be purchased from a number of vendors such as Harrick Scientific (Pleasantville, NY), Buck Scientific (Norwalk, CT), Spectral Systems (Hopewell Junction, NY).

- Prepare a supported lipid bilayer on an internal reflection element (IRE) such as a germanium plate (5 × 2 × 0.1 cm, cut at 45 degree at the 2 cm edges) following one of the two methods described below:
- The first method works as follows. Clean the IRE plate with chloroform/methanol (2:1 by volume) and process in an argon plasma cleaner (Harrick Scientific) immediately before use.
- Immerse the plate vertically into an aqueous buffer (10 m*M* Tris–acetic acid, pH 5.0) in a Langmuir trough (e.g., model 611, Nima, Coventry, UK).
- Spread solution of PC in hexane/ethanol mixture (9:1 by volume) on the surface of the buffer and adjust the surface pressure to 32 mN/m, which corresponds to lipid packing density similar to that in biomembranes.
- Allow 5–10 min for the solvent to evaporate.
- Pull the Ge plate out of the buffer through the lipid monolayer at ~2 mm/min while maintaining the surface pressure constant via a feedback mechanism programmed in the trough software.

- Make sure the decrease in the surface area of the monolayer corresponds to the total surface area of IRE withdrawn.
- The plate will be covered with PC monolayer, which is stable in air for at least an hour.
- Assemble the plate with the deposited monolayer in an ATR flow-through cell.
- Prepare lipid vesicles by processing the lipid suspension with ultrasonic radiation, using any lipid composition, as appropriate, at a lipid concentration around 1 m*M*.
- Inject the sonicated vesicles into the cell and incubate for 1.0–1.5 h. During this time the vesicles will spread on the monolayer, generating a supported bilayer.
- Flush the cell with ~10 volumes of a D_2O-based buffer to remove excess lipid.
- The second method is simpler and works as follows. Prepare sonicated vesicles, containing at least 20% anionic lipid, in a buffer containing ~5 m*M* $CaCl_2$ and inject into the ATR cell with a Ge plate. A lipid bilayer at both surfaces of the plate will form spontaneously.
- Flush the cell with ~10 volumes of a D_2O-based buffer and measure transmission spectra at *p* and *s* polarizations, i.e., when the light is polarized parallel or perpendicular to the plane of incidence, respectively.
- Calculate absorbance spectra using transmission spectra of the blank buffer in the ATR cell with the IRE, measured in advance.
- Prepare protein solution and inject into the ATR cell that contains the IRE covered with a lipid bilayer, starting with the lowest protein concentration.
- Allow the protein to bind to the membrane for 15–30 min. Protein binding will be monitored by increasing amide I band intensity, so the wait time will be dictated by at least ~90% saturation of binding.
- Measure the spectra at both polarizations, inject the next, higher concentration of the protein, measure spectra again. Continue until the titration is complete.
- In an ATR-FTIR experiment, the incident light enters the IRE through the 45 degree beveled edges and travels in it via a series of internal reflections. At each reflection, an evanescent wave is created at the IRE surface that exponentially decays into the optically rarer medium with a decay length of approximately 400 nm. Molecules within this distance from the IRE surface absorb light. Thus, due to the surface-sensitive nature of ATR-FTIR spectroscopy, only the membrane-bound protein molecules will contribute to the spectra.

- Use spectra obtained at *p* and *s* polarizations at each protein concentration to calculate a "polarization-independent" spectrum: $A = A_p + GA_s$, where A_p and A_s are the absorbance intensities at *p* and *s* polarizations of the light, respectively, and the scaling factor $G = 0.8$ for a single bilayer membrane (thin film approximation). When a "thick" (≥ 0.5 μm) lipid layer is used, which can be obtained by depositing lipid solution on the IRE, drying, and hydrating, then $G = 1.44$ (Marsh, 1999).
- Determine the protein/lipid (*P*/*L*) molar ratios at each protein concentration using the ratio of integrated intensities of the protein amide I (~1700–1600 cm^{-1}) and the lipid methylene stretching bands (~2980–2820 cm^{-1}), using the "polarization-independent" spectra, corrected for corresponding molar extinction coefficients:

$$\frac{P}{L} = \frac{A_{\text{protein}} \varepsilon_{\text{lipid}} n_{\text{lipid}}}{A_{\text{lipid}} \varepsilon_{\text{protein}} n_{\text{protein}}} \tag{4}$$

where ε is the respective molar extinction coefficient, n_{protein} is the number of peptide bonds in the protein, and n_{lipid} is the number of methylene groups in the lipid hydrocarbon chains. A value of $\varepsilon_{\text{lipid}} = 4.7 \times 10^6$ cm/mol per CH_2 group of the lipid has been used (Fringeli, Apell, Fringeli, & Läuger, 1989). The amide I molar extinction coefficients of proteins depend on their secondary structure. In a D_2O-based buffer, the integrated extinction coefficients (based on band area rather than height) are $\varepsilon_{\alpha\text{-helix}} = 5.1 \times 10^7$ cm/mol, $\varepsilon_{\beta\text{-sheet}} = 7.0 \times 10^7$ cm/mol, $\varepsilon'_{\beta\text{-sheet}} = 0.6 \times 10^7$ cm/mol (for the higher frequency component of antiparallel β-sheet), $\varepsilon_{\text{unordered}} = 5.0 \times 10^7$ cm/mol, $\varepsilon_{\text{turn}} = (1.5 \pm 0.5) \times 10^7$ cm/mol (various turns have distinct extinction coefficients). In H_2O-based buffers, $\varepsilon_{\alpha\text{-helix}} = 7.6 \times 10^7$ cm/mol and those of other structures are the same as in D_2O-based buffers (Vedantham, Sparks, Sane, Tzannis, & Przybycien, 2000; Venyaminov & Kalnin, 1990).
- Determine the surface density of membrane-bound protein using the protein-to-lipid molar ratio, *P*/*L*, and the cross-sectional area per lipid, S_L, taking into account the bilayer arrangement of the lipids in the membrane:

$$n = \frac{2}{S_L} \frac{P}{L} \tag{5}$$

- The binding data, i.e., the plot of n against protein concentration, C, are described by a Langmuir-type adsorption isotherm involving a Hill cooperativity coefficient, α_H:

$$n = \frac{N_s C^{\alpha_H} K^{\alpha_H}}{1 + C^{\alpha_H} K^{\alpha_H}} \tag{6}$$

Here, N_s is the number of binding sites per unit area, and K is the binding constant. N_s can be found from extrapolated intercepts of the n/C vs n plots with horizontal axis, i.e., the $n/C=0$ line ($n/C=0$ at $C \to \infty$, when $n = N_s$). Eq. (6) can be rearranged to:

$$\ln\left(\frac{N_s}{n} - 1\right) = -\alpha_H \ln C - \alpha_H \ln K \tag{7}$$

- As seen from Eq. (7), α_H and K can be evaluated from the slope and the horizontal axis intercept of the $\ln(N_s/n - 1)$ vs ln C plots [at $C = 1/K$, $n = N_s/2$, and $\ln(N_s/n - 1) = 0$].

This ATR-FTIR-based method was utilized to characterize membrane binding of a cationic, group IIA PLA_2 (Tatulian, 2001). Both the binding constant of PLA_2 to anionic membranes and the Hill cooperativity coefficient were ionic strength-dependent and changed from $K = 4.3 \times 10^5\ M^{-1}$ to $K = 2.8 \times 10^4\ M^{-1}$ and from $\alpha_H = 1.85$ to $\alpha_H = 0.93$ upon an increase of the ionic strength from 8 mM to 1.0 M.

2.4 Determination of Orientation by Polarized Infrared Spectroscopy

2.4.1 A Helical Peptide

Before describing how the orientation of a membrane-bound protein can be determined, it is worthwhile to consider a simpler case, a membrane-bound α-helical peptide.

- Prepare a supported membrane on an IRE, assemble the IRE in an ATR cell, and measure transmission spectra at p and s polarizations, as described in Section 2.3.2.
- Inject the peptide and let it bind to the membrane.
- Measure transmission spectra at p and s polarizations and calculate the absorbance spectra using the transmission spectra of the bare membrane at respective polarizations as reference.
- Calculate the ATR dichroic ratio, $R^{ATR} = A_p/A_s$, where A_p and A_s are the amide I areas at respective polarizations.

- Use the dichroic ratio to determine the order parameter of the helix (Axelsen & Citra, 1996; Goormaghtigh, Raussens, & Ruysschaert, 1999; Tatulian, 2013):

$$S = \frac{2\left(E_x^2 - R^{\mathrm{ATR}} E_y^2 + E_z^2\right)}{(3\langle \cos^2\alpha \rangle - 1)\left(E_x^2 - R^{\mathrm{ATR}} E_y^2 - 2E_z^2\right)} \quad (8)$$

- In Eq. (8), E_x, E_y, and E_z are the electric field components of the evanescent wave at the IRE surface, normalized for the incident light amplitude, α is the angle between the amide I vibrational transition dipole moment and the helical axis, and has been determined to be 39 degree ± 1 degree (Marsh, Müller, & Schmitt, 2000; Marsh & Páli, 2001). For an incidence angle of 45 degree, $E_x = 1.40$, $E_y = 1.51$, $E_z = 1.375$ for a single bilayer. In case of a "thick" lipid layer (≥ 0.5 μm), $E_z = 1.621$ and E_x and E_y are the same.
- The orientation of the helix relative to the membrane, i.e., the average angle θ between the membrane normal and the helical axis, is then determined from:

$$S = \frac{1}{2}\left(3\langle \cos^2\theta \rangle - 1\right). \quad (9)$$

The method is schematically shown in Fig. 2.

2.4.2 Orientation of a Membrane-Bound Protein

For a protein which has multiple helices, the orientations of two or more helices should be identified to estimate the orientation of the whole protein molecule. Human pancreatic PLA_2 has three α-helices, an N-terminal 10-amino acid residue helix and two internal helices. The two internal helices are nearly antiparallel to each other, with an interhelical angle of 173.6 degree, hence a common order parameter can be assigned to them. Thus, the task narrows down to determination of the orientations of the N-terminal helix and the two internal helices. This has been achieved by creating a semisynthetic, segmentally isotope-labeled protein in which the first 10 residues are not labeled, while the rest is uniformly ^{13}C-labeled. ^{13}C-labeling results in FTIR spectral downshift and resolution of the labeled segment and allows determination of the dichroic ratios and orientations of the labeled and unlabeled helices, which then leads to the identification of the orientation of the whole protein molecule (Tatulian et al., 2005). The N-terminal

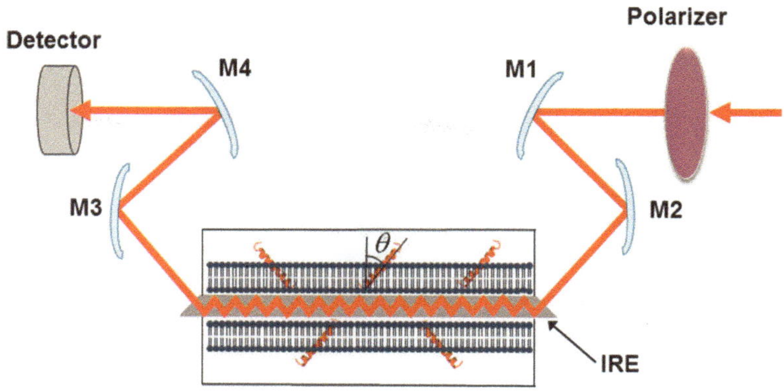

Fig. 2 Schematics of an ATR-FTIR experiment on a membrane embedded α-helix. Lipid bilayers are deposited on both surfaces of the internal reflection element (IRE), shown as a *gray* trapezoid. The helix is tilted by an angle θ with respect to the membrane normal. The infrared light beam, shown as a *red line*, undergoes plane-polarization and is directed by mirrors M1 and M2 into the IRE through a beveled edge. Inside the IRE, the beam reaches the surface at an incidence angle greater than the critical angle for internal reflection, travels along the IRE via a series of internal reflections, exits the IRE, and reaches the detector directed by mirrors M3 and M4. At each internal reflection, evanescent wave is created outside of IRE close to its surface. The membrane and the protein (in this case, a helical peptide) absorb part of the light intensity, resulting in a transmission spectrum, which is converted to absorbance by coupling it with a transmission spectrum of a bare IRE.

peptide was synthesized with a thioester group at its C-terminus, which was ligated to the N-terminal cysteine of the ^{13}C-labeled Cys^{11}-Ser^{126} fragment of the protein (Fig. 3). Production of a segmentally ^{13}C-labeled protein and determination of its orientation on a membrane surface by polarized ATR-FTIR spectroscopy are described below:

- Produce the N-terminal peptide with a C-terminal thioester by chemical synthesis. This can be done by custom synthesis using one of many companies offering peptide synthesis services.
- Use a prokaryotic expression system to generate a ^{13}C-labeled fragment of the protein with an N-terminal cysteine, using a minimal medium containing 0.2% uniformly ^{13}C-labeled D-glucose (Cambridge Isotope Laboratories, Tewksbury, MA) as a sole metabolic source of carbon. Mammalian expression systems also can be used but that will require special media, such as a Bioexpress-6000 (Mammalian) medium (Cambridge Isotope Laboratories) that contains ^{13}C-labeled nutrients and essential amino acids, which makes this choice significantly more expensive.

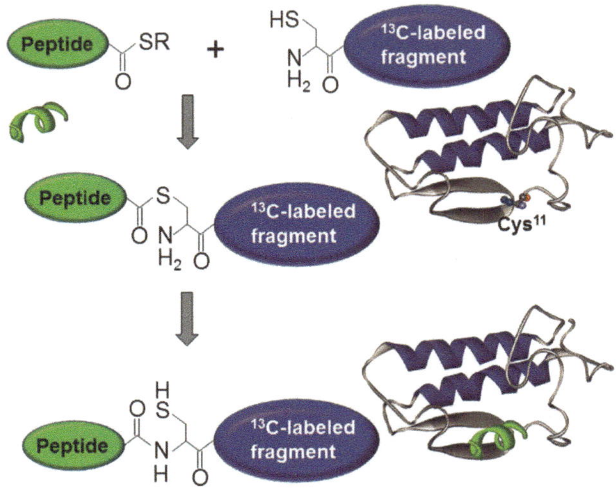

Fig. 3 Depiction of thioester-to-cysteine ligation. The peptide (*green oval*) corresponds to the N-terminal part of the protein and has a thioester group at its C-terminus ($R=CH_2COOH$). The protein fragment (*blue oval*) has a native or engineered cysteine at its N-terminus. The peptide's thioester reacts with the thiol group of the fragment's N-terminal cysteine, resulting in a new thioester linkage between them. The presence of a nearby amino group triggers *S*-to-*N* acyl shift, converting the thioester linkage to a native peptide bond.

- If the protein fragment to be expressed does not have a native cysteine at its N-terminus, it can be engineered using one of the several approaches described in the literature (Gentle, De Souza, & Baca, 2004; Liu, Xu, & Cowburn, 2009; Liu, Xu, Dutta, & Cowburn, 2008; Muralidharan & Muir, 2006). Care should be taken not to mutate a functionally important amino acid.
- Incubate the peptide with the purified ^{13}C-labeled fragment at a 1:1 molar ratio for 6 h at 37°C in 6 *M* guanidinum-HCl, 100 m*M* Na-phosphate, 5% β-mercaptoethanol, 1 m*M* EDTA, 4% thiophenol, and 4% benzyl mercaptan, pH 7.4.
- Thioester-to-cysteine ligation will occur, resulting in formation of a native peptide bond between the peptide and the fragment (Camarero & Muir, 1999; Liu et al., 2009; Muir, 2008; Muralidharan & Muir, 2006; Tam, Yu, & Miao, 1999) (Fig. 3).
- Purify the protein by ion-exchange and size exclusion chromatography.
- Verify the identity of the protein using at least two of the following: N-terminal sequencing, mass-spectrometry, Western blot, enzymatic activity assay.

- Inject the semisynthetic, segmentally ^{13}C-labeled protein into the ATR cell containing the supported lipid bilayer and measure FTIR absorbance spectra at *p* and *s* polarizations using the transmission spectra of the bare membrane as reference that should be measured in advance.
- Conduct curve-fitting (aka peak-fitting) of spectra at both polarizations to identify the amide I components generated by labeled and unlabeled helices (Fig. 4).

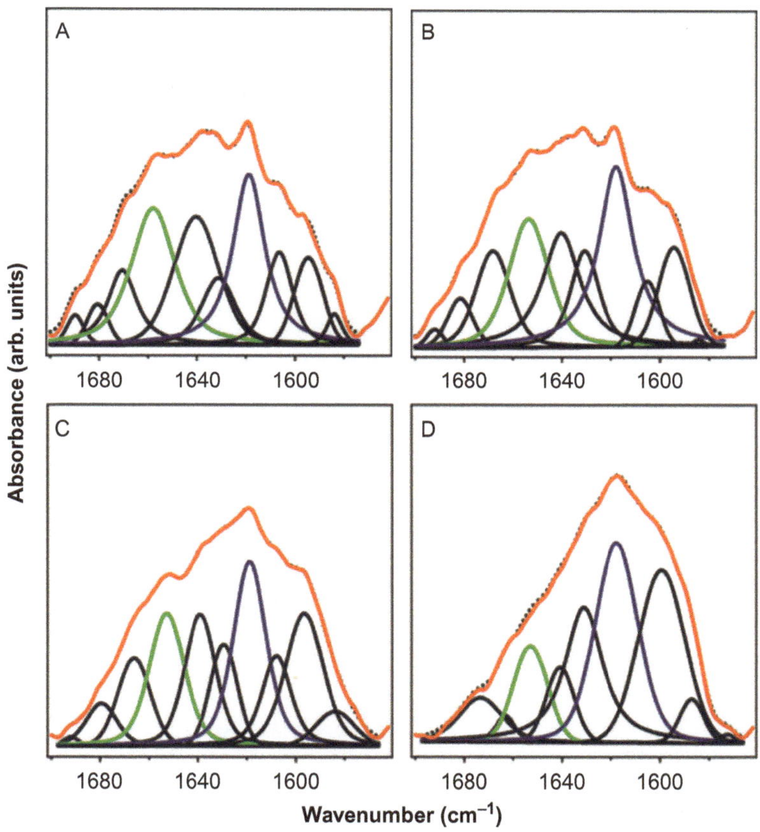

Fig. 4 ATR-FTIR spectra of human pancreatic PLA_2 with an unlabeled N-terminal helix ligated to a uniformly ^{13}C-labeled fragment bound to a POPC/POPG (4:1) membrane, exposed to a D_2O-based buffer for 15, 30, 40, and 60 min (panels (A) through (D), respectively). Amide I components have been obtained by curve fitting. The *red line* in each panel is the experimental spectrum and the *dotted line* is the sum of all components. (In some cases the *dotted line* is barely visible because it is masked by the experimental spectrum, indicating perfect fitting.) The *green* and *blue* components are assigned to the unlabeled and ^{13}C-labeled helices, respectively.

- Determine the orientations of the labeled and unlabeled helices (angles θ between helical axes and membrane normal) following procedures described in Section 2.4.1.
- Identify the orientations of the helices relative to the protein coordinate system (X_p, Y_p, Z_p) using the C_α atom coordinates, as described (Tatulian, 2008).
- Use the angles between each helical axes and the X_p, Y_p, Z_p axes along with the information on the θ angles of each helix relative to the membrane normal (Z_m) to evaluate the angles between X_p, Y_p, Z_p axes and the Z_m axis.
- Since the membrane-bound protein has a rotational freedom about Z_m, the azimuthal angles can be selected arbitrarily (observing the laws of direction cosines).
- Finally, the angles between the axes of the protein and membrane coordinate systems can be used to obtain the protein atom coordinates in the membrane coordinate system, x_m, y_m, and z_m. This will provide the three-dimensional angular orientation of the protein relative to the membrane.

This method determines the orientation of the protein but not its location along the membrane normal, i.e., the degree of membrane insertion. Determination of depth of membrane insertion of proteins is described in the next section.

2.5 Determination of Membrane Insertion

Membrane insertion of peripheral proteins can be evaluated by fluorescence quenching techniques, utilizing the following procedures.

- Prepare unilamellar vesicles containing 10 mol% of dibromo-PC (Br_2PC) brominated at 6–7, 9–10, or 11–12 positions of the *sn*-2 acyl chain (Avanti Polar Lipids). More than 10 mol% of Br_2PCs may cause membrane structure perturbations.
- Prepare identical vesicles in the absence of Br_2PCs.
- Incubate the protein with the vesicles at a ratio when most of the protein (>90%) is membrane-bound. This will require membrane-binding studies conducted separately (see Section 2.3).
- Measure Trp fluorescence intensity of the protein bound to vesicles without Br_2PC (F_0).
- Incubate the protein at the same ratio with vesicles containing each of the three Br_2PCs and measure Trp fluorescence (F_{Br}).

- Calculate the values of $\ln(F_0/F_{Br})$ for each Br_2PC and plot against the distance of the bromine atoms from the membrane center: $h=11.0$, 8.3, and 6.5 Å for 6–7, 9–10, and 11–12 Br_2PCs, respectively (McIntosh & Holloway, 1987).
- Stronger quenching by a certain brominated lipid will indicate a closer location of the Trp residue to the bromines of that lipid. Quantitatively, a Gaussian distribution for the Trp location along the membrane normal is used to fit the experimental data (Ladokhin, 1997; London & Ladokhin, 2002; Pande, Qin, Nemec, He, & Tatulian, 2006):

$$\ln\frac{F_0}{F_{Br}}=\frac{S}{\sigma\sqrt{2\pi}}\exp\left[-\frac{(h-h_m)^2}{2\sigma^2}\right] \tag{10}$$

  In Eq. (10), σ is the dispersion of the distribution curve, S is the area under the distribution curve, h is the distance from the membrane center, and h_m is the most probable location of the fluorophore with respect to the membrane center.
- The location of the fluorophore along the Z_m axis is determined by the best-fit value of h_m.

Membrane insertion depths of human group IB and a V3W mutant of IIA PLA_2, as well as bee venom PLA_2, have also been evaluated by this method (Pande et al., 2006). Trp[3] of structurally similar group IB and V3W-IIA PLA_2 was located at ~9 Å and at ~13 Å from the center of fluid and solid membranes, respectively, which correlated with higher enzymatic activity against fluid membranes. Membrane insertion of human pancreatic (IB) PLA_2 is schematically shown in Fig. 5. These data underscored the importance of membrane fluidity for an optimal membrane anchoring and activity of secreted PLA_2s. The structurally divergent bee venom PLA_2 did not penetrate into membranes as no fluorescence quenching of its tryptophans was detected, suggesting distinct membrane binding, substrate accession, and product release modes of these PLA_2s.

The mode of membrane binding of pancreatic PLA_2 (Protein Data Bank entry 1ysk) has been confirmed by both MD and OPM results (Lomize et al., 2007; Wee et al., 2008). In addition to hydrophobic protein–membrane interactions due to membrane insertion of Trp[3], Phe[19], and Leu[20] side chains of pancreatic PLA_2, salt bridging between lysines 121 and 122 and lipid phosphate groups, as well as H-bonding of Arg[6], Lys[7], Lys[10], and Lys[116] with lipid carbonyls were identified. The crevice of substrate-binding pocket was turned to the membrane with a slight tilt, consistent with earlier

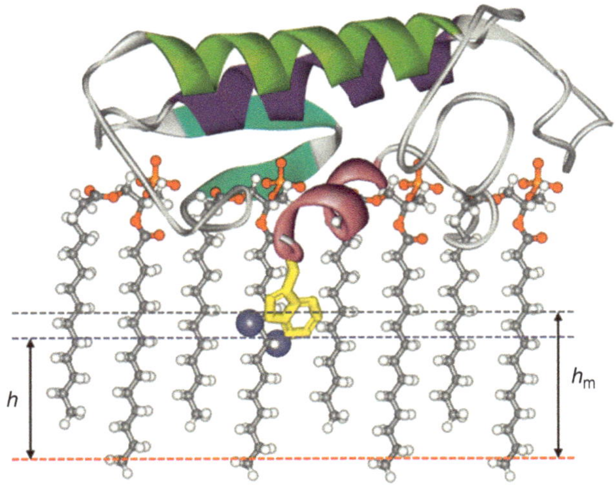

Fig. 5 Determination of membrane insertion of human pancreatic PLA_2. The protein is shown in *ribbon format*, the N-terminal helix colored *magenta*, and the two internal helices *green* and *blue*. The minor β-sheet structure is shown in *cyan*. The protein has a single tryptophan at position 3, shown in *yellow*. The lipids are shown in *ball* and *stick* format, colored according to atom type: carbons *gray*, hydrogens *white*, oxygens *red*, and phosphorus *orange*. One lipid is brominated at 6–7 positions of the *sn*-2 acyl chain, shown as *blue spheres*. Physical proximity of the tryptophan to the bromines results in fluorescence quenching. The *red dashed line* marks the membrane center. The *black* and *blue dashed lines* show the location of the tryptophan and the average location of the bromine atoms, respectively.

data on the membrane-binding mode of group I/II secreted PLA_2s (Canaan et al., 2002; Yu, Janssen, Verheij, & Jain, 2000). The distance between the catalytic histidine from the phospholipid's *sn*-2 ester bond was about 7 Å, implying that the lipid needs to travel only ~7 Å to reach the catalytic site, as opposed to earlier estimates of 15 Å (Berg, Gelb, Tsai, & Jain, 2001).

This method works well for proteins that have a single fluorophore, such as a native tryptophan. Single-Trp proteins can be created by site-directed mutagenesis, either by inserting a Trp or by replacing all tryptophans by Phe and leaving only one Trp. However, extensive mutagenesis can affect the protein structure and its membrane binding. An alternative approach is to create a set of single-Cys proteins and dansyl-label with a thiol-specific probe such as 5-IAEDANS (5-2-iodoacetylaminoethylaminonaphthalene-1-sulfonate; Teilum, Maki, Kragelund, Poulsen, & Roder, 2002). Each fluorescent-labeled protein can be used in membrane-binding experiments, where dansyl fluorescence quenching by several Br_2PCs is measured. In addition to Br_2PCs, a lipid nitroxide spin-labeled at the headgroup, such as 1-palmitoyl-2-oleoyl-phospho-TEMPO-choline (POPTC), can be used to identify the sites of the protein close to the membrane–water interface. Analysis of data of fluorescence quenching through Eq. (10) will provide the locations of a number of protein sites within the membrane, i.e., the mode of membrane binding.

Insertion of human 5-LO into solid (dipalmitoyl-phosphatidylcholine, DPPC) and fluid (1-palmitoyl-2-arachidonoyl-phosphatidylcholine, PAPC) membranes has been evaluated qualitatively. Significant Trp fluorescence quenching occurred by Br_2PCs in fluid PAPC membrane, but there was no quenching in case of solid DPPC membranes (Pande et al., 2005). In both cases, Trp fluorescence was strongly quenched by POPTC. Together with a significantly (fourfold) higher enzymatic activity of 5-LO bound to PAPC vs DPPC membranes, these findings indicated the importance of the presence of unsaturated lipids in membranes for efficient activity of 5-LO. Abundance of arachidonic acid (AA) containing lipids in the nuclear envelope might contribute to localization of 5-LO on the nuclear membrane.

2.6 Measuring PLA_2 and LO Activities

2.6.1 PLA_2

A number of methods have been utilized to measure the catalytic activity of secreted PLA_2s. The pH-stat method uses a pH electrode and a feedback device connected to an automated titrator. Phospholipid hydrolysis by PLA_2 takes place in an aqueous medium adjusted by NaOH to mildly alkaline pH, such as pH 8.0. The liberated fatty acid releases protons triggering acidification of the medium. This is sensed by the feedback mechanism, which activates the titrator and pumps certain amount of 10 m*M* NaOH into the reaction mixture to maintain the pH at the initial level. A time scan of the added volume of NaOH reflects the reaction kinetics (Bayburt et al., 1993; Ghomashchi et al., 1998; Karray, Gargouri, Verger, & Bezzine, 2012; Singh, Ranganathan, & Hajdu, 2010; Yu et al., 2000). At certain lipid and enzyme concentrations, enzyme kinetics and activity parameters can be measured quantitatively as each added NaOH molecule corresponds to a hydrolyzed lipid molecule. This method is easy and quantitative, but the absence of buffer can make the measured signal unstable because of pH fluctuations due to atmospheric CO_2, even when the sample is kept under a stream of nitrogen.

Another widely used method is based on environment-dependent changes in fluorescence emission intensity of a dansylated fatty acid analog, such as 11-(dansylamino)undecanoic acid (DAUDA). DAUDA has high affinity for rat liver fatty acid-binding protein (FABP), and binding to FABP results in strong enhancement of its fluorescence intensity ($\lambda_{ex} = 350$ nm, $\lambda_{em} = 500$ nm). Phospholipid hydrolysis by PLA_2 releases long chain fatty acids, which competitively displace DAUDA from FABP, resulting in reduction in fluorescence intensity. Time-dependent changes in DAUDA fluorescence thus allow monitoring of PLA_2 activity (Kinkaid & Wilton, 1991; Wilton, 1990). Modified FABP-based methods (Richieri & Kleinfeld, 1995), as well as alternative fluorescence-based approaches, such as changes in fluorescence of membrane-bound 8-anilino-1-naphthalenesulfonate during PLA_2-catalyzed membrane hydrolysis (Vivek et al., 2014), have been described.

One of the commercially available $sPLA_2$ activity kits is that of Cayman Chemical Co., which uses diheptanoyl-thio-phosphatidylcholine (DHTPC) micelles as PLA_2 substrate. The assay buffer contains DHTPC/Triton X-100 comicelles (1.5 and 0.3 m*M*, respectively) and 0.4 m*M* 5,5′-dithio-bis-(2-nitrobenzoic acid) (DTNB). Absorption at 414 nm is monitored in the time-scan mode. Following a stable baseline, hydrolytic reaction is initiated by adding PLA_2 to a known concentration. Hydrolysis of the *sn*-2 thioester bond of DHTPC results in exposure of free thiols and conversion of DTNB to 5-thio-nitrobenzoic acid, which is detected by its characteristic absorption at 414 nm. The initial slope of the time scan provides the specific PLA_2 activity, using an extinction coefficient of 5-thio-nitrobenzoic acid $\varepsilon_{414} = 13{,}600\ M^{-1}\ c^{-1}$ (Nemec et al., 2006; Qin et al., 2005, 2004; Tatulian et al., 2005).

A PLA_2 activity assay, effectively used in this laboratory, is based on the fluorescence properties of 1,2-bis-(1-pyrenedecanoyl)-*sn*-glycero-3-phosphocholine (bisPyPC), i.e., a phosphatidylcholine that is pyrenylated at the termini of both acyl chains (Fig. 6). The protocol is described below:

- Prepare unilamellar vesicles of desired lipid composition, containing 5 mol% bisPyPC, at a total lipid concentration of 0.5 m*M*, by extrusion through 100-nm pore-size polycarbonate membranes, as described in Section 2.3.1.
- The assay buffer should contain $CaCl_2$ at a low millimolar concentration (typically 1–5 m*M*), adjusted to pH between 7.0 and 8.5.
- Measure fluorescence spectra between 370 and 490 nm, with excitation at 347 nm.

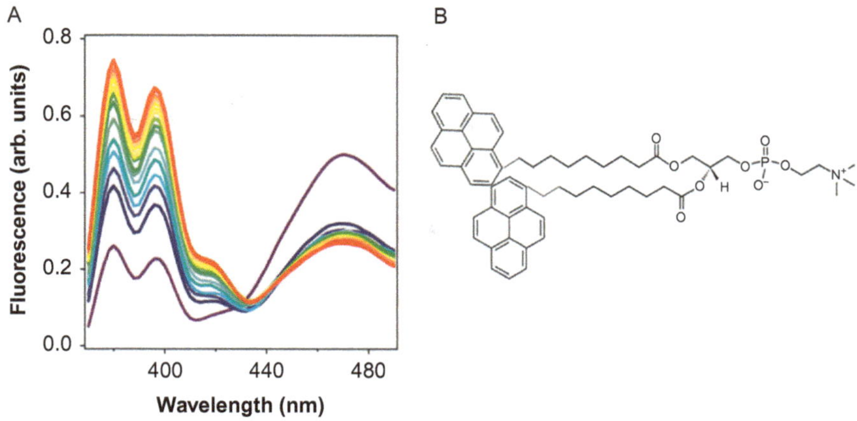

Fig. 6 Lipid hydrolysis by human group IIA PLA_2 measured by bysPyPC method (A) and the structure of bisPyPC (B). (A) Fluorescence spectra are shown for unilamellar vesicles (100 nm in diameter) composed of 75% POPC, 20% POPG, and 5% bisPyPC in 100 m*M* NaCl, 2 m*M* $CaCl_2$, 20 m*M* Hepes, pH 8.5, before (*blue*) and after addition of 2.0 μ*M* PLA_2 (*blue* to *red*) measured with 32 s periodicity. Excitation was at 347 nm, and total lipid concentration was 0.5 m*M*.

- Initiate the lipolytic reaction by adding PLA_2.
- Pyrene emission spectrum has two peaks generated by the monomeric state of the fluorophore (around 378 and 396 nm) and a peak around 470 nm that is generated by the excited-state dimers or excimers (Lakowicz, 1999).
- Initially there will be relatively weak fluorescence intensity in the 370–400 nm region and strong excimer fluorescence around 470 nm because of close proximity of pyrene moieties attached to the *sn*-1 and *sn*-2 chains of bisPyPC.
- Following addition of PLA_2, hydrolysis of bisPyPC will result in separation of the *sn*-1 and *sn*-2 chains from each other, resulting in a decrease in the excimer signal at 470 nm and an increase in the monomer signal at 378 nm (Fig. 6).
- Conduct consecutive measurements of emission spectra and determine the fluorescence intensity ratios $R_t = F_{378}/F_{470}$ at successive time points t after PLA_2 addition.
- Plot $(R_t/R_0 - 1)$ vs time, where R_0 is R_t before addition of PLA_2.
- The catalytic rate of PLA_2 can be obtained from the initial slope of the curve, i.e., change in $(R_t/R_0 - 1)$ per second immediately following addition of PLA_2 to lipid vesicles. It is not uncommon to detect relatively

long lag periods before "burst" activation of PLA_2 (Ray, Scott, & Tatulian, 2007).

2.6.2 Lipoxygenase

Plant lipoxygenase-catalyzed biosynthesis of plant hormone jasmonic acid is a multistep process and involves production of conjugated dienes (the primary products of polyunsaturated fatty acid hydroperoxidation) and ketodienes (the secondary products). Linoleic acid is a good substrate for soybean LO and has been used as such either by itself or incorporated in lipid vesicle membranes (Tatulian, Steczko, & Minor, 1998). The activity assay of soybean LO is described below:

- Transfer appropriate volume of 10 m*M* linoleic acid (Sigma-Aldrich) in ethanol into a glass vial and remove the solvent by desiccation.
- Add buffer (50 m*M* NaCl, 10 m*M* Hepes, 10 m*M* Capso, pH 8.2) to reach a final linoleic acid concentration of 0.1–0.5 m*M* and vortex for 2–3 min.
- The buffer should contain around 200 μ*M* dissolved oxygen to avoid anaerobic ketodiene production (Garssen, Vliegenthart, & Boldingh, 1971). If a dissolved oxygen meter is not available, just bubbling the buffer with air will do.
- $CaCl_2$ can be added to the buffer at 0.1–2 m*M* concentration. At low concentration, it will promote and at high concentration will inhibit conjugated diene production. At higher linoleic acid concentration (e.g., 2 m*M*), Ca^{2+} will have the opposite effect, however (Tatulian, Steczko, et al., 1998).
- Measure time-scan of absorption of linoleic acid in a quartz cuvette at 234 nm. If time mode is not available, take consecutive measurements. An identical sample can be used as reference to offset light-scattering effects.
- When a stable baseline is established, add LO to 50 n*M* final concentration.
- Determine the kinetics of production of conjugated diene from the slope of time dependence of A_{234} and quantitate using $\varepsilon_{234} = 25{,}000\ M^{-1}\ \mathrm{cm}^{-1}$ (Axelrod, Cheesbrough, & Laakso, 1981; Garssen et al., 1971; Steczko, Donoho, Dixon, Sugimoto, & Axelrod, 1991).
- Production of conjugated ketodienes can be measured by recording A_{280} and using $\varepsilon_{280} = 22{,}000\ M^{-1}\ \mathrm{cm}^{-1}$.
- Soybean LO activity can also be measured using linoleic acid in combination with lipid vesicles, as described (Tatulian, Steczko, et al., 1998).

5-LO-catalyzed conversion of AA to 5-hydroperoxyeicosatetraenoic acid (5-HPETE) can be measured using a simple spectrophotometric assay, as described below:

- Prepare 5-LO activity buffer containing 31.64 mM Na_2HPO_4, 5.4 mM KH_2PO_4, 0.2 mM ATP, 0.1 mM dithiothreitol, 0.1 mM EDTA, and 0.3 mM $CaCl_2$ (pH 7.5).
- Prepare unilamellar lipid vesicles at a final lipid concentration of 0.35 mM in the activity buffer. PC containing a palmitoyl residue at the *sn*-1 position and a fatty acyl chain with 18–20 carbons and 1–4 *cis* double bonds at the *sn*-2 position can be used.
- Transfer an appropriate volume of AA dissolved in ethanol into a glass vial and remove solvent by desiccation. (Make sure AA is not supplemented with antioxidants to protect against autoxidation.)
- Add the vesicle suspension to dry AA and gently vortex. The final AA concentration may range from 10 to 50 μM. Higher AA concentration may cause substrate inhibition (Pande et al., 2004).
- Place equal volumes of AA-containing and blank (i.e., without AA) vesicles in two quartz cuvettes and measure time-scan of absorption at 238 nm on a double-beam spectrophotometer.
- Initiate the AA oxygenation reaction by adding 5-LO into the sample cuvette to a final concentration between 1 and 10 μg/mL.
- Evaluate the specific activity of 5-LO from the initial slope of time-dependence of A_{238} nm, using an extinction coefficient of 23,000 M^{-1} cm^{-1} (Percival, 1991; Skorey & Gresser, 1998).

3. STRENGTHS AND LIMITATIONS

3.1 Strengths

Interfacial enzymes bind to cellular membranes transiently, using membrane-binding modules that in most cases cannot be readily identified based on their atomic structures. On the other hand, the understanding of the mechanism of action of these enzymes is not complete until their membrane-binding mode is elucidated. The major strength of the methods described here is that they provide this crucial piece of information and hence present a molecular description of the real functional unit, the protein–membrane complex. As described in the preceding sections, a variety of approaches have been developed to tackle this challenging task, often combining molecular biology with chemistry, biophysics, and computation. One of the approaches allowing identification of the orientation and membrane insertion of interfacial enzyme is polarized FTIR coupled with membrane depth-dependent fluorescence quenching, as described earlier. The protein is studied when bound to phospholipid membranes under near-physiological conditions, which is advantageous compared to methods that use micelles because of molecular size limitations. FTIR does not require labeling with relatively bulky probes such as nitroxide or fluorescent moieties, which can adversely affect the membrane and/or protein structure. ^{13}C- or other stable isotope labeling does not cause structural perturbations yet provides valuable information on local structure and orientation of proteins. Moreover, the overall structure of the protein and conformational changes upon membrane binding, as well as the structure and the order of membrane lipids, can be directly measured by polarized FTIR. Finally, light-scattering problems encountered in UV spectroscopy are practically absent in the infrared region.

Ideally, the structure of a membrane protein, i.e., the atom coordinates, should be presented in a membrane, rather than arbitrary, coordinate system. Then, the atom coordinates of each amino acid residue will tell its location with respect to the membrane, while the overall structure of the protein–membrane complex will display the geometric positioning of the active site cleft and the catalytic machinery of the enzyme. While FTIR spectroscopy cannot determine the atomic structure of proteins, structural and orientational information derived from polarized FTIR measurements can be combined with molecular modeling and computational methods to achieve atomic structures of proteins within a membrane coordinate system. This has been demonstrated for human pancreatic PLA_2 and a chimera of IB and IIA PLA_2s (Qin et al., 2005; Tatulian et al., 2005). As noted by Berrade and Camarero (2009), "This approach is in principle quite general, and it is likely to become a fundamental tool for determination and analysis of the structure of membrane proteins, which will undoubtedly provide valuable information on the molecular mechanisms of this important class of proteins."

3.2 Limitations

Determination of membrane binding, orientation, and membrane insertion of interfacial enzymes by fluorescence and FTIR methods is associated with certain shortcomings. These methods cannot determine the atomic resolution structure of the protein. They can only be applied to proteins for which the structure is available from X-ray crystallographic or NMR experiments or by homology modeling. Even when the protein structure is known, still the question remains as to whether the protein maintains its structure or undergoes conformational changes upon membrane binding. These issues can be addressed by measuring the structural changes in the protein caused by membrane binding, and ATR-FTIR spectroscopy is well suited to achieve this goal.

Membrane insertion determination using Trp fluorescence quenching by bromolipids may not be applicable to proteins that have no tryptophan or more than one tryptophan. On the other hand, bromolipids may perturb the global or local membrane structure and hence become a source of artifact. Therefore, monitoring the membrane structure by additional methods would be an advisable addition of these experiments. A good solution of this issue is measurements of the lipid order parameter by polarized ATR-FTIR before and after membrane binding of the protein (Tatulian, 2013; Tatulian, Jones, Reddy, Stokes, & Tamm, 1995; Tatulian, Cortes, & Perozo, 1998).

In assessment of membrane binding of interfacial enzymes by ATR-FTIR, membrane-proximal but not bound protein molecules may contribute to the spectra. This potential drawback can be addressed by minimizing the evanescence wave decay length using larger incidence angles of the incoming infrared light. Also, the ATR cell may be gently flushed with buffer to remove unbound or loosely bound proteins before the measurements.

4. CONCLUSIONS

Interfacial enzymes are biomedically important as they catalyze the biosynthesis of lipid-derived mediators involved in a plethora of cell signaling events. They also represent a unique and intriguing class of proteins because they operate in a complex environment of membrane–water interface. During the last 2–3 decades we have witnessed tremendous progress in understanding the molecular mechanisms of interfacial enzymes, including the physical forces governing protein–membrane interactions, the mode of membrane binding, including the orientation and membrane insertion, and the role of these factors in the mode of action of the enzyme such as substrate acquisition and product release. Multiple experimental approaches, e.g., advanced versions of vibrational and electronic spectroscopy and magnetic resonance, as well as computational methods, some of which have been developed into web-based servers, have provided unprecedented insight into our understanding of interfacial enzymes. Still much is left to be clarified. The details of the dynamic and coordinated conformational changes in both enzymes and lipids during interfacial activation, the specific roles in functionally important amino acids in membrane binding and the catalytic turnover are waiting to be elucidated (Lambeau & Gelb, 2008). With recent advances in multiscale models for complex chemical systems (Karplus, 2014), researchers will conquer new territory of knowledge and will combine classical and quantum physics approaches to take a closer glance at the workings of interfacial enzymes at atomic and subatomic levels.

REFERENCES

Axelrod, B., Cheesbrough, T. M., & Laakso, S. (1981). Lipoxygenase from soybeans. *Methods in Enzymology, 71*, 441–451.

Axelsen, P. H., & Citra, M. J. (1996). Orientational order determination by internal reflection infrared spectroscopy. *Progress in Biophysics and Molecular Biology, 66*, 227–253.

Bayburt, T., Yu, B. Z., Lin, H. K., Browning, J., Jain, M. K., & Gelb, M. H. (1993). Human nonpancreatic secreted phospholipase A_2: Interfacial parameters, substrate specificities, and competitive inhibitors. *Biochemistry, 32*, 573–582.

Berg, O. G., Gelb, M. H., Tsai, M. D., & Jain, M. K. (2001). Interfacial enzymology: The secreted phospholipase A_2-paradigm. *Chemical Reviews, 101*, 2613–2654.

Berrade, L., & Camarero, J. A. (2009). Expressed protein ligation: A resourceful tool to study protein structure and function. *Cellular and Molecular Life Sciences, 66*, 3909–3922.

Brunecky, R., Lee, S., Rzepecki, P. W., Overduin, M., Prestwich, G. D., Kutateladze, A. G., et al. (2005). Investigation of the binding geometry of a peripheral membrane protein. *Biochemistry, 44*, 16064–16071.

Camarero, J. A., & Muir, T. W. (1999). Native chemical ligation of polypeptides. *Current Protocols in Protein Science*, 18.4.1–18.4.21.

Canaan, S., Nielsen, R., Ghomashchi, F., Robinson, B. H., & Gelb, M. H. (2002). Unusual mode of binding of human group IIA secreted phospholipase A_2 to anionic interfaces as studied by continuous wave and time domain electron paramagnetic resonance spectroscopy. *The Journal of Biological Chemistry, 277*, 30984–30990.

Cheng, X., Kim, J. K., Kim, Y., Bowie, J. U., & Im, W. (2016). Molecular dynamics simulation strategies for protein-micelle complexes. *Biochimica et Biophysica Acta, 1858*, 1566–1572.

Cook, G. A., Dawson, L. A., Tian, Y., & Opella, S. J. (2013). Three-dimensional structure and interaction studies of hepatitis C virus p7 in 1,2-dihexanoyl-sn-glycero-3-phosphocholine by solution nuclear magnetic resonance. *Biochemistry, 52*, 5295–5303.

Damodaran, S. (2015). Water at biological phase boundaries: Its role in interfacial activation of enzymes and metabolic pathways. *Sub-Cellular Biochemistry, 71*, 233–261.

Emptage, R. P., Tonthat, N. K., York, J. D., Schumacher, M. A., & Zhou, P. (2014). Structural basis of lipid binding for the membrane-embedded tetraacyldisaccharide-1-phosphate 4'-kinase LpxK. *The Journal of Biological Chemistry, 289*, 24059–24068.

Fringeli, U. P., Apell, H.-J., Fringeli, M., & Läuger, P. (1989). Polarized infrared absorption of Na^+/K^+-ATPase studied by attenuated total reflection spectroscopy. *Biochimica et Biophysica Acta, 984*, 301–312.

Garssen, G. J., Vliegenthart, J. F., & Boldingh, J. (1971). An anaerobic reaction between lipoxygenase, linoleic acid and its hydroperoxides. *The Biochemical Journal, 122*, 327–332.

Gelb, M. H., Cho, W., & Wilton, D. C. (1999). Interfacial binding of secreted phospholipases A_2: More than electrostatics and a major role for tryptophan. *Current Opinion in Structural Biology, 9*, 428–432.

Gelb, M. H., Min, J.-H., & Jain, M. K. (2000). Do membrane-bound enzymes access their substrates from the membrane or aqueous phase: Interfacial versus non-interfacial enzymes. *Biochimica et Biophysica Acta, 1488*, 20–27.

Gentle, I. E., De Souza, D. P., & Baca, M. (2004). Direct production of proteins with N-terminal cysteine for site-specific conjugation. *Bioconjugate Chemistry, 15*, 658–663.

Ghomashchi, F., Lin, Y., Hixon, M. S., Yu, B.-Z., Annand, R., Jain, M. K., et al. (1998). Interfacial recognition by bee venom phospholipase A_2: Insights into nonelectrostatic molecular determinants by charge reversal mutagenesis. *Biochemistry, 37*, 6697–6710.

Goormaghtigh, E., Raussens, V., & Ruysschaert, J. M. (1999). Attenuated total reflection infrared spectroscopy of proteins and lipids in biological membranes. *Biochimica et Biophysica Acta, 1422*, 105–185.

Jain, M. K., DeHaas, G. H., Marecek, J. F., & Ramirez, F. (1986). The affinity of phospholipase A_2 for the interface of the substrate and analogs. *Biochimica et Biophysica Acta, 860*, 475–483.

Jo, S., Kim, T., Iyer, V. G., & Im, W. (2008). CHARMM-GUI: A web-based graphical user interface for CHARMM. *Journal of Computational Chemistry, 29*, 1859–1865.

Karplus, M. (2014). Development of multiscale models for complex chemical systems: From $H+H_2$ to biomolecules (Nobel Lecture). *Angewandte Chemie (International Ed. in English), 53*, 9992–10005.

Karray, A., Gargouri, Y., Verger, R., & Bezzine, S. (2012). Phospholipase A_2 purification and characterization: A case study. *Methods in Molecular Biology, 861*, 283–297.

Kimmett, T., Smith, N., Witham, S., Petukh, M., Sarkar, S., & Alexov, E. (2014). ProBLM web server: Protein and membrane placement and orientation package. *Computational and Mathematical Methods in Medicine, 2014*, 838259.

Kinkaid, A., & Wilton, D. C. (1991). Comparison of the catalytic properties of phospholipase A_2 from pancreas and venom using a continuous fluorescence displacement assay. *The Biochemical Journal, 278*, 843–848.

Kutateladze, T. G. (2012). Molecular analysis of protein-phosphoinositide interactions. *Current Topics in Microbiology and Immunology, 362*, 111–126.

Kutateladze, T. G., Capelluto, D. G. S., Ferguson, C. G., Cheever, M. L., Kutateladze, A. G., Prestwich, G. D., et al. (2004). Multivalent mechanism of membrane insertion by the FYVE domain. *The Journal of Biological Chemistry, 279*, 3050–3057.

Ladokhin, A. S. (1997). Distribution analysis of depth-dependent fluorescence quenching in membranes: A practical guide. *Methods in Enzymology, 278*, 462–473.

Lakowicz, J. R. (1999). *Principles of fluorescence spectroscopy* (2nd ed.). . New York: Kluwer Academic/Plenum Publishers.

Lambeau, G., & Gelb, M. H. (2008). Biochemistry and physiology of mammalian secreted phospholipases A_2. *Annual Review of Biochemistry, 77*, 495–520.

Li, Y., Li, Q., Wong, Y. L., Liew, L. S., & Kang, C. (2015). Membrane topology of NS2B of dengue virus revealed by NMR spectroscopy. *Biochimica et Biophysica Acta, 1848*, 2244–2252.

Lin, Y., Nielsen, R., Murray, D., Hubbell, W. L., Mailer, C., Robinson, B. H., et al. (1998). Docking phospholipase A_2 on membranes using potential-modulated spin relaxation magnetic resonance. *Science, 279*, 1925–1929.

Liu, D., Xu, R., & Cowburn, D. (2009). Segmental isotopic labeling of proteins for nuclear magnetic resonance. *Methods in Enzymology, 462*, 151–175.

Liu, D., Xu, R., Dutta, K., & Cowburn, D. (2008). N-terminal cysteinyl proteins can be prepared using thrombin cleavage. *FEBS Letters, 582*, 1163–1167.

Lomize, M. A., Lomize, A. L., Pogozheva, I. D., & Mosberg, H. I. (2006). OPM: Orientations of proteins in membranes database. *Bioinformatics, 22*, 623–625.

Lomize, M. A., Pogozheva, I. D., Joo, H., Mosberg, H. I., & Lomize, A. L. (2012). OPM database and PPM web server: Resources for positioning of proteins in membranes. *Nucleic Acids Research*, *40*, D370–D376.

Lomize, A. L., Pogozheva, I. D., Lomize, M. A., & Mosberg, H. I. (2006). Positioning of proteins in membranes: A computational approach. *Protein Science*, *15*, 1318–1333.

Lomize, A. L., Pogozheva, I. D., Lomize, M. A., & Mosberg, H. I. (2007). The role of hydrophobic interactions in positioning of peripheral proteins in membranes. *BMC Structural Biology*, *7*, 44.

London, E., & Ladokhin, A. S. (2002). Measuring the depth of amino acid residues in membrane-inserted peptides by fluorescence quenching. *Current Topics in Membranes*, *52*, 89–115.

Marsh, D. (1999). Quantitation of secondary structure in ATR infrared spectroscopy. *Biophysical Journal*, *77*, 2630–2637.

Marsh, D., Müller, M., & Schmitt, F.-J. (2000). Orientation of the infrared transition moments for an α-helix. *Biophysical Journal*, *78*, 2499–2510.

Marsh, D., & Páli, T. (2001). Infrared dichroism from the X-ray structure of bacteriorhodopsin. *Biophysical Journal*, *80*, 305–312.

McIntosh, T. J., & Holloway, P. W. (1987). Determination of the depth of bromine atoms in bilayers formed from bromolipid probes. *Biochemistry*, *26*, 1783–1788.

Muir, T. W. (2008). Studying protein structure and function using semisynthesis. *Biopolymers*, *90*, 743–750.

Muralidharan, V., & Muir, T. W. (2006). Protein ligation: An enabling technology for the biophysical analysis of proteins. *Nature Methods*, *3*, 429–438.

Nemec, K. N., Pande, A. H., Qin, S., Bieber Urbauer, R. J., Tan, S., Moe, D., et al. (2006). Structural and functional effects of tryptophans inserted into the membrane-binding and substrate-binding sites of human group IIA phospholipase A_2. *Biochemistry*, *45*, 12448–12460.

Newie, J., Andreou, A., Neumann, P., Einsle, O., Feussner, I., & Ficner, R. (2016). Crystal structure of a lipoxygenase from *Cyanothece* sp. may reveal novel features for substrate acquisition. *Journal of Lipid Research*, *57*, 276–287.

Nielsen, R. D., Che, K., Gelb, M. H., & Robinson, B. H. (2005). A ruler for determining the position of proteins in membranes. *Journal of the American Chemical Society*, *127*, 6430–6442.

Pande, A. H., Moe, D., Nemec, K. N., Qin, S., Tan, S., & Tatulian, S. A. (2004). Modulation of human 5-lipoxygenase activity by membrane lipids. *Biochemistry*, *43*, 14653–14666.

Pande, A. H., Qin, S., Nemec, K. N., He, X., & Tatulian, S. A. (2006). Isoform-specific membrane insertion of secretory phospholipase A_2 and functional implications. *Biochemistry*, *45*, 12436–12447.

Pande, A. H., Qin, S., & Tatulian, S. A. (2005). Membrane fluidity is a key modulator of membrane binding, insertion, and activity of 5-lipoxygenase. *Biophysical Journal*, *88*, 4084–4094.

Percival, D. M. (1991). Human 5-lipoxygenase contains an essential iron. *The Journal of Biological Chemistry*, *266*, 10058–10061.

Prior, S. H., Fulcher, Y. G., Koppisetti, R. K., Jurkevich, A., & Van Doren, S. R. (2015). Charge-triggered membrane insertion of matrix metalloproteinase-7, supporter of innate immunity and tumors. *Structure*, *23*, 2099–2110.

Qin, S., Pande, A. H., Nemec, K. N., He, X., & Tatulian, S. A. (2005). Evidence for the regulatory role of the N-terminal helix of secretory phospholipase A_2 from studies on native and chimeric proteins. *The Journal of Biological Chemistry*, *280*, 36773–36783.

Qin, S., Pande, A. H., Nemec, K. N., & Tatulian, S. A. (2004). The N-terminal α-helix of pancreatic phospholipase A_2 determines productive-mode orientation of the enzyme at the membrane surface. *Journal of Molecular Biology*, *344*, 71–89.

Ray, S., Scott, J. L., & Tatulian, S. A. (2007). Effects of lipid phase transition and membrane surface charge on the interfacial activation of phospholipase A_2. *Biochemistry, 46*, 13089–13100.

Richieri, G. V., & Kleinfeld, A. M. (1995). Continuous measurement of phospholipase A_2 activity using the fluorescent probe ADIFAB. *Analytical Biochemistry, 229*, 256–263.

Sambrook, J., & Russell, D. W. (2001). *Molecular cloning. A laboratory manual:* Vol. 2 (3rd ed.) Cold Spring Harbor, NY: Cold Spring Harbor Laboratory Press.

Scott, J. L., Musselman, C. A., Adu-Gyamfi, E., Kutateladze, T. G., & Stahelin, R. V. (2012). Emerging methodologies to investigate lipid-protein interactions. *Integrative Biology (Camb), 4*, 247–258.

Singh, J., Ranganathan, R., & Hajdu, J. (2010). Surface dilution kinetics using substrate analog-enantiomers as diluents: Enzymatic lipolysis by bee-venom phospholipase A_2. *Analytical Biochemistry, 407*, 253–260.

Skorey, K. I., & Gresser, M. J. (1998). Calcium is not required for 5-lipoxygenase activity at high phosphatidylcholine vesicle concentrations. *Biochemistry, 37*, 8027–8034.

Steczko, J., Donoho, G. A., Dixon, J. E., Sugimoto, T., & Axelrod, B. (1991). Effect of ethanol and low-temperature culture on expression of soybean lipoxygenase L-1 in *Escherichia coli*. *Protein Expression and Purification, 2*, 221–227.

Tam, J. P., Yu, Q., & Miao, Z. (1999). Orthogonal ligation strategies for peptide and protein. *Biopolymers, 51*, 311–332.

Tatulian, S. A. (2001). Toward understanding interfacial activation of secretory phospholipase A_2 (PLA_2): Membrane surface properties and membrane-induced structural changes in the enzyme contribute synergistically to PLA_2 activation. *Biophysical Journal, 80*, 789–800.

Tatulian, S. A. (2003). Structural effects of covalent inhibition of phospholipase A_2 suggest allosteric coupling between membrane binding and catalytic sites. *Biophysical Journal, 84*, 1773–1783.

Tatulian, S. A. (2008). Determination of helix orientations in proteins. *Computational Biology and Chemistry, 32*, 370–374.

Tatulian, S. A. (2013). Structural characterization of membrane proteins and peptides by FTIR and ATR-FTIR spectroscopy. *Methods in Molecular Biology, 974*, 177–218.

Tatulian, S. A., Cortes, D. M., & Perozo, E. (1998). Structural dynamics of the *Streptomyces lividans* K^+ channel (SKC1): Secondary structure characterization from FTIR spectroscopy. *FEBS Letters, 423*, 205–212.

Tatulian, S. A., Jones, L. R., Reddy, L. G., Stokes, D. L., & Tamm, L. K. (1995). Secondary structure and orientation of phospholamban reconstituted in supported bilayers from polarized attenuated total reflection FTIR spectroscopy. *Biochemistry, 34*, 4448–4456.

Tatulian, S. A., Qin, S., Pande, A. H., & He, X. (2005). Positioning membrane proteins by novel protein engineering and biophysical approaches. *Journal of Molecular Biology, 351*, 939–947.

Tatulian, S. A., Steczko, J., & Minor, W. (1998). Uncovering a calcium-regulated membrane binding mechanism for soybean lipoxygenase-1. *Biochemistry, 37*, 15481–15490.

Teilum, K., Maki, K., Kragelund, B. B., Poulsen, F. M., & Roder, H. (2002). Early kinetic intermediate in the folding of acyl-CoA binding protein detected by fluorescence labeling and ultrarapid mixing. *Proceedings of the National Academy of Sciences of the United States of America, 99*, 9807–9812.

Vedantham, G., Sparks, H. G., Sane, S. U., Tzannis, S., & Przybycien, T. M. (2000). A holistic approach for protein secondary structure estimation from infrared spectra in H_2O solutions. *Analytical Biochemistry, 285*, 33–49.

Venyaminov, S. Y., & Kalnin, N. N. (1990). Quantitative IR spectrophotometry of peptide compounds in water (H_2O) solutions. II. Amide absorption bands of polypeptides and fibrous proteins in α-, β-, and random coil conformations. *Biopolymers, 30*, 1259–1271.

Vivek, H. K., Swamy, S. G., Priya, B. S., Sethi, G., Rangappa, K. S., & Swamy, S. N. (2014). A facile assay to monitor secretory phospholipase A_2 using 8-anilino-1-naphthalenesulfonic acid. *Analytical Biochemistry*, *461*, 27–35.

Wee, C. L., Balali-Mood, K., Gavaghan, D., & Sansom, M. S. (2008). The interaction of phospholipase A2 with a phospholipid bilayer: Coarse-grained molecular dynamics simulations. *Biophysical Journal*, *95*, 1649–1657.

Weiss, J. P. (2015). Molecular determinants of bacterial sensitivity and resistance to mammalian Group IIA phospholipase A_2. *Biochimica et Biophysica Acta*, *1848*, 3072–3077.

Wilton, D. C. (1990). A continuous fluorescence displacement assay for the measurement of phospholipase A_2 and other lipases that release long-chain fatty acids. *The Biochemical Journal*, *266*, 435–439.

Wu, E. L., Cheng, X., Jo, S., Rui, H., Song, K. C., Dávila-Contreras, E. M., et al. (2014). CHARMM-GUI membrane builder toward realistic biological membrane simulations. *Journal of Computational Chemistry*, *35*, 1997–2004.

Yu, B. Z., Janssen, M. J. W., Verheij, H. M., & Jain, M. K. (2000). Control of the chemical step by leucine-31 of pancreatic phospholipase A_2. *Biochemistry*, *39*, 5702–5711.

Zhang, Y.-Y., Rådmark, O., & Samuelsson, B. (1992). Mutagenesis of some conserved residues in human 5-lipoxygenase: Effects on enzyme activity. *Proceedings of the National Academy of Sciences of the United States of America*, *89*, 485–489.

CHAPTER TEN

Measuring Diacylglycerol Kinase-θ Activity and Binding

B. Tu-Sekine, D.M. Raben[1]
Johns Hopkins University, Baltimore, MD, United States
[1]Corresponding author: e-mail address: draben@jhmi.edu

Contents

Abstract

This section provides detailed protocols for the analysis of a mammalian diacylglycerol kinase, DGKθ, including an activity assay, a kinetic analysis, preparation of small unilamellar vesicles, and a vesicle pulldown assay. The goal of this section is to provide an overview of the unique challenges inherent in the study of an interfacial enzyme such as DGKθ and to outline methods useful for analysis. We include a short tutorial on selecting lipids for forming the interface since this is critical for a successful

Methods in Enzymology, Volume 583
ISSN 0076-6879
http://dx.doi.org/10.1016/bs.mie.2016.09.002

in vitro assay, and lipids are important regulators of this enzyme. The general principles can be applied to the study of other interfacial enzymes.

1. THEORY

1.1 Properties of Interfacial Enzymes

Interfacial enzymes are soluble proteins that metabolize insoluble substrates (lipids) in membranes or lipid aggregates. Analysis of these enzymes can be challenging for several reasons: first, because these enzymes have hydrophobic regions that facilitate membrane association, they tend to aggregate and are difficult to purify in large quantities. Second, the mechanism for interfacial binding varies between proteins, necessitating empirical studies to identify appropriate interfaces. Third, the vesicle residence time (processivity), is dependent on the composition of the interface and can be strongly impacted by mutations, complicating kinetic analyses. In the simplest case (kinetically speaking), the enzyme binds tightly to the membrane and consumes the available substrate without detaching. However, DGKθ repeatedly binds and detaches from membranes under all *in vitro* conditions tested to date, and is therefore categorized as a hopping or quasi-scooting enzyme. This chapter includes some general information about interfacial enzymes, with specific protocols for the analysis of the hopping interfacial enzyme DGKθ.

1.2 The Effect of Interfacial Binding on Enzyme Kinetics

One reaction cycle of DGKθ (and all interfacial enzymes) comprises four steps:

(1) membrane binding
(2) substrate binding
(3) catalysis
(4) product release

Highly processive enzymes (termed "scooters") bind to membranes in an essentially irreversible manner and complete step 1 only once (Jain & Gelb, 1991). This is ideal for determining the kinetic parameters of an enzyme. However, peripheral interfacial enzymes like DGKθ bind and release vesicles multiple times during a reaction and complete an unknown number of substrate turnovers during each interfacial binding event, greatly complicating the kinetic profile. This means that only *apparent* kinetic parameters can be measured, as the K_M will be inflated due to the repeated binding events. Jain and Gelb use a variable termed E^* (see Fig. 1) to clarify the difference between total enzyme (E_{tot}) and enzyme bound to the interface (E^*), which is the only form that can access the insoluble substrate. This is a critical difference between enzymes that bind soluble substrates and enzymes that bind insoluble substrates. The former has access to all available substrate in the assay, while the latter can only access substrate once it has bound to an interface (such as a vesicle, micelle, or membrane). This requires a shift in thinking since the substrate essentially occupies a two-dimensional space (the membrane) and must be considered in relative terms (i.e., mole percent or mole fraction of a vesicle) rather than a three-dimensional space (i.e., aqueous solution).

Fortunately, the kinetics of hopping interfacial enzymes can produce nearly hyperbolic curves, allowing the application of Michaelis–Menten and Briggs–Haldane formalisms. While this is useful for interpretive purposes, the calculated K_M value does *not* reflect the intrinsic substrate affinity of the enzyme and must be reported as a K_M^{app}. Similarly, the measured V_{MAX} is likely to be an underestimate of the intrinsic turnover rate since

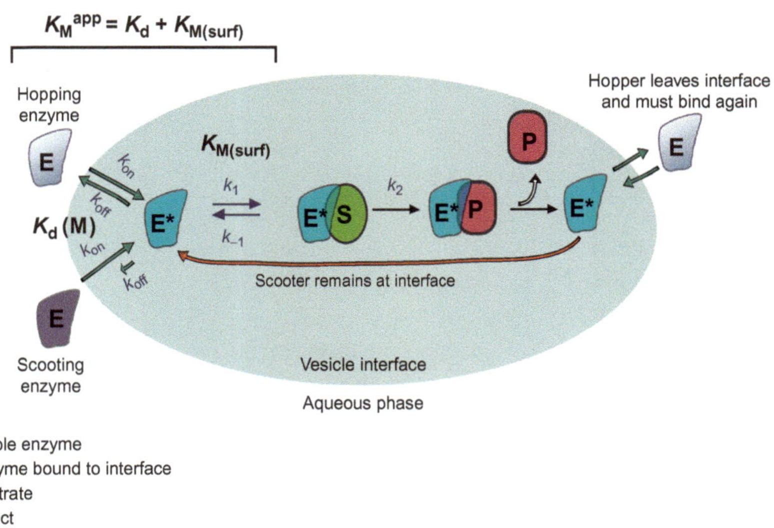

Fig. 1 A visual representation of hopping and scooting at the interface. A fundamental difference between hopping and scooting behavior lies in the enzyme's affinity for the interface. "Scooting" enzymes bind tightly to membranes (they have a low interfacial K_d), and as a result they perform hundreds to thousands of turnovers prior to dissociating. "Hopping" enzymes often have a high interfacial K_d and may perform only a few substrate turnovers before dissociating.

$[E^*] < [E_{tot}]$ by an unknown amount. In other words, some part of the total enzyme is *not* bound to an interface during the assay. For these (and other) reasons, while the Michaelis–Menten paradigm can be applied to the study of DGKθ activity, the observed kinetic values reflect the intrinsic enzyme activity *plus* the effects of repeated interfacial binding and must be reported as apparent values (i.e.,: K_M^{app}, K_{MS}^{app}, and V_{MAX}^{app}).

2. EQUIPMENT

Lipid extruder (Avanti Lipids 610020 or t10000)
Bath sonicator (e.g., Branson CPX Ultrasonic cleaning bath)
Centrifuge
Ultracentrifuge (e.g., Beckman Coulter Optima MAX with TLA-55 rotor)
N_2 evaporator (e.g., RapidVap by Labconco)
Hamilton syringes (gas tight 10, 25, 50, 250, 500, and/or 1000 μL)
TLC developing chamber (e.g., VWR 21432)
Fluorescence spectrophotometer with filter for NBD detection (*Ex*/*Em*:460 nm/535 nm)

3. MATERIALS

Extruder membranes, 0.1 μm (61005—Avanti Polar Lipids, Alabaster, AL)
Membrane filter supports (610014—Avanti Polar Lipids)
POPC (850457C—Avanti Polar Lipids)
POPE (850757C—Avanti Polar Lipids)
DOG—1,2-dioleoyl-*sn*-glycerol (800811C—Avanti Polar Lipids)
POPS (840034C—Avanti Polar Lipids)
POPA (840857C—Avanti Polar Lipids)
LMW poly-L-lysine (P7890—Sigma) or Histone H1 (14-155—EMD Millipore, Billerica, MA)
Triton X-100 (X100—Sigma-Aldrich, St. Louis, MO)
Hepes, pH 7.5 (15630080, 1 *M* solution—ThermoFisher Scientific, Waltham, MA)
ATP (45-001-322—GE Healthcare Life Sciences, Pittsburg, PA)[a]

[a] Acidic and alkaline ATP solutions are less stable than neutral ATP solutions. Store solutions frozen (−20°C or −80°C) at neutral pH in single use aliquots for best results.

$MgCl_2$ (AM9530G, 1 *M* solution—ThermoFisher Scientific)
(γ-^{32}P)ATP, 3000 Ci/mmol 10 mCi/mL EasyTide 1 mCi (BLU502A001MC—Perkin Elmer, Waltham, MA)

Note: (γ-^{32}P)ATP is also available in 100, 250, and 500 μCi amounts)

n-Dodecyl β-D-maltoside (DDM) (89902—ThermoFisher Scientific)
Silica gel 60 TLC plates with preconcentration zone (105582—EMD Millipore)
CytoScint (88245305—MP Biomedicals, Santa Ana, CA)
Sigmacote (SL2-100mL—Sigma-Aldrich)
$CHCl_3$, HPLC grade (C607-4—Fisher Chemical, Waltham, MA)
MeOH, HPLC grade (A452-1—Fisher Chemical)
1 *M* NaCl and/or 1% perchloric acid (60–62%, w/w%, Fisher Chemical)
Acetone (CAS 67-64-1 (Certified ACS)—Fisher Chemical)
Glacial acetic acid (CAS 64-19-7 (Certified ACS)—Fisher Chemical)

4. SOLUTIONS AND BUFFERS

- *DGK75 Assay buffer*: 50 m*M* Hepes, pH 7.5, 0.1 *M* NaCl, 1 m*M* DTT
- *DGK-S buffer*: DGK75 buffer with 170 m*M* sucrose
- *DGK-D buffer* (dilution buffer for purified DGKθ): 25 m*M* Tris–HCl pH 8.0 (RT), 1 m*M* DTT or 3 m*M* BMe, 0.005% DDM, 0.01% Triton X-100
- *Polybasic activator* (PBA): 10 μ*M* stock of LMW poly-L-lysine or Histone H1 in 25 m*M* Hepes, pH 7.5
- *20 mM ATP solution*: 20 m*M* ATP in 25 m*M* Hepes, pH 7.0
- *10× [γ-^{32}P]ATP/$MgCl_2$ mix*: 10 m*M* ATP solution containing [γ-^{32}P] ATP *specific activity* = *2.5* × *10^5 cpm/nmol on supplier's calibration date*

5. ESTIMATED DURATION

Day 1
- Prepare silanized test tubes: *15 min hands on, overnight drying*
- Prepare 10 × [γ-^{32}P]ATP/$MgCl_2$ mix: *0.5 h*

Day 2
- Aliquot and dry down lipids: *1.5 h*
- Prepare vesicles—*2 h*
- Conduct activity assay—*1 h assembly* + *chosen incubation time*

Day 2 or Day 3
- Extract ^{32}PtdOH and TLC separation—*3 to 4 h*
- Product measurement—*2 to 3 h*

6. DGKθ ACTIVITY ASSAY

The following protocol is broken down into sections to facilitate planning. The complete assay can be done in a single day once tubes are silanized and stock buffers are prepared. Where possible we have indicated useful stopping points. The protocol for purifying DGKθ can be found elsewhere (Tu-Sekine & Raben, 2012).

Day 1

6.1 Preparation of Test Tubes

Lipids bind to glass test tubes through hydrogen bonding and electrostatic interactions. To minimize loss of lipids and increase reproducibility between experiments, glass tubes used for making vesicles should be pretreated with a charge-neutralizing reagent such as Sigmacote (Sigma-Aldrich) before using. *We do not recommend drying lipids in plastic tubes due to their poor resistance to organic solvents.*

1. Add liquid Sigmacote (1–5 mL) directly to a test tube, tilting and rotating the tube to coat the inside. Residual Sigmacote can be poured into the next test tube to repeat the process until the desired numbers of tubes are treated.
2. Store treated test tubes upside down in a test tube rack lined with Kimwipes and allow to dry overnight. Store long term at RT; keep inverted to prevent dust accumulation.

6.2 Preparation of 10× AT^{32}P

To create a 10 m*M* ATP solution with a specific activity of 2.5×10^5 cpm/nmol, 1 mCi of [γ-^{32}P]ATP will contain approximately 8.8 mol of ATP. *(Note that the actual concentration of ATP in the* [γ-^{32}P]ATP *from Perkin Elmer is very low, and can be ignored).*

1. Prepare a 20 m*M* ATP solution (e.g., 121 mg ATP in 9 mL ATP Buffer)
2. Measure the A_{259} of a 500-fold dilution of the 20 m*M* ATP solution using a spectrophotometer to determine the actual concentration. The $\varepsilon_{\mathrm{ATP}} = 15.4 \times 10^3\ \mathrm{M}^{-1}$, therefore:

$$[\mathrm{ATP}] = A_{259}/15.4 \times 500\,(\text{dilution factor})$$

3. Adjust the ATP solution to achieve a final concentration of 20 m*M*.

4. To prepare the 10 × $AT^{32}P/MgCl_2$ solution (S.A. of 2.5×10^5 cpm/nmol, 10 m*M* ATP, 15 m*M* $MgCl_2$) add the following directly to the 100 μL of isotope in its container:
   450 μL 20 m*M* ATP solution
   316.9 μL ddH_20
   13.2 μL $MgCl_2$
5. Store at −20°C in suitable protective container.

Day 2–3

Aliquot and dry down lipids

1. Aliquot and combine the desired amount of lipids predissolved in chloroform or other suitable organic solvent(s) into the bottom of a Sigmacoted glass test tube and dry under a gentle stream of nitrogen.
   The following table provides an example of the lipids required for a triplicate reaction of two vesicle compositions, and assumes a 50 μL loss to extruder dead space.
2. Loosely wrap test tubes containing the lipid film in foil and store under vacuum in a desiccation chamber for at least 1 h (RT), or overnight (4°C) to remove residual solvent such as $CHCl_3$ that may inhibit enzyme activity.

6.3 Preparation of Vesicles

Rehydrate the lipid film

1. Add 35 μL DGK75 assay buffer per reaction (175 μL total for lipids in Table 1) to the lipid films to create a 1.5 × concentration of lipid (3 m*M*) and incubate for 30 min in a 37–40°C water bath.

Table 1 Vesicle Lipids for DGKθ Assay

Lipid	Single Reaction Without POPS (mg)	# of Reactions	Total Lipid (mg)	Single Reaction With 10% POPS	# of Reactions	Total Lipid (mg)
POPC	0.023	*5*	0.114	0.020	*5*	0.101
POPE	0.043	*5*	0.215	0.038	*5*	0.191
DOG	0.006	*5*	0.031	0.006	*5*	0.031
POPS	0	*5*	0	0.008	*5*	0.039

The types and amounts of lipids required for the DGKθ activity assay are listed. The lipid components are indicated on the left, and the mass amount of each lipid is noted both per individual reaction, and for a total of 5 reactions. Lipid for 5 reactions is sufficient to assay one sample in triplicate with a final bulk concentration of 2 m*M* lipid per 50 μl reaction, and assumes approximately 50 μl of lipid solution is lost during extrusion. These values can be readily scaled for larger assays.

2. Several times during hydration, vortex on medium speed and sonicate for 3 s in a bath sonicator (approximately 50 W) to completely disperse the lipid aggregates. *Note: do not oversonicate the lipid as this may produce some small liposomes and increase vesicle polydispersity.*

Form vesicles by extrusion

1. Prepare the mini-extruder by adding 1 filter support prewet with DGK75 buffer, within the o-ring on each half of the Teflon extruder insert. Add the 0.1 μm nitrocellulose membrane over the o-ring on one side of the insert and carefully reassemble the extruder and finger tighten the metal housing.
2. If using, preheat the extruder support to 37–40°C (above the T_M of the lipids used). Failure to heat the lipid may result in lipid phase separation and unequal distribution of lipid in vesicles (see Section 9.2)
3. Using a gastight Hamilton syringe, pass 0.2–0.4 mL DGK assay buffer through the extruder to wet the nitrocellulose membrane and test for leaks. Remove and discard this buffer solution from the extruder.
4. Load the hydrated lipid solution into a Hamilton syringe and pass through the prepared mini-extruder for 15 single passes.[b] The solution must be removed from the syringe on the side of the extruder opposite to the loading syringe to avoid contaminants or lipid aggregates that may remain in the loading syringe.
5. Using the gradations on the syringe, note the recovered volume of newly made vesicles prior to transfering the solution into a new silanized test tube or Eppendorf tube.
6. Just prior to use, supplement the vesicle solution to 1 m*M* DTT using freshly prepared DTT. Use immediately in enzyme assay.

6.4 Activity Assay

6.4.1 Assemble Master Mixes

Enzyme Master Mix—5 μL/reaction

1. Dilute purified DGKθ to 0.02 ng/μL using cold Dilution buffer and mix gently
2. Incubate on ice for 10–15 min prior to using in assay

    Note: DGKθ overexpressed in lysates (5–10 μg) can be used in place of purified enzyme. In this case, the PBA can be replaced with DGK75

[b] Vesicles containing anionic phospholipids such as PtdSer and PtdOH become very difficult to extrude at pH values >8.0, and some species of PI(4,5)P_2 aggregate under low-salt conditions. Salt concentrations between 50 and 100 m*M* are commonly used, and NaCl is frequently chosen.

buffer in the Vesicle Reaction Mix, since endogenous polybasic proteins present in lysates activate the enzyme.

Vesicle Reaction Master Mix—45 μL/reaction

1. 35 μL vesicles in DGK75 buffer
2. 5 μL of 0.5 μ*M* PBA (dilute 10 μ*M* stock 1:20) (or DGK75 buffer for reactions without protein activator)
3. 5 μL 10× $AT^{32}P$ mix.

   Note: Create separate master mixes for each different lipid composition and activator combination

6.4.2 Complete the Activity Assay

- Place labeled glass or polypropylene 13 × 100 mm test tubes in a rack on wet ice.
- Add 5 μL of diluted DGKθ enzyme (0.02 ng/μL) to the bottom of each sample tube.
- Initiate the reaction by adding 45 μL of Vesicle Reaction Mix directly on top of the enzyme solution.
- Tap tubes to mix and incubate for 15–30 min in a 37°C water bath.
- Stop the reaction by transferring the tubes to wet ice and adding:
  (a) 0.75 mL 1 *M* NaCl
  (b) 2 mL MeOH

Continue to ^{32}PtdOH extraction (Section 6.5), or store covered with foil at −20°C.

6.5 Extraction and Quantitation of ^{32}PtdOH

Thaw reactions from previous step if stored at −20°C. Continue extraction as follows:

1. Add 1 mL $CHCl_3$ to each tube and vortex to mix. *Solution should be a clear single phase. Use extreme caution when vortexing as the solution contains ^{32}P.*
2. Add 1 mL $CHCl_3$ to each tube.
3. Add 1 mL of 1 *M* NaCl (or 1% $HClO_4$) and vortex to mix. *Solution should separate into 2 phases—a lower cloudy phase (lipid) and a clear upper phase (aqueous).*
4. Centrifuge at approximately 500 × *g* for 5 min to accelerate phase separation. *Both upper and lower phases should be clear and may be separated by a white precipitate at the interface.*
5. Carefully aspirate the upper (aqueous) phase without disturbing the lower (organic) phase.
6. Wash the lower, organic phase with 2 mL 1 *M* NaCl (or 1% $HClO_4$).

7. Vortex well and centrifuge at $500 \times g$ for 5 min.
8. Carefully transfer the lower organic phase to a new glass test tube using a 9″ glass, plugged Pasteur pipette, taking care to avoid transferring any aqueous phase.

   Note: long glass pipettes extract more residual aqueous ^{32}P from the organic phase during transfer, resulting in lower background on TLC.
9. Dry samples completely under N_2 with low heat (approximately 40–45°C).

Store dried lipids, covered, at −20°C or proceed immediately to TLC separation.

TLC separation of ^{32}PtdOH

1. Resuspend dried lipid in 30 μL 95:5 $CHCl_3$:MeOH, vortex, and spot entire sample[c] on prescored silica gel 60 TLC plate.,[d e] Include a control sample of POPA (20 μg) in one lane of each TLC to verify the identity of radiolabelled product.
2. Develop plates in $CHCl_3$:acetone:MeOH:acetic acid:H_2O (5:4:3:2:1).[f]
3. Remove the TLC plate from the tank when the solvent is approximately 1–2 cm from top and mark the solvent front with a pencil to facilitate taking R_f measurements if desired.
4. Dry plates completely in a hood behind a screen.
5. Incubate the dried plates in a chamber containing crystals of iodine to develop the nonlabeled POPA control. Outline the spot with a pencil.
6. Develop the plate using film or a phosphoimager. Be sure to note the orientation of the film. It can be useful to mark the upper and lower corners of the undeveloped film on one side of the plate prior to developing it to avoid later confusion about film orientation.
7. If using a scintillation counter, align the exposed film with TLC plate and mark the upper and lower boundary of the ^{32}PtdOH region in pencil on the film (for documentation) and on the TLC plate. Cut out the desired region from the aluminum TLC plate and transfer the piece directly to a scintillation vial.
8. Add scintillation fluid to the samples and count.

---

[c] If counts are very high by Geiger counter (>10,000 cpm), a portion of the sample can be spotted, and the remainder saved or discarded.

[d] $CHCl_3$ evaporates rapidly. Add solvent to 12 tubes or less at one time and spot the samples on the TLC plate before preparing the next group.

[e] Passing warm air (e.g., from a portable hair dryer) *gently* over the TLC plate during spotting speeds drying and minimizes solvent spread.

[f] Solvent mix for TLC run should be clear and homogeneous with no phase separation apparent. After the solvent is added to the developing chamber, it should be allowed to equilibrate for 30–60 min prior to adding plates.

7. SURFACE DILUTION EXPERIMENT: DETERMINING THE K_D^{APP}, K_{MS}^{app} AND V_{MAX}^{app} OF DGKθ

A surface dilution experiment not only provides data on the kinetic values of an enzyme, but also reveals information about its processivity, or scooting behavior. For example, enzymes like PLA_2 are insensitive to changes in bulk DMPM, a lipid to which it binds very tightly, while a quasi-scooting enzyme like DGKθ shows a clear dependence on both surface and bulk concentrations (see Fig. 2).

While this experiment may at first appear somewhat overwhelming due to the number of samples, the amount of data obtained is well worth the effort. It is extremely useful for comparative studies, such as determining the effect of point mutations or lipid composition on the affinity of the enzyme for the vesicle interface and substrate(s). Importantly, kinetic parameters can only be extracted from saturated kinetic curves, and the range of lipid has been chosen to satisfy this requirement for purified DGKθ.

The following experiment covers a bulk range of lipid from 0.2 to 20 m*M*, with surface concentrations of DOG ranging from 0% to 10%. In this experiment eight master sets of vesicles are made, each containing a different surface concentration of substrate, resulting in eight tubes of dried lipid. Once the vesicles are prepared by extrusion, each set of vesicles is sequentially diluted to produce a series of eight bulk concentrations, resulting in a total of 64 vesicle stock solutions containing sufficient volume for triplicate reactions.

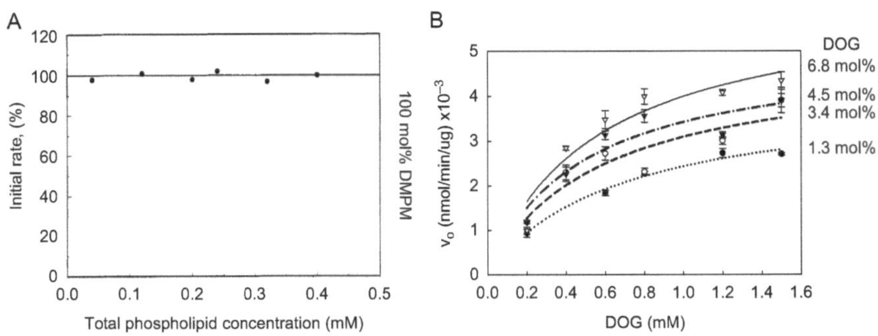

Fig. 2 The effect of increasing bulk substrate at a constant surface concentrations on the rate profiles of hopping and scooting enzymes. (A) An obligate scooter such as PLA_2 on DMPM vesicles is insensitive to increasing bulk concentrations of substrate, while (B) a hopping enzyme such as DGKθ on OG micelles is sensitive to changes in both bulk and surface concentrations of substrate.

Table 2 Vesicle Lipids for Surface Dilution Experiment

Mol% DOG	DOG (mg)	POPS (mg)	POPC (mg)	POPE (mg)
0	0	0.79	2.3	4.3
1	0.06	0.79	2.3	4.3
2.0	0.12	0.79	2.2	4.2
3.0	0.19	0.79	2.2	4.2
4.0	0.25	0.79	2.2	4.1
5.0	0.31	0.79	2.2	4.1
7.5	0.47	0.79	2.1	3.9
10	0.62	0.785	2.0	3.8

The mass amount of lipids required for a surface dilution experiment ranging from 0 to 10 mol% DOG. Excess lipid (10 reactions) is used to ensure sufficient recovery.

Protocol

1. Aliquot lipids for eight sets of vesicles containing 10 mol% POPS and increasing surface concentrations of DOG with a final concentration of 20 m*M* total lipid per 50 μL reaction (see Table 2).[g]
2. Add 400 μL DGK75 assay buffer to each tube of dried lipids. Form vesicles as indicated in Section 6.3.
3. Transfer 300 μL of each stock (25 m*M*) to a new Eppendorf tube.
4. Complete a twofold serial dilution by transferring 150 μL of each 25 m*M* lipid stock to a new Eppendorf tube and adding 150 μL DGK75 buffer. Mix and repeat to create final lipid concentrations of twofold using DGK75 assay buffer to create a series of 1.25 × vesicle stocks with lipid concentrations of 25, 12.5, 6.25, 3.13, 156, 0.844, 0.423, and 0.211 m*M* for each surface concentration.
5. Complete the DGK assay as indicated in Section 6.4.

   NB: DGKθ enzyme can be added directly to the Vesicle Reaction Mix, but should be added last (after addition of an activator and ATP). Incubation of purified DGKθ with PBA has been seen to deactivate the enzyme (unpublished observations).
6. Extract and measure ^{32}PtdOH as indicated in Sections 6.4 and 6.5.
7. Plot initial velocity vs bulk substrate to determine the K_d^{app}, and plot initial velocity vs surface substrate concentration to determine the K_M^{app} and

[g] To increase the surface concentration of DOG, POPC and POPE are decreased simultaneously to maintain a proportion of 2:1 (POPC:POPE).

V_{MAX}^{app} using the methods of Ranganathan, Tcacenco, Rosseto, and Hajdu (2006), Carman, Deems, and Dennis (1995), or Jain, Yuan, and Gelb (1989). These plots typically produce rectangular hyperbolae for DGKθ under the conditions given.

8. VESICLE PULLDOWN TO DETERMINE INTERFACIAL K_D

Vesicle pulldown assays can be a simple and effective way to determine the binding affinity of an enzyme under specific conditions and are particularly useful when minimal amounts of purified enzyme are available for study. A binding curve can be readily constructed by serial diluting a single batch of sucrose-loaded vesicles. Spiking vesicles with a small percentage of a fluorescent lipid allows easy quantization of losses during centrifugation steps. In this experiment, DGKθ (up to 1:1 enzyme:vesicle at the highest vesicle concentration chosen) is incubated with each concentration of vesicles under standard assay conditions. The sucrose-loaded vesicles are recovered by high-speed centrifugation, and the amount of enzyme remaining in the supernatant is quantified by densitometry of silver stain or western blot images to construct a binding curve. Note that determining the amount of enzyme associated with vesicles is more direct, but high levels of lipid produce streaking and smearing during electrophoresis and can interfere with gel analysis. As always, the mass amount of enzyme necessary for detection should be confirmed in advance to guide experimental setup (e.g., 10–20 ng purified DGKθ is required for silver stain, while 5–10 ng is sufficient for western blot).

Controls and Calculations

It is important to include two basic control reactions to normalize the results:

1. enzyme without vesicles.
2. vesicles without enzyme.

These reactions are used to account for variations in protein solubility or vesicle recovery. For example, with DGKθ we have found it necessary to maintain the purified enzyme in a solution of 0.01% DDM (DGK-D dilution buffer) in the control reaction to prevent enzyme loss due to aggregation or adhesion to the ultracentrifuge Eppendorf tubes which occurs over the time required for the high-speed centrifugation step. The following equation can be used to determine the amount of *soluble* enzyme that pellets with the vesicles. This calculation assumes that all NBD signal comes from a vesicle that has the potential to bind enzyme.

$$E_{bound} = \frac{(E_{sol}*E_{pell}) + (E_{sol} - 1)*E_{total}}{\text{Lipid}_{pell} + (E_{sol} - 1)} \quad (1)$$

where:

E_{sol} = fraction of enzyme in the supernatant of the no vesicles control
E_{pell} = fraction of enzyme in the pellet of a reaction containing vesicles
E_{total} = total amount of enzyme added to the reaction
Lipid_{pell} = fraction of lipid (or NBD signal) in the pellet

8.1 Preparation of Sucrose-Loaded Vesicles

The following is derived from a protocol designed by the Newton lab to measure the binding constant of PKCα for a phorbol ester (Mosior & Newton, 1995) and has been optimized for DGKθ.

1. Form vesicles essentially as indicated in Section 6.3 *except*:
   - **(a)** Replace 0.5–0.7% of the POPE with NBD-POPE to track vesicle recovery
   - **(b)** Determine the amount of required lipid based on a 100 μL final volume (not a 50 μL final volume, as in the activity assay).
   - **(c)** Increase the number of reactions by 6–8 to provide vesicles for NBD measurements.
   - **(d)** Use DGK-S buffer to hydrate the lipid film.
2. Dilute the extruded vesicle preparation in DGK-S buffer with three parts DGK75 buffer (*no* sucrose) and transfer to ultracentrifuge tubes. Centrifuge at 100,000 × *g* for 1 h at 25°C.
3. Remove and save the supernatant for fluorescence readings. Like supernatants should be combined for accuracy.
4. Resuspend the pellet containing the sucrose-loaded vesicles in DGK75 buffer (*without* sucrose) to approximately 1/3rd the original extrusion volume.
5. Calculate the percent recovery of vesicles by comparing the NBD-PE signal (*Ex*/*Em*: 460 nm/535 nm) of the vesicle solution before and after centrifugation. The NBD count of the vesicle supernatant from the 100,000 × *g* spin can be used to verify the calculation (NBDsupe + NBDpellet = NBDinput).
6. Increase the volume of the recovered vesicle pellet to create a 2 × final concentration of vesicles for the pulldown assay.

8.2 Vesicle Pulldown Assay

1. To an ultracentrifuge Eppendorf tube add:
   a. 50 μL of 2 × sucrose-loaded vesicles (or no vesicles)
   b. 40 μL of DGK75 containing 2 m*M* ATP (nonradioactive), 3 m*M* MgCl2
   c. 10 μL (20 ng) purified DGKθ in DGK-D buffer (or DGK-D buffer alone)
2. Incubate reactions at 25°C for 15–30 min.
3. Recover sucrose-loaded vesicles by centrifugation at 100,000 × *g* for 1 h at 25°C. Carefully remove the supernatant from the pellet and transfer to a new tube *(a protein gel tip works well)*.
4. Gently and carefully rinse the pellet with 100–200 μL DGK75 buffer to remove any residual supernatant and resuspend pellet in 100 μL of DGK75 buffer.
5. Remove aliquots of supernatant and pellet for NBD-PE fluorescence readings. These values represent the percent distribution of vesicles in each fraction and will be used to normalize the amount of bound DGKθ.
6. Separate aliquots of supernatant and pellet by reducing SDS PAGE and measure the amount of DGKθ using band densitometry.
7. Normalize results using Eq. (1).

9. TIPS FOR NONSPECIALISTS

9.1 How to Choose a Lipid Composition

The activity of an enzyme depends in large part on the suitability of the interface in which the substrate resides. Both detergent micelles and liposomes have been shown to be suitable interfaces for many DGKs. However, liposomes have the distinct advantage of limiting exchange of components (both substrate and enzyme), which is critical for experiments that use substrate depletion as a readout for evaluating the scooting behavior of an enzyme.

Determining an appropriate lipid composition for vesicles can be daunting. A reasonable starting point is to use a simplified version of the cellular membrane composition where the enzyme under investigation (e.g., DGKθ has been detected at the plasma membrane, Golgi, and neuronal synapse and would be expected to have different activities at these diverse membranes). To facilitate experimental design, Table 3 lists known or estimated

Table 3 Major Lipid Components of Subcellular Membranes

Organelle	Phospholipid (mol%)	Cholesterol: Phospholipid Ratio	References
Plasma membrane—extracellular leaflet	PtdCho (25–35%) PtdEth (<15%) SM (25–35%)	1.0	Lomize, Pogozheva, Joo, Mosberg, and Lomize (2012), van Meer, Voelker, and Feigenson (2008), Simons (2016), Ingólfsson et al. (2014), and Pankov, Markovska, Antonov, Ivanova, and Momchilova (2006)
Plasma membrane—cytoplasmic leaflet	PtdCho (15–30%) PtdEth (25–40%) SM (10%) PtdSer (10–20%) PtdIns (5%) $PI(4,5)P_2$ (<1%) $PI(3,4,5)P_3$ (trace)	1.0	Lomize et al. (2012), Virtanen, Cheng, and Somerharju (1998), and Ingólfsson et al. (2014)
Endoplasmic reticulum—cytoplasmic leaflet	PtdCho (50%) PtdEth (20%) SM PtdSer (5–10%) PtdIns (5–15%)	0.15	Lomize et al. (2012)
Endoplasmic reticulum—luminal leaflet	PtdCho (50%) PtdEth (20%) SM PtdSer (5–10%) PtdIns (5–15%)	0.15	Lomize et al. (2012)

Cis-Golgi	PtdCho (30%) PtdEth (30%) SM (10%) PtdIns (5–15%)	0.2	Duran et al. (2012) and van Meer et al. (2008)
Trans-Golgi	PtdCho (30%) PtdEth (30%) Sphingolipids (15%) PtdIns (15%) PI(4)P (unknown)	0.2	Klemm et al. (2009) and van Meer and de Kroon (2011)
Early endosome	PtdCho (50%) PtdEth (20%) SM (9%) PtdSer (5–10%) PtdIns (5–10%) PI(3)P (unknown)	0.5	Hullin-Matsuda, Taguchi, Greimel, and Kobayashi (2014) and Kobayashi et al. (2002)
Late endosome	PtdCho (33%) PtdEth (18%) SM (5–9%) PtdSer (3–5%) PtdIns (4–9%) BMP/LBPA (11–15%) PI(3,5)P_2 (unknown)	Low	Kobayashi et al. (2002, 1998)
Synaptic vesicle	PtdCho (20–25%) PtdEth (12–15%) SM (2–3%) PtdSer (4–6%) PtdIns (1–2%)	0.5	Deutsch and Kelly (1981) and Takamori et al. (2006)

Continued

Table 3 Major Lipid Components of Subcellular Membranes—cont'd

Organelle	Phospholipid (mol%)	Cholesterol: Phospholipid Ratio	References
Nuclear inner membrane	PtdCho (44%) PtdEth (27%) PtdIns (15%)	Unknown	Zinser et al. (1991)
Mitochondrion—outer membrane	PtdCho (45–50%) PtdEth (30%) PtdIns (10%) CL (<1%)	0.1	Horvath and Daum (2013) and van Meer et al. (2008)
Mitochondrion—inner membrane	PtdCho (35%) PtdEth (35–40%) PtdIns (5%) CL (15–20%)	0.1	Horvath and Daum (2013) and van Meer et al. (2008)

This table contains phospholipid compositions of subcellular membranes derived from a variety of resources. Lipidomic studies of organelles are still scarce; therefore, the data represent composites from multiple cell types (including yeast), resulting in ranges for some lipid classes. Totals do not necessarily add up to 100 mol%. Lipids that are characteristic markers of an organelle have been included even when quantified data was not available. Cholesterol/phospholipid ratios have been modified from van Meer et al. (van Meer et al., 2008) to reflect the indicated membrane region where possible and are approximations.

membrane compositions from several organelles in mammalian and yeast cells. Note that membrane composition can vary widely between tissues and organisms, and the following table serves only as a general guide.

9.2 Considerations When Making Liposomes

The following points should be considered when choosing lipids for an experiment:

1. *Avoid mismatches in the length of the fatty acids.* For example, combining short-chain lipids (<16 carbons/fatty acid) with long-chain lipids (>16 carbons/fatty acid) can result in lipid partitioning (an uneven distribution of the shorter chain FAs on the inside of the vesicle due to packing constraints), and/or phase separation between liquid ordered ($l_{o)}$ and liquid disordered ($l_{d)}$ phases. This is common in vesicles containing sphingomyelin and cholesterol and can be visualized in giant unilamelar vesicles using various fluorescent phase-sensitive probes (Bagatolli, Sanchez, Hazlett, & Gratton, 2003). Note that the majority of plasma membrane lipids have 0–1 elements of fatty acid unsaturation, with a higher concentration of lipids having 3–6 elements of unsaturation in the inner bilayer than the outer bilayer (Ingólfsson et al., 2014).
2. *Where possible, use lipids with a low melting point.* Vesicles should be formed under conditions that ensure they are in the liquid disordered state to avoid formation of nonvesicular structures and/or exclusion of the high T_M lipid from formed vesicles. Note that the operational temperature limit for a handheld extruder is typically around 55°C.
3. *Be aware of the physical properties of the lipids used.* For example, long-chain species of $PI(4,5)P_2$ require acidification of the head group to facilitate solubilization, but storage under acidic conditions can cause hydrolysis of the fatty acid chains. Where possible, lipids such as these should be purchased in solution, and an aliquot saved for comparison with new batches or when purchasing from different sources.
4. *Whenever possible, determine the average size and polydispersity of vesicles.* This is most easily accomplished using dynamic light scattering (DLS). This information can be particularly useful when determining the effect of lipid composition on enzyme activity, or when troubleshooting assays. Extrusion of lipids through a polycarbonate membrane with a pore size of 100 nm results in the formation of approximately 100 nm unilamellar vesicles with a narrow distribution of vesicle size. Membrane pore sizes

below 100 nm tend to produce vesicles larger than membrane pore size, while membranes with pore sizes above 100 nm tend to produce vesicles smaller than the pore size used ((Hunter & Frisken, 1998). Note that vesicle composition and temperature (above the gel phase) do not appear to impact vesicle size, though extrusion force does have some effect.

5. *There is a limit to the practical range of DOG concentration.* This depends on the ability of the lipids to form a stable interface, and seems to occur around 15–18% DOG under our assay conditions. LUV solutions are clear to slightly hazy with a bluish tinge, while aggregates appear more opaque and white in color.
6. *Make vesicles on the day of use.* Note that the lipid substrate, sn-1,2-diacyglycerol, has been shown to isomerize to sn-1,3-diacyglycerol in a temperature and time-dependent manner (Hamilton, Bhamidipati, Kodali, & Small, 1991). If vesicles will not be used on the same day that they are made tests should be run to determine the effect of storage on the size polydispersity (by DLS) and lipid integrity (e.g., by TLC). For example, we have observed an increase in polydispersity after overnight 4°C storage of PC:PE:PS:DOG (27:54:9:10) vesicles (unpublished data).

9.3 A Note on Using Detergent Micelles

DGKθ activity can be determined in detergent-solubilized lipid mixed micelles, and we have used such micelles to determine the apparent kinetic parameters of DGKθ (Tu-Sekine, Ostroski, & Raben, 2007). However, we recommend that liposomes be used where possible to achieve better control over the size and polydispersity of the interface, and to avoid unexpected detergent effects on enzyme activity, which can complicate interpretation of kinetic profiles. In most cases mixed micelles should be reserved for assays that confirm enzyme activity or to quickly evaluate the effects of various additives.

9.4 Calculating the Number of Vesicles or Micelles

It is useful to estimate the number of vesicles in order to more accurately determine the vesicle:enzyme stoichiometry in an experiment. To simplify calculations in this example the average phospholipid headgroup is estimated to be 70 A^2 (the approximate headgroup area of hydrated PtdCho Thurmond, Dodd, & Brown, 1991) and the inner and outer lipid layers are assumed to contain the same number of lipids. (For more accurate calculations, use cross-sectional areas from the lipids of interest if available).

Using the formula for the surface area of a sphere ($4\pi r^2$), the phospholipid contained in a vesicle with a 100 nm diameter (500 A radius) can be calculated as follows:

$$4\pi \times 500A^2 = \frac{3.14 \times 10^6 A^2}{70A^2 \text{ phospholipid}^{-1}} = 44,857 \text{phospholipids}$$

$$\frac{44,857 \text{ phospholipids}}{\text{layer}} \times \frac{2\text{layers}}{\text{vesicle}} = \frac{89,714\text{phospholipids}}{\text{vesicle}(100\text{nm})}$$

As long as the reaction volume and lipid concentration are known, it is a simple matter to determine the number of vesicles per reaction. For example, in 100 μL of 1 m*M* solution of 100 nm vesicles, there are approximately

$$1 \times 10^{-7}\text{moles} * \frac{6.02 \times 10^{23} \text{ lipids}}{\text{mole}} * \frac{\text{vesicle}(100\ nm)}{89,714 \text{ lipids}} = 6.7 \times 10^{11}\text{vesicles}$$

Example Calculation for Micelle number: To determine the mol% lipid in micelles, it is important to know the critical micelle concentration (CMC) of the chosen detergent under the chosen assay conditions. By definition the detergent concentration present at concentrations above the CMC will spontaneously form micelles, while the detergent at a concentration below the CMC will be present as monomers. For example, Triton X-100 (Sigma-Aldrich) has a CMC of approximately 0.2 m*M*, meaning that the first 0.2 m*M* Triton X-100 is present as monomers in water. To make a 1 m*M* solution of Triton X-100 micelles containing 10% diglyceride, the final mix should contain 1.1 m*M* Triton X-100 and 0.1 m*M* diglyceride. Theoretically, 0.2 m*M* of the Triton X-100 will be present as monomers and 0.9 m*M* will form micelles with 0.1 m*M* diglyceride. To calculate the concentration of micelles, subtract the monomeric detergent from the bulk detergent and divide by the number of detergent molecules per micelle (the aggregation number):

$$[\text{micelles}]\frac{(C_s - CMC)}{N}$$

where C_s is the bulk molar concentration of detergent and N is the mean aggregation number. For example, a solution containing 1.1 m*M* of Triton X-100 (avg. M.W. = 625, N = 100) in water will have $\frac{(1.1 - 0.2)}{100}$ or 9 μ*M* micelles.

Recall that the aggregation number reported for detergents refers to the pure detergent under the reported conditions only, and is not directly related to detergent–lipid mixed micelles, which form larger aggregates due to the presence of phospholipid. It is worthwhile to search for information relevant to the effects of detergents on substrate lipids as this may help with detergent selection. Note also that the detergent:lipid ratio can effect the CMC and the micellar aggregation number of a detergent (Eidelman, Blumenthal, & Walter, 1988), as can the presence of salts (Bhairi, 1997).

REFERENCES

Bagatolli, L. A., Sanchez, S. A., Hazlett, T., & Gratton, E. (2003). Giant vesicles, laurdan, and two-photon fluorescence microscopy: Evidence of lipid lateral separation in bilayers A2. In G. Marriott & I. Parker (Eds.), *Enzyme kinetics and mechanism part D: Developments in enzyme dynamics: Vol. 360*. (pp. 481–500): Academic Press.

Bhairi, S. M. (1997). A guide the properties and uses of detergents in biology and biochemistry. Calbiochem, Doc No. CB0068-401 Copyright updated 2001 by Calbiochem-Novabiochem Corporation.

Carman, G. M., Deems, R. A., & Dennis, E. A. (1995). Lipid signaling enzymes and surface dilution kinetics. *Journal of Biological Chemistry*, *270*(32), 18711.

Deutsch, J. W., & Kelly, R. B. (1981). Lipids of synaptic vesicles: Relevance to the mechanism of membrane fusion. *Biochemistry*, *20*(2), 378–385.

Duran, J. M., Campelo, F., van Galen, J., Sachsenheimer, T., Sot, J., Egorov, M. V., et al. (2012). Sphingomyelin organization is required for vesicle biogenesis at the Golgi complex. *The EMBO Journal*, *31*(24), 4535–4546. http://doi.org/10.1038/emboj.2012.317.

Eidelman, O., Blumenthal, R., & Walter, A. (1988). Composition of octyl glucoside-phosphatidylcholine mixed micelles. *Biochemistry*, *27*(8), 2839–2846.

Hamilton, J. A., Bhamidipati, S. P., Kodali, D. R., & Small, D. M. (1991). The interfacial conformation and transbilayer movement of diacylglycerols in phospholipid bilayers. *Journal of Biological Chemistry*, *266*(2), 1177–1186.

Horvath, S. E., & Daum, G. (2013). Lipids of mitochondria. *Progress in Lipid Research*, *52*(4), 590–614. http://doi.org/10.1016/j.plipres.2013.07.002.

Hullin-Matsuda, F., Taguchi, T., Greimel, P., & Kobayashi, T. (2014). Lipid compartmentalization in the endosome system. *Seminars in Cell & Developmental Biology*, *31*, 48–56. http://doi.org/10.1016/j.semcdb.2014.04.010.

Hunter, D. G., & Frisken, B. J. (1998). Effect of extrusion pressure and lipid properties on the size and polydispersity of lipid vesicles. *Biophysical Journal*, *74*(6), 2996–3002. http://doi.org/10.1016/S0006-3495(98)78006-3.

Ingólfsson, H. I., Melo, M. N., van Eerden, F. J., Arnarez, C., Lopez, C. A., Wassenaar, T. A., et al. (2014). Lipid organization of the plasma membrane. *Journal of the American Chemical Society*, *136*(41), 14554–14559. http://doi.org/10.1021/ja507832e.

Jain, M. K., & Gelb, M. H. (1991). Phospholipase A2-catalyzed hydrolysis of vesicles: Uses of interfacial catalysis in the scooting mode. *197. Phospholipases* (pp. 112–125). Elsevier. http://doi.org/10.1016/0076-6879(91)97138-O.

Jain, M. K., Yuan, W., & Gelb, M. H. (1989). Competitive inhibition of phospholipase A2 in vesicles. *Biochemistry*, *28*(10), 4135–4139.

Klemm, R. W., Ejsing, C. S., Surma, M. A., Kaiser, H.-J., Gerl, M. J., Sampaio, J. L., et al. (2009). Segregation of sphingolipids and sterols during formation of secretory vesicles at the trans-Golgi network. *The Journal of Cell Biology*, *185*(4), 601–612. http://doi.org/10.1083/jcb.200901145.

Kobayashi, T., Beuchat, M.-H., Chevallier, J., Makino, A., Mayran, N., Escola, J.-M., et al. (2002). Separation and characterization of late endosomal membrane domains. *Journal of Biological Chemistry*, *277*(35), 32157–32164. http://doi.org/10.1074/jbc.M202838200.

Kobayashi, T., Stang, E., Fang, K. S., de Moerloose, P., Parton, R. G., & Gruenberg, J. (1998). A lipid associated with the antiphospholipid syndrome regulates endosome structure and function. *Nature*, *392*(6672), 193–197. http://doi.org/10.1038/32440.

Lomize, M. A., Pogozheva, I. D., Joo, H., Mosberg, H. I., & Lomize, A. L. (2012). OPM database and PPM web server: Resources for positioning of proteins in membranes. *Nucleic Acids Research*, *40*(Database issue), D370–D376. http://doi.org/10.1093/nar/gkr703.

Mosior, M., & Newton, A. C. (1995). Mechanism of interaction of protein kinase C with phorbol esters. Reversibility and nature of membrane association. *The Journal of Biological Chemistry*, *270*(43), 25526–25533.

Pankov, R., Markovska, T., Antonov, P., Ivanova, L., & Momchilova, A. (2006). The plasma membrane lipid composition affects fusion between cells and model membranes. *Chemico-Biological Interactions*, *164*(3), 167–173.

Ranganathan, R., Tcacenco, C. M., Rosseto, R., & Hajdu, J. (2006). Characterization of the kinetics of phospholipase C activity toward mixed micelles of sodium deoxycholate and dimyristoylphosphatidylcholine. *Biophysical Chemistry*, *122*(2), 79–89. http://doi.org/10.1016/j.bpc.2006.02.012.

Simons, K. (2016). Cell membranes: A subjective perspective. *Biochimica et Biophysica Acta*, *1858*, 2569–2572. http://doi.org/10.1016/j.bbamem.2016.01.023.

Takamori, S., Holt, M., Stenius, K., Lemke, E. A., Grønborg, M., Riedel, D., et al. (2006). Molecular anatomy of a trafficking organelle. *Cell*, *127*(4), 831–846. http://doi.org/10.1016/j.cell.2006.10.030.

Thurmond, R. L., Dodd, S. W., & Brown, M. F. (1991). Molecular areas of phospholipids as determined by 2H NMR spectroscopy. Comparison of phosphatidylethanolamines and phosphatidylcholines. *Biophysical Journal*, *59*, 108–113. http://doi.org/10.1016/S0006-3495(91)82203-2.

Tu-Sekine, B., Ostroski, M., & Raben, D. M. (2007). Modulation of diacylglycerol kinase theta activity by alpha-thrombin and phospholipids. *Journal of Cellular Physiology*, *46*, 924–932.

Tu-Sekine, B., & Raben, D. M. (2012). Dual regulation of diacylglycerol kinase (DGK)-θ: Polybasic proteins promote activation by phospholipids and increase substrate affinity. *Journal of Biological Chemistry*, *287*, 41619–41627.

van Meer, G., & de Kroon, A. I. (2011). Lipid map of the mammalian cell. *Journal of Cell Science*, *124*(Pt. 1), 5–8. http://doi.org/10.1242/jcs.071233.

van Meer, G., Voelker, D. R., & Feigenson, G. W. (2008). Membrane lipids: Where they are and how they behave. *Nature Reviews. Molecular Cell Biology*, *9*(2), 112–124. http://doi.org/10.1038/nrm2330.

Virtanen, J. A., Cheng, K. H., & Somerharju, P. (1998). Phospholipid composition of the mammalian red cell membrane can be rationalized by a superlattice model. *Proceedings of the National Academy of Sciences of the United States of America*, *95*(9), 4964–4969.

Zinser, E., Sperka-Gottlieb, C. D., Fasch, E. V., Kohlwein, S. D., Paltauf, F., & Daum, G. (1991). Phospholipid synthesis and lipid composition of subcellular membranes in the unicellular eukaryote Saccharomyces cerevisiae. *Journal of Bacteriology*, *173*(6), 2026–2034.

CHAPTER ELEVEN

Studying Gastric Lipase Adsorption Onto Phospholipid Monolayers by Surface Tensiometry, Ellipsometry, and Atomic Force Microscopy

A. Bénarouche*, L. Sams*, C. Bourlieu†, V. Vié‡, V. Point*, J.F. Cavalier*, F. Carrière*[,1]

*CNRS, Aix Marseille Université, Enzymologie Interfaciale et de Physiologie de la Lipolyse, Marseille, France
†INRA CIRAD, UMR1208 IATE, Montpellier, France
‡Université Rennes 1, Institut de Physique de Rennes, UMR CNRS 6251, Rennes cedex, France
[1]Corresponding author: e-mail address: carriere@imm.cnrs.fr

Contents

Methods in Enzymology, Volume 583
ISSN 0076-6879
http://dx.doi.org/10.1016/bs.mie.2016.09.039

Abstract

The access to kinetic parameters of lipolytic enzyme adsorption onto lipids is essential for a better understanding of the overall catalytic process carried out by these interfacial enzymes. Gastric lipase, for instance, shows an apparent optimum activity on triglycerides (TAG) at acidic pH, which is controlled by its pH-dependent adsorption at lipid–water interfaces. Since gastric lipase acts on TAG droplets covered by phospholipids, but does not hydrolyze these lipids, phospholipid monolayers spread at the air–water interfaces can be used as biomimetic interfaces to study lipase adsorption and penetration through the phospholipid layer, independently from the catalytic activity. The adsorption of recombinant dog gastric lipase (rDGL) onto 1,2-dilauroyl-*sn*-glycero-3-phosphocholine (DLPC) monolayers can be monitored by surface tensiometry at various enzyme concentrations, pHs, and surface pressures (Π). These experimental data and the use of Langmuir adsorption isotherm and Verger–de Haas' lipase kinetics models further allow estimating various parameters including the adsorption equilibrium constant (K_{Ads}), the interfacial concentration (Γ_{E^*}), the molar fraction $\left(\Phi_{E^*(\%)}, \text{mol\%}\right)$ ($\Phi_{E^*(\%)}$, mol%), and the molecular area (A_{E^*}) of rDGL adsorbed onto the DLPC monolayer under various conditions. Additional insight into rDGL adsorption/insertion on phospholipid monolayers can be obtained by combining ellipsometry, Langmuir–Blodgett film transfer, and atomic force microscopy. When using multicomponent phospholipid monolayers with phase separation, these techniques allow to visualizing how rDGL preferentially partitions toward liquid expanded phase and at phase boundaries, gets adsorbed at various levels of insertion and impacts on the lateral organization of lipids.

1. INTRODUCTION

Lipases or triacylglycerol hydrolases (EC 3.1.1.3) are water-soluble enzymes acting on a water-insoluble substrate, at the surface of lipid droplets. These interfacial enzymes have therefore to bind first at the water–lipid interface prior to catalyze the hydrolysis of triglycerides (TAG) (Aloulou et al., 2006). The kinetic model developed by Verger and de Haas for describing the mode of action of lipases includes an adsorption step inspired from Langmuir adsorption isotherm, followed by the formation of an interfacial enzyme–substrate complex analogs to a Michaelis–Menten complex in a two-dimensional space (Verger & de Haas, 1973). According to this model, the rate of substrate hydrolysis depends on interfacial Michaelis–Menten (K_{M^*}) and catalytic rate (k_{cat^*}) constants, but also on the adsorption (k_a) and desorption (k_d) rate constants characterizing the enzyme binding at the water–lipid interface (Fig. 1A).

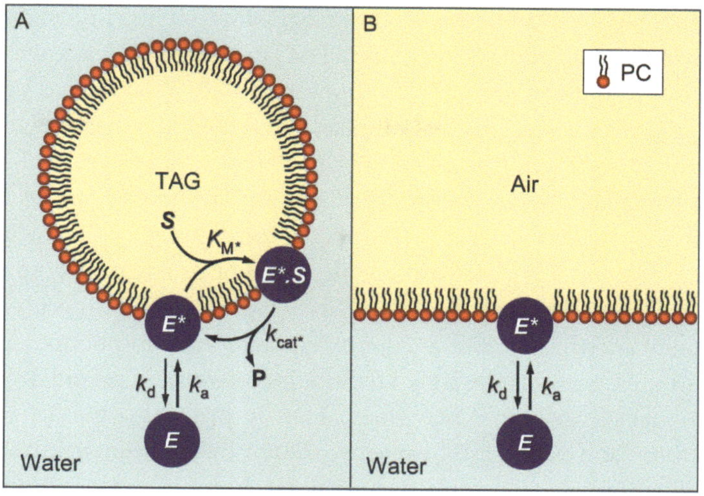

Fig. 1 Schematic representation of lipase interaction with a triglyceride (TAG) droplet covered by phospholipids (panel A) and a phospholipid monolayer spread at the air–water interface (panel B). The picture in panel A is inspired from the Verger–de Haas' kinetic model for the action of a soluble enzyme (*E*) on an insoluble substrate (*S*) (Verger, Mieras, & de Haas, 1973). The lipase has to bind at the lipid–water interface (E^*) before forming an interfacial Michaelis–Menten complex with the TAG substrate ($E^*{\cdot}S$) and catalyzing the conversion of TAG into lipolysis products (*P*). K_{M^*} and k_{cat^*} are the interfacial Michaelis–Menten constant and catalytic rate constant, respectively. A PC monolayer spread at the air–water (panel B) can be used as a biomimetic interface for studying lipase adsorption independently from enzyme activity, on the conditions that the lipase, like rDGL, is not active on PC. k_a, adsorption rate constant; k_d, desorption rate constant; *PC*, phosphatidylcholine.

The reaction of enzymatic lipolysis strongly depends on various interfacial phenomena, including the structural organization of lipids and amphiphiles (proteins, surfactants) presents at the surface of TAG droplets (Aloulou et al., 2006). On the way to reaching its TAG substrate, a lipase has often to cross an interfacial barrier made of polar phospholipids and/or amphiphilic proteins stabilizing the lipid droplet dispersed in an aqueous environment. This is the case for digestive lipases acting on oil-in-water (O/W) emulsions in the gastrointestinal tract (Staggers, Hernell, Stafford, & Carey, 1990), but also for intracellular lipases acting on lipid bodies in the aqueous cytosol of adipocytes (Thiam, Farese, & Walther, 2013). In both cases, the composition of the phospholipid layer at the surface of lipid droplets is crucial for the lipid droplet stability, protein targeting to their surfaces and lipolysis (Gargouri, Pieroni, Riviere, Lowe, et al., 1986; Thiam et al., 2013).

Gastric lipase is an attractive model enzyme for studying lipase adsorption and activity on phospholipid-stabilized TAG emulsions: it is known to bind and hydrolyze milk fat globules (Bourlieu et al., 2015), as well as vegetable oil emulsions made with egg lecithins like Intralipid™ (Carrière et al., 1991; Chahinian et al., 2006; Gargouri, Piéroni, Rivière, Saunière, et al., 1986), whereas pancreatic lipase has no direct activity on these substrates unless a prehydrolysis by gastric lipase (Bernbäck, Bläckberg, & Hernell, 1989; Gargouri, Pieroni, Riviere, Lowe, et al., 1986) or pancreatic phospholipase A2 (Borgstrom, 1980) triggers its activity. Gastric lipase has no phospholipase activity and therefore, it is able to penetrate the phospholipid layer (or trilayer in the case of native milk fat globules) surrounding lipid droplets in order to access to the TAG core. This is probably due to its high tensioactivity and penetration capacity, shown by experiments with phospholipid monomolecular films (Benarouche, Point, Carriere, & Cavalier, 2014; Benarouche, Point, Parsiegla, Carriere, & Cavalier, 2013; De la Fournière, Ivanova, Blond, Carrière, & Verger, 1994).

Phospholipid monomolecular films spread at the air–water interface can be used as biomimetic interfaces reproducing under controlled conditions the phospholipid monolayer surrounding various lipid bodies and emulsions, as well as a "half membrane." These films have been extensively used for studying the adsorption and the tensioactive properties of lipases, phospholipases, and various other proteins (Brezesinski & Mohwald, 2003; Brockman, 1999; Ransac, Moreau, Rivière, & Verger, 1991; Verger, 1980). In the case of lipases, phospholipid monolayers allow studying their adsorption step independently from substrate hydrolysis. In this setup, the gas phase (air) is considered as the hydrophobic phase at the surface of which the lipase binds (Fig. 1B). The use of these model membranes allows monitoring several physical parameters, such as the pH of the subphase, the lipid composition of the monolayer, and the density of lipids at the interface. The adsorption of a protein onto a phospholipid monolayer can be estimated from the increase in surface pressure ($\Delta\Pi$) which is directly connected to the variation of lipid molecular area (i.e., lipid packing). By varying the initial surface pressure (Π_i), it is possible to analyze the influence of lipid packing on the interfacial binding of proteins.

The biochemical characterization of recombinant dog gastric lipase (rDGL) provides several examples on the use of phospholipid monolayers for studying lipase adsorption. This chapter describes how adsorption kinetics monitored by surface tensiometry and data-fitting to Langmuir adsorption model allow the determination of kinetic constants of rDGL adsorption

(k_a, k_d, K_{Ads}) and related parameters (i.e., interfacial lipase concentration and molecular area) under various condition of pH and surface pressure. The use of additional techniques for characterizing phospholipid monolayers and lipase adsorption, like ellipsometry and atomic force microscopy (AFM), is also illustrated by studies performed with multicomponent phospholipid monolayers with phase separation. Ellipsometry combined with surface tension measurements brings insight about events occurring just beneath the interface, such as the presence of lipase molecules that do not necessarily induce changes in surface pressure. AFM imaging performed after Langmuir–Blodgett film transfer reveals the separation of liquid expanded (LE) and liquid condensed (LC) phases within complex phospholipid monolayers, as well as the preferential binding of rDGL to LE phase and at phase boundaries, with various levels of insertion into the phospholipid monolayer.

2. STUDY OF pH-DEPENDENT ADSORPTION OF GASTRIC LIPASE ONTO PHOSPHOLIPID MONOLAYERS USING SURFACE TENSIOMETRY

2.1 General Methodology

The Wilhelmy method (Wilhelmy, 1863) is used to measure the difference of the surface tension existing between the surface of the aqueous phase ($\gamma_0 \cong 72.7$ mN m^{-1} for pure water at 20°C) and that of the same aqueous phase covered by the phospholipid film (γ). This difference corresponds to the surface pressure ($\Pi = \gamma_0 - \gamma$), directly monitored by a Langmuir film balance equipped with a Wilhelmy platinum blade. Experiments reported here are performed at 25°C using various Teflon troughs and a Langmuir monomolecular film apparatus like the KSV5000 system (KSV Nima/Biolin Scientific, Stockholm, Sweden) equipped with an electromicrobalance, a mobile barrier for compression isotherm experiments and a temperature sensor probe surface pressure, barrier movement, and temperature are monitored by the KSV Device Server Software v.1.20.

Before each experiment, the Teflon trough has to be carefully cleaned and gently brushed with tap water, distilled ethanol, and Milli-Q™ water to get rid of residual surface-active impurities (Piéroni & Verger, 1983). The aqueous subphase containing 100 mN NaCl, 21 mM $CaCl_2$, and 1 mM EDTA is prepared with Milli-Q™ water and various buffers to adjust the pH value: 50 mM glycine–HCl for pH 2.0 and 3.0; 10 mM sodium acetate buffer for pH 4.0, 5.0, and 6.0; 10 mM Tris buffer for pH 7.0.

The phospholipid monolayer is prepared by spreading a few microliters of 1,2-dilauroyl-*sn*-glycero-3-phosphocholine (DLPC; Avanti Polar Lipids Inc., cat. number 850335) solution in chloroform (1 mg mL^{-1}) until the desired initial surface pressure (Π_i) is reached. The waiting time for solvent evaporation and for the film to reach equilibrium varies from 10 to 20 min depending on the spreading volume, Π_i, and pH of the aqueous subphase.

2.2 Phospholipid Monolayers and Surface Pressure-Area Isotherms

As a prerequisite to study the pH-dependent adsorption of a protein onto phospholipid monolayers, the stability of the phospholipid spread at the air/water interface has to be checked. Monolayers of DLPC spread at various Π_i and pH values ranging from 2.0 to 7.0 form stable films with only minor changes in surface pressure (± 0.2 mN m^{-1}) over 60 min. After forming the DLPC monolayer over the surface of a rectangular Teflon trough (volume 126.3 mL; surface area, 210.5 cm^2) containing the aqueous phase, the surface pressure (Π)–molecular area (A_{DLPC}) isotherms are recorded by monitoring the surface pressure increase ($\Delta\Pi$) during film compression by the mobile barrier (Fig. 2A).

2.3 Recording rDGL Adsorption Onto DLPC Monolayers

The DLPC monolayer is first formed in a round Teflon trough (volume, 18.1 mL; surface area, 13.6 cm^2) by spreading the DLPC chloroformic solution over the continuously stirred (250 rpm; 1-cm magnetic bar) aqueous phase, until the desired Π_i is reached. rDGL stock solutions are typically prepared at a concentration of 1 mg mL^{-1} in 10 m*M* MES (2-(*N*-morpholino) ethanesulfonic acid; Sigma-Aldrich, cat. number M8250), 150 m*M* NaCl, pH 6.0 buffer. rDGL is available from the authors on demand. Various volumes of rDGL stock solution (5–80 μL) are then injected with a Hamilton syringe below the monolayer to study rDGL adsorption at enzyme concentrations (C_{E0}) ranging from 5 to 90 n*M*. After injecting rDGL into the aqueous subphase, the surface pressure increase ($\Delta\Pi$) due to the adsorption of the lipase onto the DLPC monolayer is continuously recorded until the equilibrium surface pressure ($\Delta\Pi_{max}$) is reached. Fig. 2B shows the $\Delta\Pi = f(t)$ curves typically obtained upon rDGL ($C_{E0} = 20$ n*M*) adsorption onto DLPC monolayers at pH 5.0 and various Π_i.

The $\Delta\Pi_{max} = f(\Pi_i)$ plot (Fig. 2C) and linear extrapolation to zero surface pressure increase ($\Delta\Pi_{max} = 0$) allow to determine the critical surface pressure

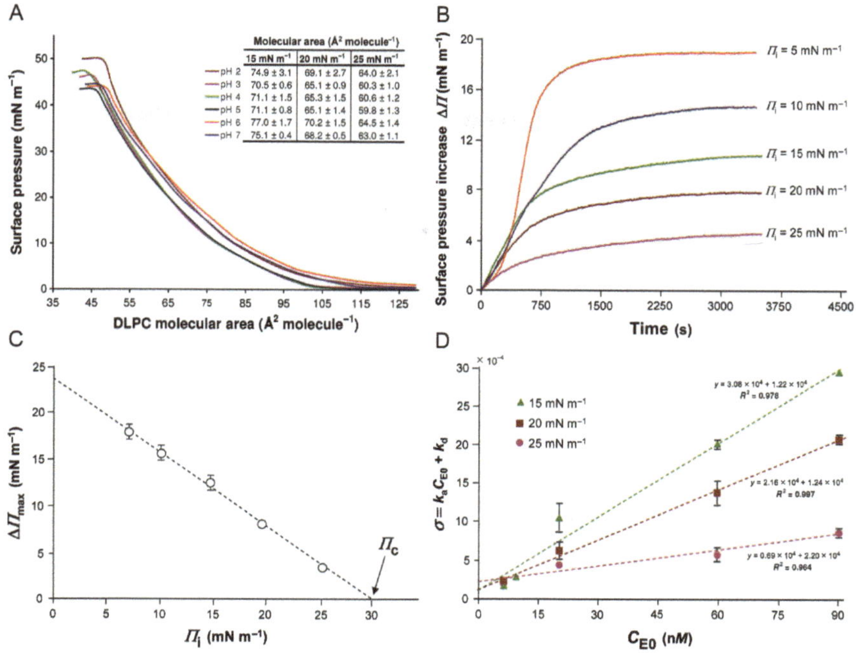

Fig. 2 Monolayer experiments for estimating rDGL adsorption/desorption parameters. Panel A: Surface pressure vs molecular area (Π–A_{DLPC}) isotherms of DLPC monomolecular films recorded at various pH upon film compression by the mobile barrier at a constant rate of 10 mm min^{-1}. The insert shows molecular areas $\pm$ SD ($n = 3$) obtained at various pH and initial surface pressures (Π_i). Panel B: Surface pressure increase ($\Delta\Pi$) with time after rDGL injection below the DLPC monomolecular film at various Π_i ranging from 5 to 25 mN m^{-1}. Experiments performed at pH 5.0 with 20 n*M* rDGL final concentration. Panel C: Plot of maximum surface pressure increase ($\Delta\Pi_{max}$) after rDGL injection as a function of the initial surface pressure (Π_i) of the DLPC monolayer. Experiments performed at pH 5.0 with 20 n*M* rDGL final concentration. Panel D: Determination of the adsorption (k_a, M^{-1} s^{-1}) and desorption (k_d, s^{-1}) rate constants from σ plot vs rDGL concentration (C_{E0}). The values of σ are obtained from curve-fitting experimental data points of surface pressure increase with time to simplified Langmuir adsorption isotherm (Eq. 2), at each pH, Π_i, and C_{E0} investigated. For each initial surface pressure, linear regression line allows to determine k_a (slope) and k_d (*y*-intercept). The data presented here are those obtained at pH 5. *Adapted from Benarouche, A., Point, V., Parsiegla, G., Carriere, F., & Cavalier, J. F. (2013). New insights into the pH-dependent interfacial adsorption of dog gastric lipase using the monolayer technique.* Colloids and Surfaces. B, Biointerfaces, 111C, *306–312.*

of insertion ($\Pi_c = 30.3$ mN m^{-1}) above which no increase in surface pressure occurs upon rDGL injection below the DLPC monolayer. The value of Π_c depends on the phospholipid molecules used for making the monolayer. With egg PC (Avanti Polar Lipids Inc., cat. number 131601), a lower Π_c

value of 21.5 mN m^{-1} is obtained for rDGL insertion into the monolayer (Benarouche et al., 2013), indicating that the longer and unsaturated acyl chains of egg PC constitute a stronger barrier for rDGL adsorption.

2.4 Determination of the Kinetic Parameters of DGL Adsorption Onto DLPC Monolayer

Experimental data (i.e., surface pressure increases with time; Fig. 2B) can be fitted to Langmuir adsorption isotherm (Langmuir, 1917) for determining the rate constants of adsorption (k_a, M^{-1} s^{-1}) and desorption (k_d, s^{-1}). The Langmuir adsorption isotherm is adapted to $\Delta\Pi$ measurements as follows (Eq. 1):

$$\Pi(t) = \Pi_i + \Delta\Pi_{max} \cdot [(k_a \cdot C_E)/(k_a \cdot C_E + k_d)] \cdot [1 - \exp(-(k_a \cdot C_E + k_d) \cdot t)] \quad (1)$$

where $\Pi(t)$ is the surface pressure measured as a function of time; Π_i is the initial surface pressure; $\Delta\Pi_{max}$ is the maximum variation of surface pressure reached upon DGL binding, and C_E (mol L^{-1}) is the enzyme concentration in the subphase of the trough.

In the monolayer system, since the ratio of surface area to volume of the trough (S_{trough}/V_{trough}) is very small and close to 1 cm^{-1}, only a small fraction of the protein injected is expected to adsorb onto the lipid film (Aoubala, Ivanova, Douchet, De Caro, & Verger, 1995; Pattus, Slotboom, & de Haas, 1979; Verger, Rietsch, van Dam-Mieras, & de Haas, 1976). Based on these findings, one can make the assumption that the enzyme concentration C_E remains close to the initial concentration of the enzyme injected into the subphase (C_{E0}). With this approximation ($C_E \cong C_{E0}$), the Langmuir adsorption isotherm (Eq. 1) can be simplified to Eq. (2):

$$\Pi(t) = \Pi_i + \Delta\Pi_{max} \cdot \theta \cdot [1 - \exp(-\sigma \cdot t)] \quad (2)$$

where

$$\sigma = k_a \cdot C_{E0} + k_d \quad (3)$$

and

$$\theta = k_a \cdot C_{E0}/(k_a \cdot C_{E0} + k_d) \quad (4)$$

θ is the fraction of the total free adsorption sites coverage.

Curve-fitting the experimental data points ($\Delta\Pi = f(t)$) to Eq. (2) allows to calculate σ values for multiple C_{E0} concentrations at each pH and Π_i

investigated (Fig. 2D). Values of k_a and k_d are then determined from the slope and y-intercept, respectively, of the linear plot of σ vs C_{E0} (Fig. 2D). The association (K_{Ads}) and dissociation (K_D) constants of the adsorption equilibrium are obtained from the ratio of the adsorption and desorption rate constants ($K_{Ads}=k_a/k_d=1/K_D$).

When these calculations are applied to rDGL adsorption onto DLPC monolayers at various pH values and a single Π_i value of 20 mN m^{-1}, k_a and k_d are found to be maximum at pH 2.0 and minimum at pH 5 ($k_a=2.16\times10^4$ M^{-1} s^{-1}; $k_d=1.24\times10^{-4}$ s^{-1}; Fig. 3A and B). The maximum adsorption equilibrium constant is obtained at pH 5.0

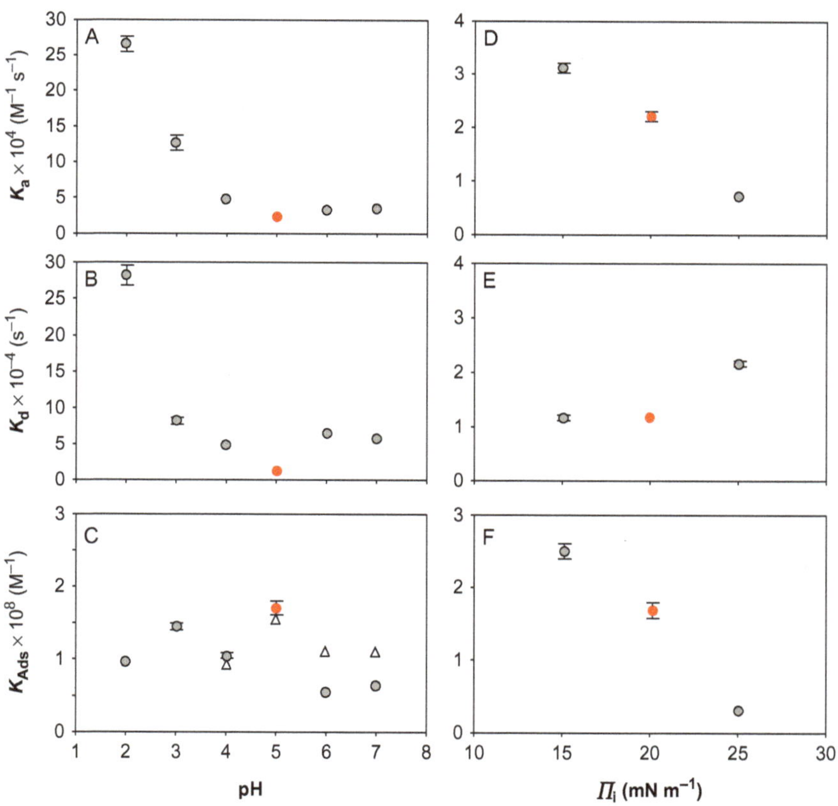

Fig. 3 Adsorption rate (k_a), desorption rate (k_d), and adsorption equilibrium (K_{Ads}) constants derived from the kinetics of rDGL adsorption onto DLPC films at various pH (panels A, B, and C) and initial surface pressure (Π_i; panels D, E, and F). *Red dots* indicate values estimated at pH 5 and $\Pi_i=20$ mN m^{-1}. *Triangles* in panel C correspond to (K_{Ads}) values estimated by TIRF upon rDGL adsorption on solid hydrophobic surface (Chahinian et al., 2006).

($K_{Ads} = 1.7 \times 10^8\ M^{-1}$; Fig. 3C), in good agreement with the fact that gastric lipase preferentially binds to interfaces at acidic pH and displays its optimum activity between pH 4 and 6 (Chahinian et al., 2006; Sams, Paume, Giallo, & Carriere, 2016).

Using the same experimental approach and calculations, one can estimate the effects of initial surface pressure Π_i on rDGL adsorption kinetics at a constant pH value. Fig. 3D–F shows the kinetic parameters estimated at pH 5.0 with Π_i varying from 15 to 25 mN m^{-1}. The adsorption rate constant (k_a) is found to decrease from $3.1 \times 10^4\ M^{-1}\ s^{-1}$ at 15 mN m^{-1} to $0.69 \times 10^4\ M^{-1}\ s^{-1}$ at 25 mN m^{-1}, whereas the desorption rate constant (k_d) increases from $1.2 \times 10^{-4}\ s^{-1}$ at 15–20 mN m^{-1} to $2.2 \times 10^{-4}\ s^{-1}$ at 25 mN m^{-1}. The resulting adsorption equilibrium constant K_{Ads} shows a decreasing dependency upon the initial surface pressure from 15 to 25 mN m^{-1}.

2.5 Estimation of rDGL Interfacial Concentration and Molecular Area

Using the equations derived from the interfacial kinetic model of Verger–de Haas (Panaiotov & Verger, 2000), it is possible to estimate the enzyme interfacial concentration (Γ_{E^*}, molecule cm^{-2}) based on Eq. (5), where the trough volume (V_{trough}) is 18.1 mL and the surface area (S_{trough}) is 13.6 cm^2:

$$\Gamma_{E^*} = C_{E0} \Big/ \left[(1 + K_D) \cdot \left(\frac{S_{trough}}{V_{trough}} \right) \right] \tag{5}$$

The molar fraction $\left(\Phi_{E^*(\%)}, \text{mol\%}\right)$ of lipase adsorbed onto the lipid monolayer can then be deduced from Eq. (6):

$$\Phi_{E^*(\%)} = \left[\Gamma_{E^*} \cdot \left(\frac{S_{trough}}{V_{trough}} \right) \right] \Big/ C_{E0} = 100/(1 + K_D) \tag{6}$$

As expected, Γ_{E^*} and $\Phi_{E^*(\%)}$ values determined for rDGL adsorption onto DLPC monolayers (Table 1) have the same pH and surface pressure dependencies as K_{Ads}. When calculating $\Phi_{E^*(\%)}$ at $\Pi_i = 20$ mN m^{-1} in a pH range from 2.0 to 7.0, the maximum fraction of absorbed DGL adsorbed on DLPC monolayer ($\Phi_{E^*(\%)} = 14.8 \pm 1.2$ mol%) is reached at pH 5.0, and the lowest one ($\Phi_{E^*(\%)} = 5.1 \pm 0.2$ mol%) at pH 6.0 (Table 1). At pH 5.0, DGL adsorption decreases with increasing surface pressures and $\Phi_{E^*(\%)}$

Table 1 Characteristic Values Related to rDGL Adsorption Onto DLPC Films at Various pH and Initial Surface Pressures (Π_i)

pH	Π_i (mN m^{-1})	$\Delta\Pi_{max}$ (mN m^{-1})	Γ_{E^*} (Molecule cm^{-2}) × 10^{12}	Φ_{E^*}(%) (mol%)	A_{E^*} (Å^2 molecule^{-1})
2	20	7.9±0.1	1.39±0.04	8.7±0.2	1084±30
3		8.6±0.3	2.01±0.07	12.6±0.4	494±1
4		8.7±0.05	1.51±0.06	9.4±0.4	771±32
5		8.2±0.2	2.37±0.06	14.8±1.2	718±19
6		7.6±0.2	0.81±0.03	5.1±0.2	1395±61
7		8.4±0.3	0.95±0.05	5.9±0.3	1498±82
5	15	12.6±0.5	3.23±0.09	20.2±0.9	793±22
	20	8.2±0.2	2.37±0.06	14.8±1.2	718±19
	25	3.6±0.03	0.49±0.01	3.1±0.2	674±12

Adapted from Benarouche, A., Point, V., Parsiegla, G., Carriere, F., & Cavalier, J. F. (2013). New insights into the pH-dependent interfacial adsorption of dog gastric lipase using the monolayer technique. *Colloids and Surfaces. B, Biointerfaces, 111C*, 306–312.

drops from 20.2±0.9 mol% at 15 mN m^{-1} to 3.1±0.2 mol% at 25 mN m^{-1} (Table 1).

The molecular area (A_{E^*}, Å^2 molecule^{-1}) of rDGL adsorbed onto DLPC monolayers can be further calculated from Eq. (7):

$$\begin{aligned} A_{E^*} &= \left[\left(S_{\text{trough}} - A_{\text{DLPC}} \cdot n_{\text{DLPC}}\right) \cdot \theta\right] / \left(\Gamma_{E^*} \cdot S_{\text{trough}}\right) \\ &= (\text{surface occupancy})/\Gamma_{E^*} \end{aligned} \quad (7)$$

in which A_{DLPC} (Å^2 molecule^{-1}) is the molecular area of DLPC deduced from the surface pressure (Π)–molecular area isotherms (Fig. 2A) at the final equilibrium surface pressure reached upon rDGL adsorption; n_{DLPC} is the number of DLPC molecules spread at the air/water interface at the working initial surface pressure (Π_i); and θ is the fractional surface coverage determined in Eq. (4).

When keeping Π_i at a constant value of 20 mN m^{-1}, A_{E^*} is found to vary significantly with pH, with a minimum at a pH value of 3 and a maximum at a pH value of 7 (Table 1). When keeping the pH at 5.0, the A_{E^*} value decreases from 793 Å^2 at 15 mN m^{-1} to 674 Å^2 molecule^{-1} at 25 mN m^{-1} (Table 1 and Fig. 4). The lowering of A_{E^*}, Γ_{E^*}, and

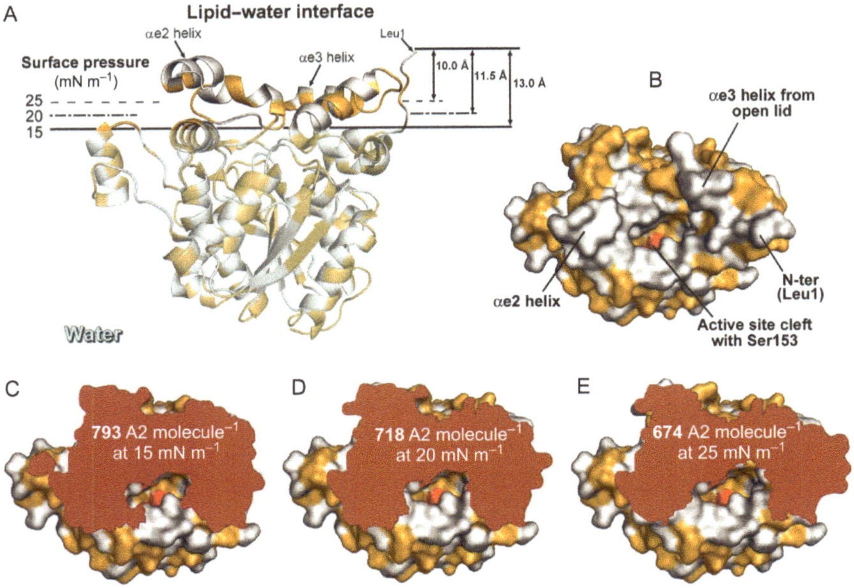

Fig. 4 Schematic representation of the surface pressure-dependent penetration of rDGL into a phospholipid monolayer spread at the air–water interface. Panel A: Side view of rDGL molecule (Cα ribbon tracing) oriented at the lipid–water interface (see Chapter "Probing conformational changes and interfacial recognition sites of lipases with surfactants and inhibitors" by Mateos-Diaz et al.). Panel B: Top view of rDGL in molecular surface representation. The hydrophobic residues, particularly those surrounding the active site entrance and forming a ring parallel to the putative lipid–water interface, are shown as *white surface* while polar residues are shown in *yellow color*. The catalytic Ser153 is colored in *red*. Panels C, D, and E: Top views of rDGL with cross-sections (*purple color*) made parallel to the interface plane. The cross-section areas correspond to DGL molecular areas deduced from rDGL adsorption onto DLPC monolayers at surface pressures of 15, 20, and 25 mN m^{-1}, respectively. The distances between these sections and the top of the DGL molecule, indicated in panel A, are indicative of rDGL penetration into the phospholipid monolayer. DGL 3D models were drawn using PyMOL Molecular Graphics System (Version 1.3, Schrödinger, LLC) and DGL X-ray structure with the lid in the open conformation (PDB ID: 1K8Q). *Adapted from Benarouche, A., Point, V., Parsiegla, G., Carriere, F., & Cavalier, J. F. (2013). New insights into the pH-dependent interfacial adsorption of dog gastric lipase using the monolayer technique.* Colloids and Surfaces. B, Biointerfaces, 111C, *306–312.*

$\Phi_{E^*(\%)}$ from 15 to 25 mN m^{-1} at pH 5.0 is in agreement with the general behavior of lipolytic enzymes showing a decreased interfacial binding capacity at high surface pressure and more compact lipid packing (Aoubala et al., 1995; Piéroni & Verger, 1979; Sugihara et al., 1986; Verger et al., 1973).

2.6 Level of Confidence of Kinetic Parameters and Molecular Areas Deduced from Monolayer Data-Fitting to Langmuir Adsorption Model

Although the Langmuir adsorption model is often used for analyzing the adsorption of biological macromolecules onto various surfaces (e.g., protein adsorption/desorption kinetics using surface plasmon resonance (Wegner et al., 2004)), this model is limited by several constraints. The use of Langmuir adsorption model implies that all adsorption sites are equivalent with the same binding energy to the surface, from the very first to the final adsorbing molecule. The adsorption capacity is strictly limited to a single monolayer of molecules binding the surface.

The surface pressure (Π)–molecular area (A_{DLPC}) compression isotherms obtained for DLPC are characteristic of a single LE state at surface pressures below the collapse (i.e., below 43 mN m^{-1}) and before reaching the gas phase (Fig. 2A). The use of DLPC therefore allows studying rDGL binding to phospholipids in a single physical state at the surface pressures (15–25 mN m^{-1}) selected for measuring adsorption kinetics. Moreover, the surface occupied by rDGL at the air–water interface, estimated from A_{E*} and Γ_{E*} values (Table 1) does not exceed 25% of the total surface of the monolayer trough (13.6 cm^2), thus precluding the formation of multilayers of adsorbed rDGL.

It is also worth noticing that the pH-dependent K_{Ads} values estimated for rDGL adsorption onto DLPC monolayers are similar to those estimated from rDGL adsorption onto a solid hydrophobic surface using total internal reflection fluorescence (TIRF) (Chahinian et al., 2006). The maximum adsorption equilibrium constant deduced from TIRF experiments, also at pH 5.0, is nearly identical ($K_{\mathrm{Ads}} = 1.54 \times 10^8\ M^{-1}$; Fig. 2C). The two sets of experiments differ, however, with respect to rDGL concentration (C_{E}) in the bulk phase. In TIRF experiments, rDGL concentration is kept constant ($C_{\mathrm{E}} = C_{\mathrm{E0}}$) during the assay using a continuous flow of 20 nM rDGL solution in the flow cell (Chahinian et al., 2006). This allows using the Langmuir adsorption equation without approximation of C_{E} value. In DLPC monolayer experiments, only the initial rDGL concentration (C_{E0}) is controlled upon enzyme injection into the subphase and it is assumed that C_{E} remains close to C_{E0} to simplify Langmuir adsorption equation (Eq. 2) and further estimate kinetic constants. The estimation of rDGL partitioning between the bulk phase and DLPC monolayer shows, however, that the molar fraction of rDGL adsorbed onto the lipid monolayer ($\Phi_{E*(\%)}$, mol%) can reach 14.8% under the most favorable conditions of adsorption (pH 5 and 25 mN m^{-1}; Table 1). The resulting variation in rDGL bulk concentration might be considered as too high for simplifying the Langmuir adsorption equation with $C_E = C_{E0}$. Nevertheless, the similarity of the kinetic parameters estimated from tensiometry (monolayer experiments) and TIRF (adsorption onto hydrophobic solid surface) suggests that the approximation of rDGL bulk concentration is acceptable. All together, these data support a preferential adsorption of rDGL at acidic pH and an important contribution of hydrophobic interactions.

Since the 3D structure of rDGL is known (PDB ID: 1K8Q) (Roussel et al., 2002), one can check whether A_{E^*} values estimated from monolayer studies are compatible with the dimensions of the enzyme and its putative orientation at the interface (see Chapter "Probing conformational changes and interfacial recognition sites of lipases with surfactants and inhibitors" by Mateos-Diaz et al.). The rDGL molecule with the open lid shows a large hydrophobic ring surrounding the entrance of the active site (Fig. 4A and B). After aligning this hydrophobic ring parallel to a putative lipid/water interface (Fig. 4A), stepwise cross-sections distant by 1.5 Å can be made parallel to this plane (Fig. 4C–E). The cross-section areas are then compared to the A_{E^*} values estimated from rDGL adsorption onto DLPC monolayers. Thus, the molecular areas estimated from DGL adsorption onto DLPC at 25, 20, and 15 mN m^{-1} are similar to the areas of the cross-sections made at 10.0, 11.5, and 13.0 Å from the top of the rDGL molecule (Fig. 4A, C–E). These findings support the estimation of rDGL molecular area by monolayer studies and reflect the decreasing penetration of rDGL into DLPC monolayers when the surface pressure is increased.

3. STUDY OF GASTRIC LIPASE ADSORPTION USING ELLIPSOMETRY

3.1 Principle of Ellipsometry

The principle of ellipsometry is based on the measurements of a polarized light when it is reflected by an interface (solid or liquid). The "null ellipsometer" configuration (see later) allows achieving pure linear polarization after beam reflection. Then, the ellipsometric angle (Δ, degrees) which is the phase difference between p- and s-polarized light can be measured and corresponds to the extinction of the reflected light (Azzam & Bashara, 1977). Indeed, the incident angle of the beam should be equal to the Brewster angle respect to the interface composition. In order to get more precision on the measurements, this angle is lifted of 1 degree and the minimum of light reflection is calculated by fitting the intensity of the reflected light during the analyzer's rotation. The ellipsometric angle is dependent of the optical properties of the interface through its reflective index (n) and its thickness (e).

In the case of liquid interface, and in order to follow interfacial changes induced by the presence of lipids or proteins or both, the reference is the water (or buffer) contained in the trough. Indeed the setup is calibrated to give a value of 0 for the bare air/water interface with a dispersion value of 0.03 degree.

The setup is described as follows: a polarized He–Ne laser beam ($\lambda = 632.8$ nm, Melles Griot, Glan–Thompson polarizer) is reflected on the surface of the trough. The incidence angle is 52.12 degree, i.e., Brewster angle minus 1 degree. After reflection on the liquid surface, the laser light passed through a $\lambda/4$ retardation plate, a Glan–Thompson analyzer, and a photomultiplier. The analyzer angle, multiplied by two, yielded the value of the ellipsometric angle (Δ), i.e., the phase difference between parallel and perpendicular polarization of the reflected light. The laser beam probed the 1 mm^2 surface with a depth in the order of 1 μm. Initial values of the ellipsometric angle (Δ_0) and surface tension of pure buffer solutions are recorded for at least half an hour and no variation allows beginning the experiments. Values of Δ are recorded every 4 s (Berge & Renault, 1993).

The surface pressure is measured simultaneously using the Wilhelmy method as described in Section 2.1, using a filter paper as plate (Whatman, Velizy-Villacoublay, France) connected to a microelectronic feedback system (Nima PS4; KSV Nima/Biolin Scientific, Stockholm, Sweden). The surface pressure (Π) is recorded every 4 s with a precision of ± 0.2 mN m^{-1}. Moreover, the trough is placed on a balance and connected to a water syringe in order to monitor and compensate water evaporation. The water flow is adjusted not to disturb the interfacial molecule behavior.

3.2 Applications of Ellipsometry to Protein Adsorption Characterization

As previously shown with phospholipid monolayers, the lipid interfacial properties can be determined by surface pressure measurements, which correspond to the evolution of the lateral interfacial force measured by the force device. The surface pressure can, however, hide possible presence of molecules either at very low surface concentration (in gas phase or low cohesive interaction) (Jacquemet, Vie, Lemiegre, Barbeau, & Benvegnu, 2010) or behind a preexisting interfacial monolayer (Desfougeres et al., 2011). It is why the use of ellipsometry is precious and gives additional information about events occurring just beneath the interface. Sometimes, the protein insertion induces new protein adsorption and the ellipsometric angle increases again while surface pressure might not change. Combination of surface pressure measurement and ellipsometry is a very good way to differentiate protein forming monolayers from multilayers. In the first case, proteins insertion at the interface is characterized by a linear evolution of surface pressure vs ellipsometric angle. In the second case (multilayer), proteins are characterized by an increase in ellipsometric angle at constant pressure (Bolanos-Garcia, Renault, & Beaufils, 2008).

3.3 Example of Results Obtained on Gastric Lipase

We have seen previously the limitations of surface tensiometry and data-fitting to Langmuir adsorption isotherm for investigating rDGL adsorption onto phospholipid monolayers (see Section 2.6). Ellipsometry allows studying rDGL adsorption onto multicomponent lipid monolayers with phase separation (Bourlieu et al., 2016). As an example, rDGL adsorption onto a mixed (45:45:10 mol/mol/mol) monolayer of 1,2-dioleoyl-*sn*-glycero-3-phosphocholine (DOPC; Avanti Polar Lipids Inc., cat. number 850375), 1,2-dipalmytoyl-*sn*-glycero-3-phosphocholine (DPPC; Avanti Polar Lipids Inc., cat. number 850355), and 1,2-dioleoyl-*sn*-glycero-3-phosphoserine (DOPS; Avanti Polar Lipids Inc., cat. number 840035) is given in Fig. 5. The initial ellipsometric angle ($\Delta_i = 7.5 \pm 0.5$) and surface pressure ($\Pi_{\mathrm{i}} = 19.5 \pm 0.2$ mN m^{-1}) are related to the properties of the phospholipid monolayer. Following lipase injection, the surface pressure increases of around 2 mN m^{-1}, before rapid reorganization and relaxation of the film followed by a steady increase in surface pressure up to 27 mN m^{-1} indicating protein insertion. Meanwhile, ellipsometric angle increases very sharply ($\delta\Delta = 10.6$ degree) suggesting that an important amount of protein is present in the first micrometer below the interface. More examples of lipase adsorption into mixed lipid monolayers characterized both by surface pressure and ellipsometry are detailed in Bourlieu et al. (2016).

4. STUDY OF GASTRIC LIPASE ADSORPTION BY AFM

4.1 Principles of AFM and Sample Preparation

In imaging configuration, AFM consists in scanning the surface of a sample using a very sharp tip fixed at the end of a cantilever. Atomic interactions

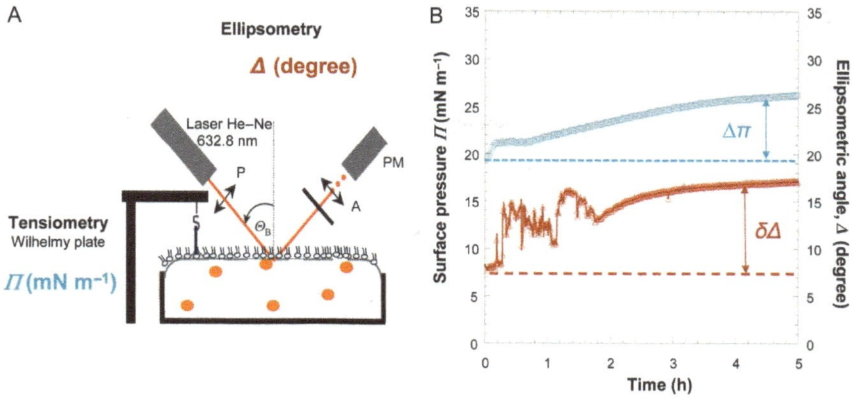

Fig. 5 Evolution of surface pressure and ellipsometric angle upon rDGL adsorption onto a mixed phospholipid monolayer. Panel A: Schematic representation of the coupling of tensiometry and ellipsometry for studying lipid monolayers spread at the air–water interface. Panel B: Variations in surface pressure (Π, mN m^{-1}; *circles*) and ellipsometric angle (Δ, degrees; *triangle*) after injection of rDGL below a mixed monolayer of DOPC/DPPC/DOPS (45:45:10 w/w/w) after the initial surface pressure was adjusted to 20 mN m^{-1}. The final subphase concentration of rDGL is 40 n*M*. Determination of $\Delta\Pi$ and $\delta\Delta$ from kinetics are indicated in Panel B. *Adapted from Bourlieu, C., Paboeuf, G., Chever, S., Pezennec, S., Cavalier, J. F., Guyomarc'h, F., et al. (2016). Adsorption of gastric lipase onto multicomponent model lipid monolayers with phase separation.* Colloids and Surfaces. B, Biointerfaces, 143, *97–106.*

between the end of the tip and the surface sample induce cantilever deflection detected by an optical detector. In order to regulate the tip/sample forces, a feedback device controls the distance between the tip and sample by varying the height either of the tip or of the sample through piezoelectric elements (depending of the commercial system). The height evolution is recorded line by line (pixel by pixel) and at the end, the reconstructed image gives a topographic view of the sample surface. During the scan, the measured forces are minimized to avoid sample damages. The lateral AFM resolution is in the order of the nanometer due to the tip convolution, in contrast the vertical resolution can be up to 0.1 nm. Such resolution is well adapted to reveal the dynamic nanoscale of imaged surfaces, for instance in the case of complex lipid mono- or bilayers and their phase separation. The AFM can be operated in air or in liquid which allows the observations of a wide variety of biomolecules under conditions close to physiological ones (Fotiadis, Scheuring, Muller, Engel, & Muller, 2002; Radmacher, Tillamnn, Fritz, & Gaub, 1992; Vié et al., 1998). Other AFM application concerns force measurements to perform force spectroscopy.

Lipase adsorption onto lipid film formed at the air/water interface can be imaged by AFM after transferring a Langmuir film onto a freshly cleaved mica plate by the LB method (Fig. 6A). It can be done following the enzyme adsorption onto nonhydrolysable phospholipid monolayers, as in the case of gastric lipase (Benarouche et al., 2013; De la Fournière et al., 1994), but

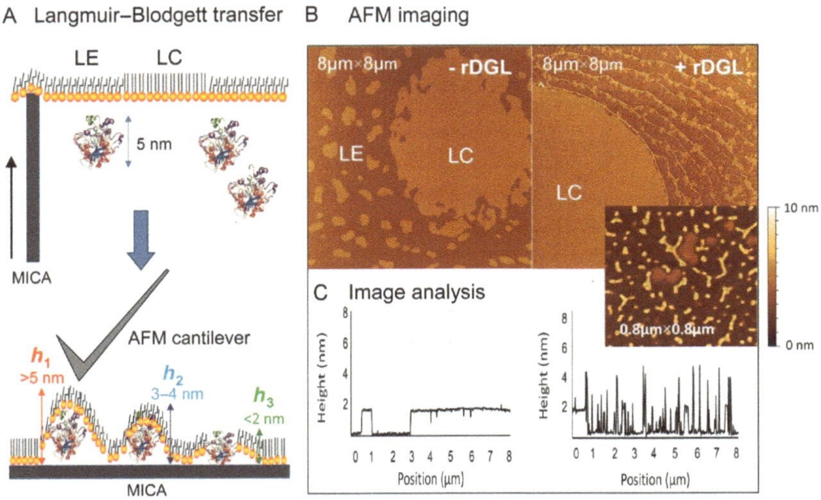

Fig. 6 Heterogeneous lipid monolayers (DOPC/DPPC/DOPS 45:45:10 w/w/w) obtained by Langmuir–Blodgett (LB) transfer and imaged by AFM in the air in the presence or in the absence of rDGL. The lipid monolayers were transferred onto mica either 40 min after lipid deposition at the air–water interface (without lipase) or 2 h after injection of the lipase below the lipid monolayer. Panel A: Principle of LB transfer onto mica after rDGL adsorption onto the lipid monolayers (*top*) and AFM imaging of the resulting rDGL distribution in heterogeneous monolayer presenting liquid expanded (LE)/liquid condensed (LC) phase coexistence at various height levels. h_1, lipase adsorbed underneath interacting with polar head only, rapid adsorption evidenced by ellipsometry; h_2, insertion of lipase on 1–2 nm depth, such insertion is coherent with the penetration depth reported by Bénarouche et al. (2013), i.e., 1.15 nm at 20 mN m^{-1} for the lipase active form; h_3, lipase totally inserted on 3–5 nm, probably denaturated at the interface and contributing to lower surface tension. Panel B: Topographic AFM 8×8 μm^2 images of LB films (initial surface pressure of 20 mN m^{-1}) before (−rDGL) or after injection of rDGL (+rDGL; 40 n*M* final subphase concentration). Lipase adsorption onto the LE phase and at boundaries of LC domains in heterogeneous monolayer results in an increase in electrostatic interactions between molecules, in small nanodomains alignments; Insert presents higher resolution 0.8×0.8 μm^2 AFM image with individual lipases or clusters and small LC domains. Panel C: Representative AFM cross-sections in the absence (*right*) and in the presence (*left*) of rDGL. *Adapted from Bourlieu, C., Paboeuf, G., Chever, S., Pezennec, S., Cavalier, J. F., Guyomarc'h, F., et al. (2016). Adsorption of gastric lipase onto multicomponent model lipid monolayers with phase separation.* Colloids and Surfaces. B, Biointerfaces, 143, *97–106.*

hybrid films based on mixtures of acylglycerols (tri-, di-, or monoacylglycerols) and phospholipids have also been used for studying both enzyme adsorption and activity of microbial lipases and phospholipase A2 (Balashev, Jensen, Kjaer, & Bjornholm, 2001).

The LB transfer is done at constant pressure and very low speed (0.5 mm min^{-1}) to limit film perturbation. Imaging is then typically carried out in contact mode in air (20°C) with standard silicon cantilever (spring constant of 0.06 or 0.12 N m^{-1}, SNL, Bruker, France) and at a scan rate of 1 Hz. However, imaging in liquid cell has been advised to preserve the native structure and activity of some biomolecules (Balashev et al., 2001). In this case supported bilayers are used. They can be prepared either (i) by collapsing liposomes obtained by sonication or extrusion or (ii) by successive depositions of two monolayers by the LB method. Comparing the two methods, Balashev et al. (2001) concluded that bilayers obtained by the LB deposition method present less defects than bilayers obtained by collapsing liposomes. Since lipase will preferentially adsorb where lipid molecular packing is lower, i.e., typically at the edge of defect zones, such defects must be avoided during lipid bilayer preparation.

For LB-obtained lipid bilayers, the bottom layer is generally based on 1,2-dipalmitoyl-*sn*-glycero-3-phosphocholine (DPPC), compressed to 30–35 mN m^{-1} and transferred onto mica at very low speed to give a sound bottom layer. The second or top layer can be formed with DPPC to obtain a phospholipid bilayer, but for studying lipase activity, a monoacylglycerol layer (either monooleoylglycerol or monopalmitoylglycerol) previously compressed at 25 mN m^{-1} can be spread on top of the DPPC layer. For detailed methodologies consult (Balashev et al., 2003; Prim, Iversen, Diaz, & Bjornholm, 2006). The supported bilayer is then transferred to the liquid cell of the AFM and after equilibration (30 min) the enzyme solution is flushed into the cell.

An alternative method for AFM imaging of lipase in liquid cell consists in adding the adsorbed lipase/lipid monolayer onto hydrophobic mica support by Langmuir–Schaefer transfer before liquid cell AFM imaging (see Supplementary data of Bourlieu et al., 2016). Briefly, after lipase adsorption on a mixed lipid Langmuir film, the film is collected by lowering horizontally the hydrophobic (DPPC-coated) mica disc at very low speed (0.5 mm min^{-1}) using a dipper device (KSV Nima/Biolin Scientific, Stockholm, Sweden). After contact with the film, the sample is maintained blocked on the sealed chamber. Then, all system is reverted, and the chamber is placed on the AFM. The sample must be maintained hydrated all the time.

4.2 Imaging Lipase Adsorption and Activity on Supported Bilayers and Monolayers by AFM

In situ investigation of lipase kinetics using AFM was first proposed by Balashev et al. (2001). In homogeneous systems, they imaged successfully lipase adsorption and activity on model lipid membranes (Balashev, John DiNardo, Callisen, Svendsen, & Bjornholm, 2007; Balashev et al., 2001). Using AFM liquid cell, *Thermomyces* (formerly *Humicola*) *lanuginosus* lipase was visualized trapped in a DPPC bilayer. The lipase molecules appeared as protrusion of 2-nm height over the membrane surface and did not cluster. Using mixed bilayer of DPPC/monooleoylglycerol or DPPC/monopalmitoylglycerol exposed to lipase in liquid cell, allowed imaging the preferential adsorption of *Thermomyces lanuginosus* and *Candida rugosa* lipases onto the edge of nanoscale structural defects, as well as the enzyme activity resulting in the growth of these defects after lipase injection and monoacylglycerol hydrolysis (Prim et al., 2006). Lag phases for the hydrolysis of lipid substrate could be observed or not depending on the lipase and DPPC/monoacylglycerol mixed bilayer, suggesting that AFM imaging is well adapted for studying the effects of the substrate physical state on enzyme adsorption and activity. However, very few AFM characterizations of lipase adsorption in heterogeneous lipid system with phase coexistence have been proposed despite the obvious biological interest of such approach. Chu et al. (2010) investigated the adsorption of pancreatic lipase and colipase onto digalactosyldiacylglycerol or DPPC monolayers at the air/water interface. AFM imaging revealed that colipase and lipase selectively adsorbed onto mixed DPPC-bile salts LE regions and not onto LC DPPC domains. In LC domains, the close packing of the DPPC headgroups probably prevents lipase adsorption by steric hindrance. Further insights into the mechanisms governing lipase adsorption onto heterogeneous lipid films were provided by Bourlieu et al. (2016) using rDGL, as illustrated in the next section.

4.3 Example of AFM Imaging of Gastric Lipase Adsorption Onto Heterogeneous Lipid Monolayers

AFM observations of DOPC/DPPC/DOPS (45:45:10 mol/mol/mol) LB films in the absence or after injection of rDGL in the subphase, are displayed in Fig. 6. Topographic $8 \times 8\ \mu m^2$ images (Fig. 6B) are adequate to visualize all the objects in the systems: lipid phase separation and lipase distribution.

In the absence of rDGL (Fig. 6B, left), LB films display typical phase separation with LC (lighter zones) and LE (background) domains of variable shapes and size. After addition of rDGL, changes in the lipid film organization can be observed: the circularity of large LC domain is enhanced, while aligned smaller domains appear. rDGL appears as small grains in LE phase mainly and at LE–LC phase boundaries. These grains are clearly displayed as peaks on the cross-sections profile of the AFM images (Fig. 6C). These peaks are mainly subdivided into three categories depending on their height and frequency. Image analysis confirms distinct depths of insertion of rDGL depending on the local lateral packing of the lipids. The higher peaks ($h_1 > 5$ nm) correspond to lipase molecules adsorbed underneath the monolayer and interacting with polar head only. The intermediate depth of penetration ($h_2 = 3$–4 nm) matches with a 1–2 nm insertion of the lipase into the lipid film, which could correspond to the penetration of rDGL deduced from experiments with DLPC monolayers and surface tensiometry (Fig. 4; Benarouche et al., 2013). It also fits with previous observations made on other lipases by Balashev et al. (2001). The lower peaks ($h_3 < 2$ nm) may stand for lipase molecules totally inserted into the phospholipid monolayer and probably denaturated at the interface. Lipases are known to be fully denatured, and thus inactivated, upon adsorption at high surface tension/low surface pressure (pure oil–water interface for instance) (Aloulou et al., 2006). At higher surface pressures, such as those used here to study rDGL adsorption (15–25 mN m^{-1}), the global enzyme activity is preserved, but it may happen that some lipase molecules are still denatured due to heterogeneity of surface binding sites. Like other amphiphilic proteins that contribute to lower surface tension and preserve lipase activity (e.g., serum albumin; Gargouri, Piéroni, Lowe, Sarda, & Verger, 1986), these lipase molecules may serve as sacrificial agents changing the interface properties and preserving the folding and activity of other lipase molecules binding at the interface. This can be demonstrated by adding increasing amounts of lipase in the course of lipase assays with pure TAG dispersed in water (data not shown).

5. CONCLUSIONS

Surface tension measurement, ellipsometry, AFM, and their combinations are useful techniques for studying lipase adsorption under conditions mimicking the surface of biological lipid droplets, often covered by phospholipid monolayers (e.g., intracellular lipid bodies, food emulsions) or multilayers (e.g., milk fat globules). These biophysical approaches involving Langmuir monolayers have been particularly useful to better understand the mode of action of gastric lipase as shown here by various examples. These techniques and associated models can be applied to other lipolytic enzymes, but also to other proteins binding at the surface of lipid droplets or membranes.

REFERENCES

Aloulou, A., Rodriguez, J. A., Fernandez, S., Van Oosterhout, D., Puccinelli, D., & Carriere, F. (2006). Exploring the specific features of interfacial enzymology based on lipase studies. *Biochimica et Biophysica Acta—Molecular and Cell Biology of Lipids, 1761*, 995–1013.

Aoubala, M., Ivanova, M., Douchet, I., De Caro, A., & Verger, R. (1995). Interfacial binding of human gastric lipase to lipid monolayers, measured with an ELISA. *Biochemistry, 34*, 10786–10793.

Azzam, R. M. A., & Bashara, N. M. (1977). *Ellipsometry and polarized light*. Amsterdam: North-Holland Pub. Co., Elsevier.

Balashev, K., Gudmand, M., Iversen, L., Callisen, T. H., Svendsen, A., & Bjornholm, T. (2003). Humicola lanuginosa lipase hydrolysis of mono-oleoyl-rac-glycerol at the lipid-water interface observed by atomic force microscopy. *Biochimica et Biophysica Acta, 1615*, 93–102.

Balashev, K., Jensen, T. R., Kjaer, K., & Bjornholm, T. (2001). Novel methods for studying lipids and lipases and their mutual interaction at interfaces. Part I. Atomic force microscopy. *Biochimie, 83*, 387–397.

Balashev, K., John DiNardo, N., Callisen, T. H., Svendsen, A., & Bjornholm, T. (2007). Atomic force microscope visualization of lipid bilayer degradation due to action of phospholipase A2 and Humicola lanuginosa lipase. *Biochimica et Biophysica Acta, 1768*, 90–99.

Benarouche, A., Point, V., Carriere, F., & Cavalier, J. F. (2014). An interfacial and comparative in vitro study of gastrointestinal lipases and Yarrowia lipolytica LIP2 lipase, a candidate for enzyme replacement therapy. *Biochimie, 102*, 145–153.

Benarouche, A., Point, V., Parsiegla, G., Carriere, F., & Cavalier, J. F. (2013). New insights into the pH-dependent interfacial adsorption of dog gastric lipase using the monolayer technique. *Colloids and Surfaces. B, Biointerfaces, 111C*, 306–312.

Berge, B., & Renault, A. (1993). Ellipsometry study of 2D crystallization of 1-alcohol monolayers at the water surface. *Europhysics Letters, 21*, 773–777.

Bernbäck, S., Bläckberg, L., & Hernell, O. (1989). Fatty acids generated by gastric lipase promote human milk triacylglycerol digestion by pancreatic colipase-dependent lipase. *Biochimica et Biophysica Acta, 1001*, 286–293.

Bolanos-Garcia, V. M., Renault, A., & Beaufils, S. (2008). Surface rheology and adsorption kinetics reveal the relative amphiphilicity, interfacial activity, and stability of human exchangeable apolipoproteins. *Biophysical Journal, 94*, 1735–1745.

Borgstrom, B. (1980). Importance of phospholipids, pancreatic phospholipase A2, and fatty acid for the digestion of dietary fat: In vitro experiments with the porcine enzymes. *Gastroenterology, 78*, 954–962.

Bourlieu, C., Menard, O., De La Chevasnerie, A., Sams, L., Rousseau, F., Madec, M. N., et al. (2015). The structure of infant formulas impacts their lipolysis, proteolysis and disintegration during in vitro gastric digestion. *Food Chemistry, 182*, 224–235.

Bourlieu, C., Paboeuf, G., Chever, S., Pezennec, S., Cavalier, J. F., Guyomarc'h, F., et al. (2016). Adsorption of gastric lipase onto multicomponent model lipid monolayers with phase separation. *Colloids and Surfaces. B, Biointerfaces, 143*, 97–106.

Brezesinski, G., & Mohwald, H. (2003). Langmuir monolayers to study interactions at model membrane surfaces. *Advances in Colloid and Interface Science, 100–102*, 563–584.

Brockman, H. (1999). Lipid monolayers: Why use half a membrane to characterize protein-membrane interactions? *Current Opinion in Structural Biology, 9*, 438–443.

Carrière, F., Moreau, H., Raphel, V., Laugier, R., Bénicourt, C., Junien, J.-L., et al. (1991). Purification and biochemical characterization of dog gastric lipase. *European Journal of Biochemistry, 202*, 75–83.

Chahinian, H., Snabe, T., Attias, C., Fojan, P., Petersen, S. B., & Carrière, F. (2006). How gastric lipase, an interfacial enzyme with a Ser-His-Asp catalytic triad, acts optimally at acidic pH. *Biochemistry, 45*, 993–1001.

Chu, B. S., Gunning, A. P., Rich, G. T., Ridout, M. J., Faulks, R. M., Wickham, M. S., et al. (2010). Adsorption of bile salts and pancreatic colipase and lipase onto digalactosyldiacylglycerol and dipalmitoylphosphatidylcholine monolayers. *Langmuir, 26*, 9782–9793.

De la Fournière, L., Ivanova, M. G., Blond, J.-P., Carrière, F., & Verger, R. (1994). Surface behaviour of human pancreatic and gastric lipases. *Colloids and Surfaces. B, Biointerfaces, 2*, 585–593.

Desfougeres, Y., Saint-Jalmes, A., Salonen, A., Vie, V., Beaufils, S., Pezennec, S., et al. (2011). Strong improvement of interfacial properties can result from slight structural modifications of proteins: The case of native and dry-heated lysozyme. *Langmuir, 27*, 14947–14957.

Fotiadis, D., Scheuring, S., Muller, S. A., Engel, A., & Muller, D. J. (2002). Imaging and manipulation of biological structures with the AFM. *Micron, 33*, 385–397.

Gargouri, Y., Piéroni, G., Lowe, P. A., Sarda, L., & Verger, R. (1986). Human gastric lipase. The effect of amphiphiles. *European Journal of Biochemistry, 156*, 305–310.

Gargouri, Y., Pieroni, G., Riviere, C., Lowe, P. A., Sauniere, J. F., Sarda, L., et al. (1986). Importance of human gastric lipase for intestinal lipolysis: An in vitro study. *Biochimica et Biophysica Acta, 879*, 419–423.

Gargouri, Y., Piéroni, G., Rivière, C., Saunière, J.-F., Lowe, P. A., Sarda, L., et al. (1986). Kinetic assay of human gastric lipase on short- and long-chain triacylglycerol emulsions. *Gastroenterology, 91*, 919–925.

Jacquemet, A., Vie, V., Lemiegre, L., Barbeau, J., & Benvegnu, T. (2010). Air/water interface study of cyclopentane-containing archaeal bipolar lipid analogues. *Chemistry and Physics of Lipids, 163*, 794–799.

Langmuir, I. (1917). The constitution and fundamental properties of solids and liquids.II. Liquids. *Journal of the American Chemical Society, 39*, 1848–1906.

Panaiotov, I., & Verger, R. (2000). Enzymatic reactions at interfaces. Interfacial and temporal organisation of enzymatic lipolysis. In A. Baszkin & W. Norde (Eds.), *Physical chemistry of biological interfaces* (pp. 359–400). New York: Marcel Dekker.

Pattus, F., Slotboom, A. J., & de Haas, G. H. (1979). Regulation of phospholipase A2 activity by the lipid -water interface: A monolayer approach. *Biochemistry, 13*, 2691–2697.

Piéroni, G., & Verger, R. (1979). Hydrolysis of mixed monomolecular films of triglyceride/lecithin by pancreatic lipase. *Journal of Biological Chemistry, 254*, 10090–10094.

Piéroni, G., & Verger, R. (1983). Hydrolysis of mixed monomolecular films of phosphatidylcholine/triacylglycerol by pancreatic phospholipase A_2. *European Journal of Biochemistry, 132*, 639–644.

Prim, N., Iversen, L., Diaz, P., & Bjornholm, T. (2006). Atomic force microscope studies on the interactions of Candida rugosa lipase and supported lipidic bilayers. *Colloids and Surfaces. B, Biointerfaces, 52*, 138–142.

Radmacher, M., Tillamnn, R. W., Fritz, M., & Gaub, H. E. (1992). From molecules to cells: Imaging soft samples with the atomic force microscope. *Science, 257*, 1900–1905.

Ransac, S., Moreau, H., Rivière, C., & Verger, R. (1991). Monolayer techniques for studying phospholipase kinetics. *Methods in Enzymology, 197*, 49–65.

Roussel, A., Miled, N., Berti-Dupuis, L., Riviere, M., Spinelli, S., Berna, P., et al. (2002). Crystal structure of the open form of dog gastric lipase in complex with a phosphonate inhibitor. *The Journal of Biological Chemistry*, *277*, 2266–2274.

Sams, L., Paume, J., Giallo, J., & Carriere, F. (2016). Relevant pH and lipase for in vitro models of gastric digestion. *Food & Function*, *7*, 30–45.

Staggers, J., Hernell, O., Stafford, R. J., & Carey, M. C. (1990). Physical-chemical behavior of dietary and biliary lipids during intestinal digestion and absorption. 1. Phase behavior and aggregation states of model lipid systems patterned after aqueous duodenal contents of healthy adult human beings. *Biochemistry*, *29*, 2028–2040.

Sugihara, A., Gargouri, Y., Piéroni, G., Rivière, C., Sarda, L., & Verger, R. (1986). Activities and interfacial properties of Rhizopus delemar and porcine pancreatic lipases after treatment with phospholipids. *Biochemistry*, *25*, 3430–3434.

Thiam, A. R., Farese, R. V., Jr., & Walther, T. C. (2013). The biophysics and cell biology of lipid droplets. *Nature Reviews. Molecular Cell Biology*, *14*, 775–786.

Verger, R. (1980). Enzyme kinetics of lipolysis. *Methods in Enzymology*, *64*, 340–392.

Verger, R., & de Haas, G. H. (1973). Enzyme reactions in a membrane model. 1: A new technique to study enzyme reactions in monolayers. *Chemistry and Physics of Lipids*, *10*, 127–136.

Verger, R., Mieras, M. C. E., & de Haas, G. H. (1973). Action of phospholipase A at interfaces. *Journal of Biological Chemistry*, *248*, 4023–4034.

Verger, R., Rietsch, J., van Dam-Mieras, M. C. E., & de Haas, G. H. (1976). Comparative studies of lipase and phospholipase A2 acting on substrate monolayers. *Journal of Biological Chemistry*, *251*, 3128–3133.

Vié, V., Van Mau, N., Lesniewska, E., Goudonnet, J. P., Heitz, F., & Le Grimellec, C. (1998). Distribution of ganglioside GM1 between two-component, two-phase phosphatidylcholine monolayers. *Langmuir*, *14*, 4574–4583.

Wegner, G. J., Wark, A. W., Lee, H. J., Codner, E., Saeki, T., Fang, S., et al. (2004). Real-time surface plasmon resonance imaging measurements for the multiplexed determination of protein adsorption/desorption kinetics and surface enzymatic reactions on peptide microarrays. *Analytical Chemistry*, *76*, 5677–5684.

Wilhelmy, L. (1863). Ueber die abhängigkeit der capillaritäts-constanten des alkohols von substanz und gestalt des benetzten festen körpers. *Annalen der Physik und Chemie*, *119*, 177–217.

CHAPTER TWELVE

Probing Conformational Changes and Interfacial Recognition Site of Lipases With Surfactants and Inhibitors

E. Mateos-Diaz*, S. Amara*, A. Roussel†, S. Longhi†, C. Cambillau†, F. Carrière*[,1]

*CNRS, Aix Marseille Université, Enzymologie Interfaciale et de Physiologie de la Lipolyse, Marseille, France
†Aix-Marseille Université, CNRS, Architecture et Fonction des Macromolécules Biologiques, Marseille, France
[1]Corresponding author: e-mail address: carriere@imm.cnrs.fr

Contents

Abstract

Structural studies on lipases by X-ray crystallography have revealed conformational changes occurring in the presence of surfactants/inhibitors and the pivotal role played by a molecular "lid" of variable size and structure depending on the enzyme. Besides

Methods in Enzymology, Volume 583
ISSN 0076-6879
http://dx.doi.org/10.1016/bs.mie.2016.09.040

controlling the access to the enzyme active site, the lid is involved in lipase activation, formation of the interfacial recognition site (IRS), and substrate docking within the active site. The combined use of surfactants and inhibitors has been critical for a better understanding of lipase structure–function relationships. An overview of crystal structures of lipases in complex with surfactants and inhibitors reveals common structural features and shows how surfactants monomers interact with the lid in its open conformation. The location of surfactants, inhibitors, and hydrophobic residues exposed upon lid opening provides insights into the IRS of lipases. The mechanism by which surfactants promote the lid opening can be further investigated in solution by site-directed spin labeling of lipase coupled to electron paramagnetic resonance spectroscopy. These experimental approaches are illustrated here by results obtained with mammalian digestive lipases, fungal lipases, and cutinases.

1. INTRODUCTION

Lipases (EC 3.1.1.3; triacylglycerol hydrolases) play an important role in the uptake of fatty acids from fats and mobilization of triglycerides stored in various lipid bodies or transported by lipoproteins. They are found to occur in most organisms in the animal, plant, and microbial kingdoms (Aloulou et al., 2006). In order to catalyze the hydrolysis of ester bonds in triglycerides, they have to bind at the oil–water interface of lipid droplets, and their activity depends on the available surface (Benzonana & Desnuelle, 1965) and various interfacial parameters like surface tension (Verger, 1980). Some lipases also interact with smaller substrate aggregates like micelles and present a broader substrate specificity (monoglyceride lipase, phospholipase A1, galactolipase) (Bakala N'Goma, Amara, Dridi, Jannin, & Carriere, 2012; Eydoux et al., 2008). The biochemical characterization of lipases has always been associated with the use of surfactants, like gum Arabic to form and stabilize fine substrate emulsions (Tiss, Carriere, & Verger, 2001) or bile salts to solubilize lipolysis products (Hofmann, 1963). Surfactants also have a significant impact on the partitioning of lipase between the water phase and oil–water interface (Delorme et al., 2011). They can compete with lipases for adsorption (Bezzine et al., 1999) or promote lipase exchange between lipid droplets via lipase interaction with micelles in solution (Haiker, Lengsfeld, Hadvary, & Carrière, 2004). Last but not least, they are required for efficient lipase inhibition in solution, and these findings have led to various speculations on their interactions with lipase and/or inhibitors.

In early studies on pancreatic lipase inhibition by serine hydrolase inhibitors, it was shown that diethyl *p*-nitrophenyl phosphate (E600; Fig. 1) could

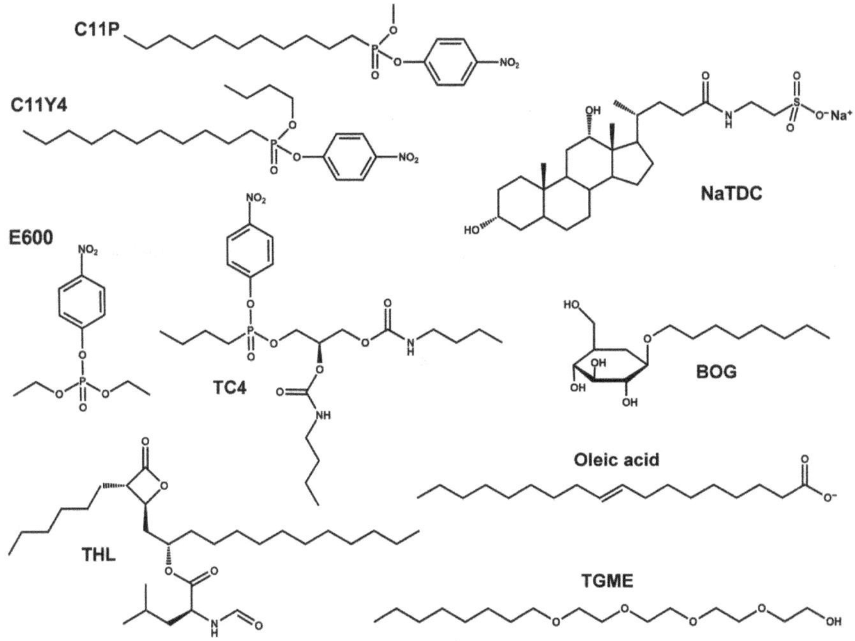

Fig. 1 Inhibitors and surfactants commonly used for biochemical and structural studies of lipases. *BOG*, β-octylglucoside; *C11P*, *O*-methyl *O*-(*p*-nitrophenyl) *n*-undecylphosphonate; *C11Y14*, (R_P)-*O*-butyl *O*-(*p*-nitrophenyl) *n*-undecylphosphonate; *E600*, diethyl *p*-nitrophenylphosphate; *NaTDC*, sodium taurodeoxycholate; *TC4*, (*R*)-1,2-dibutyl-carbamoylglycero-3-*O*-(*p*-nitrophenyl) butyl-phosphonate; *THL*, tetrahydrolipstatine (or Orlistat®); *TGME*, tetraethylene glycol monooctyl ether.

be included into bile salt micelles and that pancreatic lipase inhibition rate was strongly dependent on the bile salt concentration and micellar concentration of the organophosphate, as well as on the presence of colipase (Rouard, Sari, Nurit, Entressangles, & Desnuelle, 1978). It was proposed that the "solubilization" of the poorly water-soluble organophosphate in micelles was a prerequisite for lipase inhibition after enzyme interaction with mixed micelles. 3D structures of lipases were unknown at that time, and there was no indication that bile salts could promote lipase inhibition by other mechanisms. The use of bile salts like sodium taurodeoxycholate (NaTDC; Fig. 1) and other surfactants like β-octylglucoside (BOG; Fig. 1) and tetraethylene glycol monooctyl ether (TGME; Fig. 1) has become a classical approach to promote lipase inhibition by poorly water-soluble inhibitors (Delorme et al., 2011; Hermoso et al., 1996; Moreau, Moulin, Gargouri, Noël, & Verger, 1991) and investigate structural changes by various means, including tryptophan fluorescence spectroscopy (Bourbon-Freie, Dub,

Xiao, & Lowe, 2009; Lüthi-Peng & Winkler, 1992; Yapoudjian, Ivanova, Brzozowski, et al., 2002; Yapoudjian, Ivanova, Douchet, et al., 2002).

When the first lipase 3D structures revealed the presence of a molecular lid blocking the access to the active site (Brady et al., 1990; Winkler, d'Arcy, & Hunziker, 1990), it was rapidly envisioned to cocrystallized lipases with surfactants and covalent inhibitors to unravel conformational changes necessary for lipase activation and inhibition (Brzozowski et al., 1991; van Tilbeurgh et al., 1993). The combined use of surfactants and inhibitors has thus been critical for elucidating the mechanism of lipase activation. As illustrated in this chapter, crystal structures of several lipases in complex with surfactants and inhibitors are today available and have revealed common structural features and shown how surfactants monomers can interact with the lid and stabilize its open conformation. The location of surfactants, inhibitors, and hydrophobic residues exposed upon lid opening provides insights into the interfacial recognition site (IRS) of lipases.

X-ray crystallography has been the method of choice for studying lipase conformational changes because most lipases are too large to be studied by nuclear magnetic resonance. Data on the mechanism of lid opening are however limited and mostly deduced from molecular dynamics simulations based on known X-ray structures of lipases (Barbe et al., 2009; Bordes et al., 2010; Jensen et al., 2002; Rehm, Trodler, & Pleiss, 2010; Santini et al., 2009; Selvan et al., 2010). Another approach for investigating the conformational changes in large proteins like lipases consists in the association of site-directed spin labeling (SDSL) and electron paramagnetic resonance (EPR) spectroscopy. EPR spectroscopy allows to document protein conformational changes at the residue level and provides dynamic information on structural transitions occurring within well-characterized structured proteins for which X-ray crystallography can only provide snapshots of the initial and final conformations as illustrated with lipases. Moreover, SDSL–EPR spectroscopy allows working in solution, at room/physiological temperature and with limited amounts of enzyme/protein (20 μL samples at 40–80 μ*M* with pancreatic lipase). This approach is illustrated here by the study of pancreatic lipase lid opening in the presence of various bile salt concentrations (Belle et al., 2007).

2. EFFECTS OF SURFACTANTS ON LIPASE INHIBITION

Surfactants are required to promote the inhibition in solution of lipases having a lid controlling the access to their active site. This is illustrated in Fig. 2

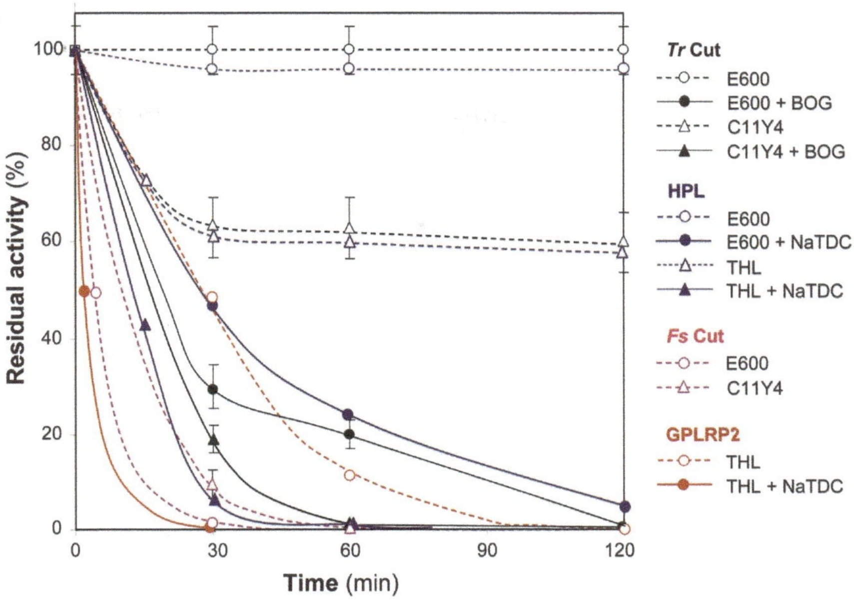

Fig. 2 Time-course inhibition of lipases in the absence and presence of surfactants. Inhibition studies were performed with two lipases having a lid, *Trichoderma reesei* cutinase (*Tr*Cut) and human pancreatic lipase (HPL), and two lipases with no lid, *Fusarium solani* cutinase (*Fs*Cut) and guinea pig pancreatic lipase-related protein 2 (GPLRP2). Residual lipase activity (*Y*-axis; % of maximum activity) was measured with the pHstat technique and tributyrin as substrate, after various times of lipase incubation with E600, C11Y4 phosphonate or THL, in the absence or presence of NaTDC or BOG. *Data for* Tr *cutinase,* Fs *cutinase, HPL, and GPLRP2 are from Roussel, A., Amara, S., Nyyssola, A., Mateos-Diaz, E., Blangy, S., Kontkanen, H., et al. (2014). A Cutinase from Trichoderma reesei with a lid-covered active site and kinetic properties of true lipases.* Journal of Molecular Biology, 426, *3757–3772; Eydoux, C., Spinelli, S., Davis, T. L., Walker, J. R., Seitova, A., Dhe-Paganon, S., et al. (2008). Structure of human pancreatic lipase-related protein 2 with the lid in an open conformation.* Biochemistry, 47, *9553–9564; Cudrey, C., van Tilbeurgh, H., Gargouri, Y. & Verger, R. (1993). Inactivation of pancreatic lipases by amphiphilic reagents 5-(dodecyldithio)-2-nitrobenzoic acid and tetrahydrolipstatin. Dependence upon partitioning between micellar and oil phases.* Biochemistry, 32, *13800–13808*

that shows the time-course inhibition of human pancreatic lipase (HPL) and *Trichoderma reesei* cutinase (*Tr*Cut), a fungal enzyme also active on triglycerides (Roussel et al., 2014). Inhibition assays are typically performed with 1–10 μ*M* enzyme and a 100-fold molar excess of inhibitor. When HPL or *Tr*Cut is preincubated with diethyl *p*-nitrophenylphosphate (E600; Sigma-Aldrich, Saint-Quentin Fallavier, France, cat. number 855790) alone, there is no inhibition of lipase activity, as measured using tributyrin (Sigma-Aldrich, cat. number

91010) as substrate. The addition of surfactants at supramicellar concentration (30 m*M* BOG (CMC = 25 m*M*; Sigma-Aldrich, cat. number O3757) for *Tr*Cut; 4 m*M* NaTDC (CMC = 2.4 m*M*; Sigma-Aldrich, cat. number T0557) for HPL) allows lipase inhibition with $\leq$30 min half-inhibition times. Complete inhibition of HPL by tetrahydrolipstatine (THL or Orlistat; Sigma-Aldrich, cat. number O4139; Fig. 1) and *Tr*Cut by (R_P)-*O*-butyl *O*-(*p*-nitrophenyl) *n*-undecylphosphonate (C11Y14; Fig. 1) also requires the addition of NaTDC and BOG, respectively, although some partial inhibition by these amphiphilic inhibitors is also observed in the absence of surfactants (Fig. 2).

Lipases devoid of a lid, like *Fusarium solani* cutinase (*Fs*Cut) and guinea pig pancreatic lipase-related protein 2 (GPLRP2), do not require the presence of surfactants to be rapidly and fully inhibited by E600, C11Y4, or THL (Fig. 2). Nevertheless, GPLRP2 inhibition by THL is accelerated in the presence of NaTDC at supramicellar concentration (4 m*M*) (Cudrey, van Tilbeurgh, Gargouri, & Verger, 1993). Similarly, the half-inactivation time for *Fs*Cut inhibition by (*R*)-1,2-dibutyl-carbamoylglycero-3-*O*-(*p*-nitrophenyl) butyl-phosphonate (TC4; Fig. 1) is reduced from 25 min to 49 s in the presence of 100 m*M* NaTDC (Longhi et al., 1997).

3. X-RAY CRYSTALLOGRAPHY FOR STUDYING LIPASE STRUCTURES, CONFORMATIONAL CHANGES, AND IRSs

3.1 General Features of Lipase Structures

Significant progress has been made in the structural characterization of lipases over the last 25 years using X-ray crystallography. Those studies revealed that lipases are serine esterases with an α/β hydrolase fold (Ollis et al., 1992) and a catalytic triad similar to that observed in serine proteases (Brady et al., 1990). In addition, many lipases have the access to their active site controlled by a lid formed by surface loops and helices of various size and structures. The crystal structures of these lipases in complex with organophosphates, organophosphonates, and/or surfactants have shown how the lid undergoes conformational changes and renders the active site accessible to solvent upon opening. Crystal structures in closed and open conformations have been obtained for several lipases, including *Rhizomucor miehei* lipase (RML) (Brady et al., 1990; Brzozowski et al., 1991; Derewenda, Brzozowski, Lawson, & Derewenda, 1992), *Thermomyces lanuginosus* lipase (TLL) (Brzozowski, 1993; Brzozowski et al., 2000), HPL

(Egloff et al., 1995; van Tilbeurgh et al., 1993; van Tilbeurgh, Sarda, Verger, & Cambillau, Cambillau, 1992; Winkler et al., 1990), porcine pancreatic lipase (PPL) (Hermoso et al., 1996), dog (DGL) and human (HGL) gastric lipases (Roussel et al., 1999, 2002), and *Tr*Cut (Roussel et al., 2014) (Fig. 3A–C). The structural differences observed between the closed and open lipase forms range from a relatively simple rigid hinge-type motion of a single helix in RML (Brzozowski et al., 1991) and TLL (Brzozowski, 1993; Brzozowski et al., 2000), to a much more complex pattern of multiple loops undergoing profound changes in their secondary structures in pancreatic and gastric lipases (Egloff et al., 1995; Hermoso et al., 1997; Roussel et al., 2014, 2002). The lid opening is not only important for enabling access

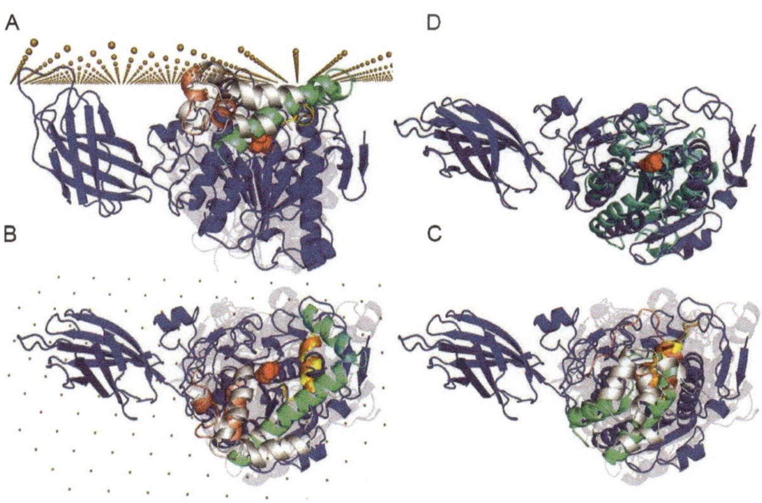

Fig. 3 Superimposition of lipase 3D structures with closed and open lid conformations. Lipases are superimposed based on their common α/β hydrolase fold. All lipase 3D structures are shown as *blue ribbon* Cα tracing with transparency, except for HPL which is in *solid blue*. The active site serine atoms are shown as *red spheres*, and the lid is colored in *pink* for HPL, *gray* for dog gastric lipase (DGL), *yellow* for *Rhizomucor miehei* lipase (RML), *orange* for *Thermomyces lanuginosus* lipase (TLL), and *cyan* for *Trichoderma reesei* cutinase (*Tr*Cut). *Side* (A) and *top* (B–D) *views* are defined according to pancreatic lipase orientation at an interface, materialized by an array of carbon atoms (*brown spheres*; see Fig. 4 legend for details) in (A) and (B). (A) *Side view* of lipases with the open lid. (B) *Top view* of lipases with the open lid, showing the accessibility of active site serine. (C) *Top view* of lipases with the closed lid and no accessibility to active site serine. (D) Superimposition of guinea pig pancreatic lipase-related protein 2 (GPLRP2; *blue ribbon* Cα tracing) and *Fusarium solani* cutinase (*Fs*Cut; *cyan ribbon* Cα tracing) showing the accessibility of active site serine in the absence of a lid.

to the active site but also for adjusting the catalytic machinery with the formation/stabilization of the oxyanion hole. In lipases without a lid, like *Fs*Cut (Martinez et al., 1994) and GPLRP2 (Withers-Martinez, Carrière, Verger, Bourgeois, & Cambillau, 1996), the active site is always accessible to the solvent, and the oxyanion hole is preformed (Fig. 3D).

The opening of the lid domain generates a large hydrophobic surface surrounding the lipase active site, and this surface represents the main IRS of lipases (Brzozowski et al., 1991; Roussel et al., 2002; van Tilbeurgh et al., 1993). In pancreatic lipases, however, the additional C-terminal domain and its distal β5′-loop provide a second site of interaction with lipids (Chahinian et al., 2002).

3.2 Building a Model for Lipase Orientation at a Lipid–Water Interface

Although pancreatic lipase is the lipase of known 3D structure with the most complex architecture (two domains; one protein cofactor), it is particularly useful for attempting at defining a common interfacial orientation of all lipases in the absence of experimental data documenting the enzyme orientation/penetration at the surface of lipid aggregates and without running molecular dynamics simulations. A general model for the orientation of lipases at a lipid–water interface can be deduced from the combined structural homologies of HPL with other lipases (common α/β-hydrolase fold) on one hand, and lipid-binding domains on the other hand.

In the two-domain 3D structure of HPL (Fig. 4A), the catalytic triad (S152–D176–H263) and the lid (peptide stretch C237–C261) are located in the large N-terminal domain (residues 1–335) harboring the typical α/β-hydrolase fold of lipases and dominated by a central parallel β-sheet (van Tilbeurgh et al., 1993; Winkler et al., 1990). The β-sandwich C-terminal domain found in HPL (residues 336–449; on the left in Fig. 4A) and other members of the pancreatic lipase gene family (Carrière et al., 1998) distinguishes these enzymes from the other lipases for which the 3D structures have been determined so far. This C-terminal domain of HPL provides a binding site for colipase, the specific protein cofactor of HPL (van Tilbeurgh et al., 1992), but it is also involved in the interaction of HPL with triglyceride emulsions, micelles, and lipid monolayers (Bezzine, Carrière, De Caro, Verger, & De Caro, 1998; Chahinian et al., 2002; Chahinian, Sias, & Carriere, 2000; Hermoso et al., 1997). It has some homology with C2 lipid-binding domains occurring in a wide range of

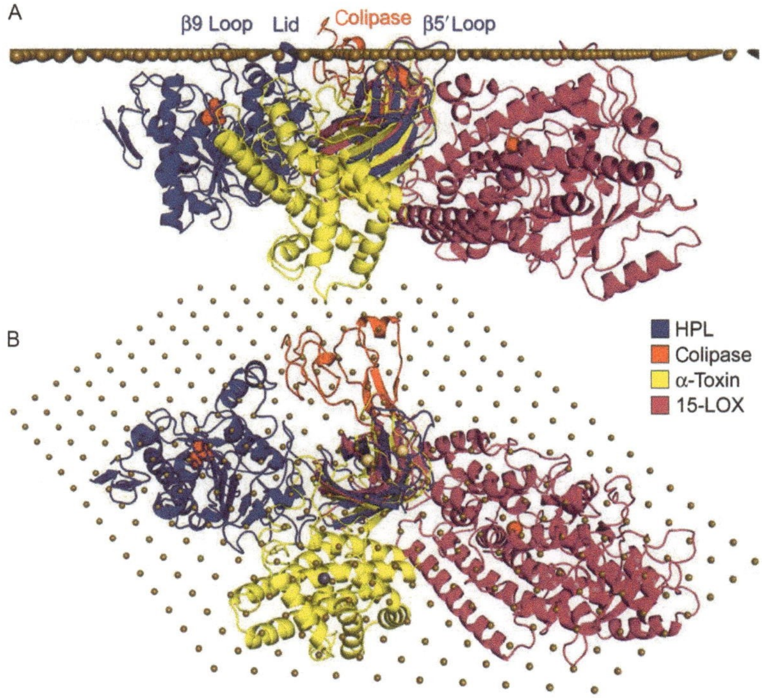

Fig. 4 3D structures of enzymes with PLAT/C2 domains aligned on a putative interface. The 3D structure of HPL (*blue ribbon* Cα tracing, open conformation) was oriented so that β9 loop, lid, and β5′ loop were in contact with the interface plane, materialized by an array of carbon atoms 6 Å apart (*brown spheres*). In this orientation, the hydrophobic moiety of colipase (*red* Cα tracing) also comes in close contact with the plane. Rabbit 15-lipoxygenase (15-LOX; *purple ribbon* Cα tracing) and *Clostridium perfringens* α-toxin/phospholipase C (*yellow ribbon* Cα tracing) were aligned onto the HPL β-sandwich C-terminal domain based on the homology of their N-terminal and C-terminal domains, respectively.

proteins involved in signal transduction (e.g., phosphoinositide-specific phospholipase C, protein kinase C, and cytosolic phospholipase A2 (cPLA2)), membrane trafficking (e.g., synaptotagmin I and rabphilin), and membrane disruption processes (e.g., perforin) (Nalefski & Falke, 1996; Rizo & Südhof, 1998). However, it differs from canonical C2 domains by a different connectivity of its β-strands and the absence of calcium ions (Chahinian et al., 2000). It is more closely related to the N-terminal domain of 15-lipoxygenase (Gillmor, Villasenor, Fletterick, Sigal, & Browner, 1997) and the C-terminal domain of *Clostridium perfringens* α-toxin/phospholipase C (Naylor et al., 1998) (Fig. 4), which are usually referred to as PLAT domains in structural databases (Bateman & Sandford, 1999). The hydrophobic β5′ loop (residues 405–414) of HPL C-terminal domain has the same topology as the CBR3 hydrophobic loop found in C2 domains (Chahinian et al., 2000). Site-directed mutagenesis studies have shown the importance of pancreatic lipase C-terminal domain, and particularly hydrophobic residues from the β5′ loop (V407, I408, L412), for lipase adsorption/penetration into phospholipid monomolecular films spread at the air–water interface and for binding to triglyceride emulsions (Bourbon Freie, Ferrato, Carriere, & Lowe, 2006; Chahinian et al., 2002). A neutron diffraction crystallography study on the PPL–colipase complex also showed that a micelle of the nonionic detergent TGME (Sigma-Aldrich, cat. number T3394; Fig. 1) can interact with colipase and with the distal part of PPL C-terminal domain, which includes several residues (I408–L412) from the β5′ loop (Hermoso et al., 1997), thus confirming the role of the β5′ loop in the interaction of pancreatic lipase with various interfaces.

A molecular model of the HPL/colipase complex at a putative interface can be built from the structural homology with cPLA2 C2 lipid-binding domain, and the orientation of this domain at the surface of liposomes deduced from tryptophan fluorescence spectroscopy (Perisic, Paterson, Mosedale, Lara-Gonzalez, & Williams, 1999) and SDSL–EPR studies (Ball, Nielsen, Gelb, & Robinson, 1999) (Fig. 4). The β5′ loop and hydrophobic loops from the N-terminal domain (lid and β9-loop) are aligned with the interface plane, as well as the N-terminal part and hydrophobic residues of colipase (Fig. 4). When the 3D structures of rabbit 15-lipoxygenase and *C. perfringens* α-toxin are superimposed with HPL based on their homologous C2/PLAT domain, one can see that their respective catalytic domains are found close to the interface plane (Fig. 4A), and the access to their catalytic site is turned toward the plane (Fig. 4B).

The model for HPL orientation at a putative lipid–water interface further allows orientating the other lipases based on their conserved α/β hydrolase fold. The superimposition of HPL, DGL, RML, TLL, and *Tr*Cut 3D structures shows how the lid in the open conformation comes in close contact with the interface plane, whatever the lipase (Fig. 3A and B) and exposes hydrophobic residues surrounding the active site entrance and facing the interface (Fig. 5). This orientation model can be further used to analyze the contribution of hydrophobic residues to the IRS of each lipase based on crystal structures obtained in the presence of surfactants and inhibitors.

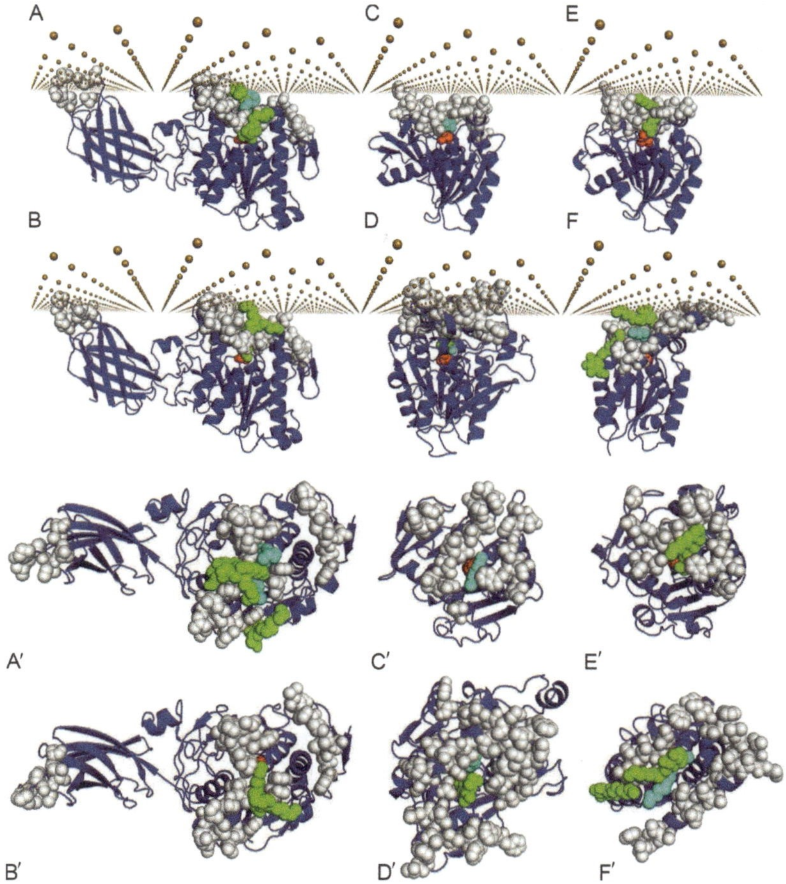

Fig. 5 3D structures of lipases in complex with inhibitors and surfactants at a putative interface. Lipase orientation is the same as in Fig. 3 with *side views* in (A)–(F) and *top views* in (A′)–(F′). All lipases are shown as *blue ribbon* Cα tracing, except the active site serine atoms that are shown as *red spheres*. Hydrophobic residues exposed upon lid opening, surfactants and covalently bound inhibitors are shown as *gray, green, and cyan spheres*, respectively. (A) A′: open HPL cocrystallized with C11 alkyl phosphonate and BOG. (B) B′: open porcine pancreatic lipase (PPL) cocrystallized with TGME. (C) C′: open RML. (D) D′: open DGL cocrystallized with C11Y4 alkyl phosphonate and BOG. (E) E′: open TLL cocrystallized with oleic acid. (F) F′: open *Tr*Cut cocrystallized with C11 alkyl phosphonate and BOG.

3.3 Crystal Structures of Pancreatic Lipases in Complex With Inhibitors and Surfactants

Crystallization of the ternary alkyl phosphonate-HPL–procolipase complex was carried out after the inhibition of HPL (160 μ*M* + 250 μ*M* colipase) by a racemic mixture of *O*-methyl *O*-(*p*-nitrophenyl) *n*-undecylphosphonate (C11P; Fig. 1), dissolved in tetrahydrofuran and added at a 100-fold molar excess with respect to lipase (Egloff et al., 1995). After 30 min of incubation, 4 μL of this solution was mixed with 4 μL of the mother liquor containing 2% PEG 8000, 0.1 *M* MES (2-(*N*-morpholino)ethanesulfonic acid; Sigma-Aldrich, cat. number M8250), and 0.4 *M* NaCl (pH 6.0) and deposited in the form of sitting drops. A BOG solution in water (1 μL at 250 m*M*) was added a few days later to the drops, to induce the formation of tetragonal crystals that grew overnight. Alternatively, BOG was added without any previous equilibration period and much more numerous but smaller crystals developed in each drop. These crystals belong to the space group $P4_22_12$ and are isomorphous with those obtained in the case of the ternary phospholipid-HPL–procolipase complex (van Tilbeurgh et al., 1993). The C11P-HPL–procolipase model built from X-ray diffraction data at 2.46 Å resolution (PDB code: 1LPB) shows the lid (peptide stretch C237–C261) in the open conformation. The C11P inhibitor (both enantiomers) is clearly observed in the active site, covalently bound to the active site serine S152. The alkyl chain establishes contacts either with the hydrophobic groove or with hydrophobic patches on the open lid, depending on the enantiomer. Several BOG molecules are also observed in the crystal structure, with some of them being located at the entrance of the active site, bound to the hydrophobic moiety of the lid (Fig. 5A and A′). The IRS contains hydrophobic residues from HPL N-terminal region (I20, P24, L25, I27, L28, W30), β9 loop (F77, I78, A207, P208, I209, V210, P211, L213, F215), lid (I241, L242, I245, V246, I251, W252, F258, A259, A260), and β5′ loop (V407, I408, P410, L412). These hydrophobic amino acids represent a total accessible surface of 2392 Å^2, with 723 Å^2 being gained upon lid opening. The hydrophobic residues of HPL lid represent only 20% of this surface (Table 1).

A structure of PPL in complex with colipase and with the lid in the open conformation was obtained after crystallization with TGME (Hermoso et al., 1996). The PPL–colipase complex was crystallized by the sitting-drop vapor diffusion method at 20°C, and suitable crystals were obtained by mixing 2 μL of PPL–colipase solution (10 mg/mL each; corresponding to a 1:5 molar ratio of lipase to colipase) with 2 μL of a reservoir solution containing 50 m*M* MES, pH 6.0, 0.5 *M* ammonium sulfate, and 15 m*M* TGME (CMC = 8 m*M* with 0.1% NaCl). Under these conditions, TGME is essential for crystallization. The crystals obtained belong to the cubic space group F23. The PPL–colipase model built from X-ray diffraction data at 2.8 Å

Table 1 Extent of Solvent Exposure of Hydrophobic Amino Acid Residues Involved in Lipase IRS Upon Lid Opening

	Total Accessible Surface of IRS Hydrophobic Residues ($Å^2$)			Lid Residues ($Å^2$)			Other IRS Residues ($Å^2$)		
Lipase	**Open Lid Conformation**	**Closed Lid Conformation**	**Difference**	**Open Lid Conformation**	**Closed Lid Conformation**	**Difference**	**Open Lid Conformation**	**Closed Lid Conformation**	**Difference**
HPL	2392	1669	+723	489 (20%)	341 (20%)	+148	1903 (80%)	1328 (80%)	+576
DGL	2383	1959	+424	1072 (45%)	618 (32%)	+454	1311 (55%)	1341 (68%)	−30
RML	2026	1407	+619	716 (35%)	326 (23%)	+391	1310 (65%)	1082 (77%)	+228
TLL	1250	638	+612	593 (47%)	145(23%)	+447	657 (53%)	493 (77%)	+164
*Tr*Cut	2209	1167	+1042	1706 (77%)	914 (78%)	+792	503 (23%)	253 (22%)	+250

Surface accessibility ($Å^2$) of lipase amino acid residues was estimated with the UCSF Chimera package (Resource for Biocomputing, Visualization, and Informatics at the University of California, San Francisco; http://www.rbvi.ucsf.edu/chimera) using the following lipase 3D structures deposited in RSCB Protein Data Bank: 1LPB (open) and 1N8S (closed) for HPL; 1K8Q (open) and 3D analogy model built from 1HGL (closed human gastric lipase) for DGL; 4TGL (open) and 3TGL (closed) for RML; 1GT6 (open) and 1DTE (closed) for TLL; 4PSE (open) and 4PSC (closed) for *Tr*Cut. Values in brackets indicate the percentage of total accessible surface of IRS hydrophobic residues in either open or closed lipase 3D structures.

resolution (PDB code: 1ETH) shows the lid in the open conformation, with a high-electron density peak in the hydrophobic canyon of the open active site. Two TGME conformers were modeled from the electron density and assigned occupancies of 0.67 and 0.33, respectively, based on electron density peak heights and temperature factor values. In both conformations of the TGME molecule, the aliphatic chain (from C-1 to C-8) is exposed to the solvent, whereas the hydrophilic moiety (from O-8 to O-16) is completely buried in the canyon of the lipase active site pocket (Fig. 5B and B′). The hydrophobic amino acids involved in the IRS are similar to those identified in HPL.

3.4 Crystal Structure of Gastric Lipase in Complex With C11Y4 Alkyl Phosphonate Inhibitor and BOG

DGL inhibition was carried out prior to crystallization in the presence of BOG (30 m*M* final concentration) and a 50-fold molar excess of C11Y4 phosphonate with respect to DGL. The C11Y4 phosphonate inhibitor was dissolved in isopropyl alcohol (4% w/w final concentration) and mixed with BOG before DGL was also added (50 μL final volume). Crystallization experiments were performed using the hanging-drop vapor diffusion method. Crystals of the DGL–C11Y4 phosphonate inhibitor complex were obtained by mixing 1 μL of a well solution (30% polyethylene glycol monomethyl ether 2000 (Sigma-Aldrich, cat. number 202509), 0.1 *M* sodium acetate, 0.2 *M* ammonium sulfate, pH 4.6) with 1 μL of protein solution at 5 mg/mL. Crystals belong to the space group $C222_1$. The DGL–C11Y4 model built from X-ray diffraction data at 2.7 Å resolution (PDB code: 1K8Q) shows the lid in the open conformation, while a molecule of C11Y4 phosphonate inhibitor and one of BOG are located in the active site (Miled et al., 2003; Roussel et al., 2002). The lid of DGL (peptide stretch Y213–N250) in its open conformation is formed by two helices H1 (F217–S228) and H2 (T231–C244), two elongated structures, and one turn. When compared to the structure of human gastric lipase (PDB code: 1HGL) with the lid in the closed conformation, the opening of the lid in gastric lipase appears more complex than the simple rotation of a rigid body observed in fungal lipases (Brzozowski et al., 1991, 2000; Derewenda et al., 1992, 1994). Lid opening in gastric lipase involves a conformational reorganization of helices and a large movement. As a result, the polar face of helices H1 and H2 becomes buried, whereas the hydrophobic part becomes exposed. The hydrophobic plateau generated upon lid opening is found to closely fit the putative interfacial plane defined by HPL orientation. In particular, the open lid of DGL is found in the same area as the open lid of HPL, and the axis of the amphiphilic helices H1 and H2 is found to run parallel to this plane (Fig. 3B).

Based on the putative interfacial orientation of DGL (Fig. 5D and D′), three peptide stretches involved in the IRS of rDGL can be identified: L1–F2 at the DGL N-terminus, Y187–F246 including most of the lid residues (213–250), and L325–L326 at the DGL C-terminus. Most hydrophobic residues surround the active site cavity, with several side chains facing the putative interface plane: L1, F2, Y187, L195, M196, L197, V198, P199, F201, L202, L205, I206, F207, I211, Y213, F217, F218, F221, L222, A223, V226, V232, L234, L235, A239, L240, I242, I243, F246, V279, L325, L326. These hydrophobic residues represent a total accessible surface of 2383 $Å^2$, with 424 $Å^2$ being gained upon lid opening. The hydrophobic residues of DGL lid represent 45% of this surface (Table 1).

3.5 Crystal Structure of RML Complexed With E600

RML has been crystallized under various conditions that resulted in the opening of the lid (S82–P96). One structure (PDB code: 5TGL) was obtained with *n*-hexylchlorophosphonate ethyl ester dissolved in DMSO (5% v/v; dimethylsulfoxide; Sigma-Aldrich, cat. number 276855) (Brzozowski et al., 1991). We describe here the similar structure (PDB code: 4TGL) obtained with the inhibitor E600 (Derewenda et al., 1992). A saturated solution of E600 was used for the solubilization of RML at a concentration of 20 mg/mL. After an overnight incubation at 37°C, precipitated protein was removed, and the remaining protein solution was used for crystallogenesis in the presence of PEG 600 (8–12% w/v). Crystals belong to the $C222_1$ space group like those obtained with the RML–*n*-hexylphosphonate complex (Brzozowski et al., 1991). The RML model built from X-ray diffraction data at 2.6 Å resolution (PDB code: 4TGL) shows the lid in the open conformation and one E600 molecule covalently bound to the active site serine (Fig. 5C and C′). From the comparison with RML structure having the lid in the closed conformation (PDB code: 3TGL), the movement of the lid (S82–P96) appears a hinge-type rigid body motion that results in profound changes in the exposed surface of RML. Several hydrophobic amino acids become significantly more exposed upon lid opening. The resulting IRS involves hydrophobic residues from the lid (I85, W88, I89, L92, F94, V95, P96) and other locations (V97, I204, V205, L208, P210, A211, A212, F213, I248, V249, P250, F251, V254, L255,

L258, L267). These hydrophobic residues represent a total accessible surface of 2026 Å^2, with 619 Å^2 being gained upon lid opening. The hydrophobic residues of RML lid represent 35% of this surface (Table 1).

3.6 Crystal Structure of TLL in Complex With Oleic Acid

TLL has been crystallized under various conditions that led to the opening of the lid (S83–D96). One structure (PDB code: 1DT5) was obtained with a supramicellar concentration (0.5% w/v) of nonionic surfactant pentaethylene glycol monooctyl ether (C8E5), but surfactant molecules are observed neither in the active site nor in any other part of the enzyme (Brzozowski et al., 2000). Another one (PDB code: 1EIN) was obtained in the presence of mixed micelles of di-dodecanoyl phosphatidylcholine and NaTDC, and a phospholipid molecule was found to occupy the active site (Brzozowski et al., 2000). We describe here the structure (PDB code: 1GT6) obtained from the mixture of an inactive variant of TLL (S146A) and oleic acid (OA). Crystallization was performed using the hanging drop technique at 291 K and at pH 8.0. OA was dissolved in isopropanol and mixed in this form with TLL S146A solution (20 mg/mL) at a 5:1 molar ratio. After an hour incubation, the resulting precipitate was removed by centrifugation, and the remaining protein was used in the crystallization experiments. NaTDC was added to the crystallization trials separately at a concentration of 10 m*M*. Good X-ray quality crystals of TLL S146A were obtained in the presence of OA in 0.1 *M* Tris/HCl pH 8.0 buffer, 10 m*M* NaTDC, 25% w/v PEG 5K MME, and 25 m*M* $MgCl_2$. Crystals of the TLL S146A variant belong to the $P2_12_12$ space group and have two molecules (A and B) in the asymmetric unit. Crystals belong to the space group $C222_1$. The X-ray diffraction data at 2.2 Å resolution (PDB code: 1GT6) allowed modeling the two TLL molecules with the open lid (S83–D96) and one OA molecule bound within the active site of TLL (Fig. 5E and E′). The two molecules of the asymmetric unit differ in their binding modes for OA. In molecule A, OA binds to TLL active site in a similar manner to *sn*-1 alkyl chain of di-dodecyl phosphatidylcholine (Brzozowski et al., 2000) and dodecyl phosphonate inhibitor (Lawson, Brzozowski, Rety, Verma, & Dodson, 1994) in complex with TLL. In molecule B, OA is trapped in the active site in an unexpected manner (referred to as anti-*sn*-1), with a rotation of ≈180 degree with respect to the main *sn*-1 alkyl-chain binding site. This alkyl chain bends and is wrapped around the W89 residue of the lid, and the C18 carbon is wedged between I255 and W89. The conformational change of the lid is similar to that observed in RML. The resulting IRS involves hydrophobic residues from the lid (I86, W89, I90, L93, F95) and other locations (I202, V203, L206, F211, I252, P253, I255, P256). These hydrophobic amino acids represent a total accessible surface of 1250 Å^2, with 612 Å^2 being gained upon lid opening. The hydrophobic residues of TLL lid represent 47% of this surface (Table 1).

3.7 Crystal Structure of *T. reesei* Cutinase in Complex With C11Y4 and BOG

T. reesei cutinase (*Tr*Cut) was covalently inhibited by C11Y4 prior to crystallization experiments (Roussel et al., 2014). *Tr*Cut (10 mg) was incubated with C11Y4 (70 μL of a 100-m*M* stock solution in isopropanol) at an inhibitor to enzyme molar excess of 20 in a total volume of 1 mL, in the presence of 30 m*M* BOG. *Tr*Cut crystallization trials were then performed using the sitting-drop vapor diffusion method in 96-well crystallization plates at 20°C using a nanodrop-dispensing robot. The *Tr*Cut in complex with C11Y4 was crystallized by mixing 300 nL of enzyme at 10 mg/mL (and 30 m*M* BOG) with 100 nL of PEG3350 (17%), Bis-Tris (0.1 *M*), pH 5.7. Crystals belong to the space group $P2_1$. The *Tr*Cut model built from X-ray diffraction data at 1.71 Å resolution (PDB code: 4PSE) shows the lid (W31–G70) in the open conformation, a C11Y4 phosphonate inhibitor bound to the catalytic serine S164 and BOG molecules (Fig. 5F and F′). The lid, made of two N-terminal helices, has a larger contribution to the IRS than in other lipases, with 19 hydrophobic residues involved (W31, P32, I34, F37, L38, L41, A42, V44, M45, P46, I47, A53, A54, L57, I58, A64, A65, L68, F69) vs 6 hydrophobic residues at other locations (I224, I225, L226, L227, P228, L230). All these hydrophobic amino acids represent a total accessible surface of 2209 Å^2, with 1042 Å^2 being gained upon lid opening. The hydrophobic residues of *Tr*Cut lid represent 77% of this surface (Table 1).

4. PROBING LIPASE LID OPENING BY SDSL AND EPR SPECTROSCOPY

4.1 Basic Principles of SDSL–EPR

EPR spectroscopy is based on the resonant absorption of a microwave radiation by paramagnetic atoms or molecules when placed in a strong magnetic field. It is well adapted for the study of molecules with unpaired electrons like free radicals, transition metal ions, or paramagnetic groups introduced into proteins via SDSL. The basic strategy of SDSL involves the covalent modification of a cysteine residue naturally present in the protein of interest or introduced by site-directed mutagenesis by a selective reagent bearing a nitroxide radical. MTSL ((1-oxyl-2,2,5,5-tetramethyl-Δ3-pyrroline-3-methyl) methanethiosulfonate; Toronto Research Chemicals, cat. number O875000) is one of the most commonly used reagent for spin labeling of proteins (Fig. 6A) (Longhi, Belle, Fournel, Guigliarelli, & Carriere, 2011). Several EPR approaches can be used to study spin-labeled proteins: continuous wave EPR spectroscopy, which consists in varying the magnetic field and holding the microwave frequency constant, is well adapted to obtain information on the side-chain mobility of the nitroxide probe

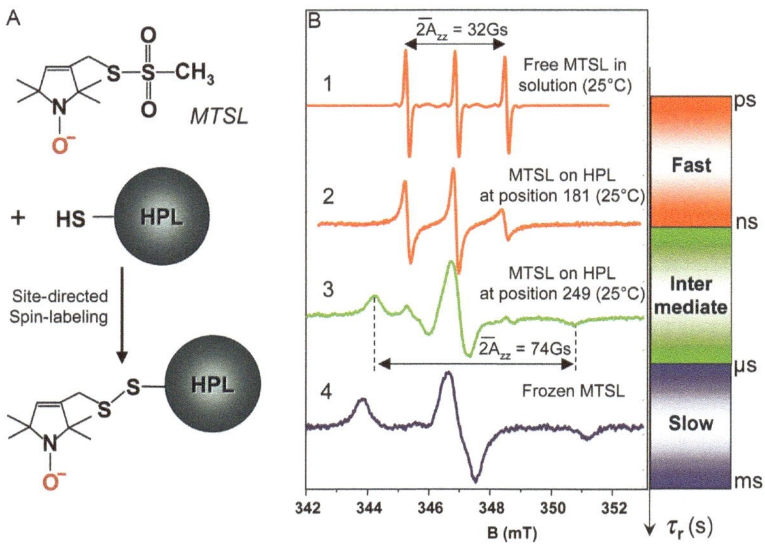

Fig. 6 Site-directed spin labeling and EPR spectra of human pancreatic lipase. (A) Chemical structure of (1-oxyl-2,2,5,5-tetramethyl-Δ3-pyrroline-3-methyl) methanethiosulfonate (MTSL) and covalent modification of the SH group of a cysteine in HPL. (B) EPR spectra of free MTSL in solution (spectrum 1), spin-labeled D249C/C181Y HPL variant (spectrum 2), HPL labeled on the natural cysteine C181 (spectrum 3), and frozen solution of MTSL (spectrum 4). These EPR spectra of the nitroxide radical correspond to different motional regimes characterized by various rotational correlation times τ_r: 1×10^{-12} s (spectrum 1, high motion); 1×10^{-9} s (spectrum 2); 1×10^{-8} s (spectrum 3); and 1×10^{-5} s (spectrum 4, slow motion). The outer line splitting $2A_{zz}$, A_{zz} being the z-component of the hyperfine tensor, is used to characterize the spectra. EPR spectra 1–3 were recorded at room temperature and 9.9 GHz with a microwave power of 10 mW and a magnetic field modulation frequency and amplitude of 100 kHz and 0.1 mT, respectively. Spectrum 4 of the frozen solution of MTSL was recorded at 9.4 GHz and 200 K with a microwave power of 4 mW and a magnetic field modulation frequency and amplitude of 100 kHz and 0.3 mT, respectively.

and its solvent accessibility (Fig. 6B). Pulsed-EPR consists in varying the microwave frequency and holding the magnetic field constant, as in the case of NMR spectroscopy. Combined with double electron-electron resonance (DEER) techniques, pulsed-EPR allows the determination of interspin distances in the range of 1.8–8.0 nm, thus covering a wide range of interest for the study of large conformational transitions (Pannier, Veit, Godt, Jeschke, & Spiess, 2000). The method is based on the introduction of two spin probes, and the distance between them is determined through spin–spin dipolar interaction analyses (Jeschke, 2002). The theoretical principles of these well-established approach have been covered in a recent volumes of Methods in Enzymology (Altenbach, Lopez, Hideg, & Hubbell, 2015; Stein, Beth, & Hustedt, 2015; Yang, Bridges, Lerch, Altenbach, & Hubbell, 2015).

4.2 Investigation of HPL Conformational Changes by SDSL–EPR

The closed and open conformations of HPL observed in crystal structures are thought to correspond to the initial and final stages of the lid opening. However, the opening of the lid can be anticipated to be a complex process probably involving several steps and implying different intermediate conformations, as in the case of the lipase from *T. lanuginosus* (Brzozowski et al., 2000). How and when the lid opens has long been a matter of discussion, although a few studies have indicated that the lid may be open in solution in the presence of bile salts and colipase (Hermoso et al., 1997; Lüthi-Peng & Winkler, 1992; Pignol et al., 2000). The presence of a lipid/water interface would not be strictly required to induce and stabilize the open conformation of the lid, with this latter being stabilized in solution by interactions with both bile salt micelles and colipase (Hermoso et al., 1997). Inhibition studies have shown, however; that the sole presence of bile salts suffices to induce a fast inactivation of pancreatic lipase, probably via the opening of the lid that normally prevents the access of inhibitors to the active site in solution (Moreau et al., 1991).

EPR spectroscopy was used to monitor HPL conformational changes after MTSL was covalently bound to the lid. For this purpose, a residue of the lid (D249) whose side chain is fully exposed to solvent in both the open and closed conformations (Fig. 7B and C) was substituted by a cysteine using site-directed mutagenesis (Belle et al., 2007). In order to ensure specific spin labeling at a single position, an accessible free cysteine (C181) naturally present in HPL was substituted by a tyrosine, as observed at the equivalent position in other

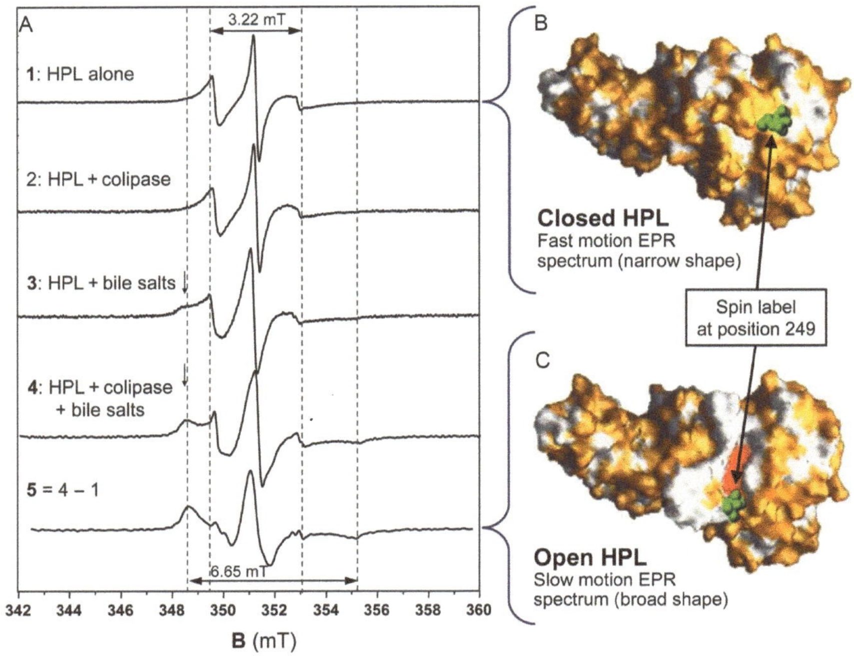

Fig. 7 Probing HPL lid opening by site-directed spin labeling and EPR spectroscopy. EPR spectra of spin-labeled D249C–C181Y HPL variant (80 μ*M*) were recorded at room temperature on an ESP 300E Bruker spectrometer equipped with an ELEXSYS super-high sensitivity resonator operating at 9.9 GHz. The microwave power was 10 mW, and the magnetic field modulation frequency and amplitude were 100 kHz and 0.1 mT, respectively. (A) The EPR spectra of the spin-labeled HPL variant were recorded with (1) the lipase alone, (2) HPL+colipase, (3) HPL+bile salts (4 m*M* NaTDC), and (4) HPL +colipase+bile salts. The difference spectrum (5) was obtained by subtracting the spectrum of HPL alone from the composite spectrum obtained in the presence of colipase and 4 m*M* NaTDC (4). The *horizontal arrows* indicate the outer line splitting (mT) of the two spectral components. (B) Molecular surface representation of HPL with the lid in closed conformation (PDB code 1N8S; van Tilbeurgh et al., 1992). Hydrophobic and polar surfaces are shown in *white* and *orange*, respectively. Atoms of residue 249 targeted for cysteine substitution and spin labeling are shown as *green* spheres. This structure corresponds to the fast-motion component of the EPR spectrum. (C) Molecular surface representation of HPL with the lid in open conformation (PDB code 1LPA; van Tilbeurgh et al., 1993). This structure corresponds to the slow-motion component of the EPR spectrum. *Panel A adapted from Belle, V., Fournel, A., Woudstra, M., Ranaldi, S., Prieri, F., Thome, V., et al. (2007). Probing the opening of the pancreatic lipase lid using site-directed spin labeling and EPR spectroscopy.* Biochemistry, *46, 2205–2214.*

pancreatic lipases (Carrière, Thirstrup, Hjorth, & Boel, 1994; Hjorth et al., 1993). The double HPL C181Y–D249C variant was produced in the yeast *Pichia pastoris* and purified to homogeneity in a single cation exchange chromatography step. A DTT reduction step prior to spin labeling by MTSL turned out to be crucial to ensure reduction of the oxidized cysteine(s) in HPL and to lead to high (approx. 80%) labeling yields (Belle et al., 2007). It was checked that the HPL variant, irrespective of the presence or absence of the spin label, had the same kinetic properties as native HPL and was inhibited by E600 only in the presence of micellar concentrations of bile salts, as native HPL (Belle et al., 2007).

The EPR spectrum of the spin-labeled HPL C181Y–D249C variant (spectrum 1 in Fig. 7A) showed a relatively narrow single-component shape, with an outer line splitting of 3.22 ± 0.02 mT, indicating a fairly fast motion of the radical grafted to a solvent-exposed amino acid residue (Mchaourab, Lietzow, Hideg, & Hubbell, 1996). While addition of a twofold molar excess of colipase did not induce any change in the EPR spectrum (spectrum 2 in Fig. 7A), the presence of 4 m*M* bile salts (NaTDC) gave rise to a composite spectrum (spectrum 3 in Fig. 7A) consisting of two components showing different mobilities of the nitroxide radical: one with a relatively narrow spectral shape with on outer line splitting of 3.22 ± 0.02 mT (referred to as the fast-motion component), and one with a very broad spectral shape with outer line splitting of 6.65 ± 0.05 mT (see vertical dotted lines in Fig. 7A) reflecting a highly restricted mobility of the radical (referred to as the slow-motion component). When both colipase and bile salts were added, the proportion of the slow-motion component increased strongly (spectrum 4 in Fig. 7A). In order to isolate the slow-motion component, the fast-motion component (spectrum 1) was subtracted from the composite spectrum 4. The resulting spectrum (spectrum 5 in Fig. 7A) is representative of a radical with reduced mobility, probably located at a site poorly accessible to the solvent (Mchaourab et al., 1996). The proportion of the slow-motion component in composite spectra was quantitatively estimated from the integrated intensities of the absorption line.

The assignment of the EPR spectrum components to the different HPL lid conformations is based on experiments with the lipase inhibitor E600 (Belle et al., 2007). When the E600 inhibitor and spin-labeled HPL C181Y–D249C variant were mixed in the absence of bile salts and colipase, the rate of lipase inhibition was extremely slow (no access of the inhibitor to the active site), and the EPR spectrum remained mainly composed of a fast-motion component. When the inhibition of spin-labeled HPL C181Y–D249C variant by E600 was performed in the presence of micellar concentrations of NaTDC (2–4 m*M*) and colipase, the inhibition rate increased significantly, and the EPR spectrum was mainly (i.e., 80%) composed of a slow-motion component. Subsequent dilution of the samples for reducing bile salt concentration below the critical micellar concentration (CMC) did not give rise to spectral changes, whereas similar dilution experiments performed in the absence of E600 allowed to restore an EPR spectrum mainly composed of a fast-motion component. Since the closed conformation of HPL lid is not compatible with the presence of E600 covalently linked to the active serine of HPL, these results allowed the assignment of the fast- and slow-motion components to the closed and open form of HPL, respectively (Fig. 7B and C). These conclusions are supported by the analysis of the 3D structures of HPL showing that the side chain of residue 249 onto which the spin label is grafted is more constrained in the open form than in the closed form of HPL (Ranaldi et al., 2009). The assignment of EPR spectrum components to HPL lid conformations was further supported by DEER experiments performed with double spin-labeled HPL and measurement of the amplitude of HPL lid opening (Ranaldi et al., 2010). One spin label was introduced at position 249 of the lid after D249C substitution, while the second one was grafted on C181, the accessible free cysteine naturally present in HPL and corresponding to a rigid position. Using pulsed EPR and the bilabeled HPL D249C variant frozen at 70 ± 5 K, DEER experiments in the absence of bile salt and colipase allowed measuring a main interspin label distance of 19 ± 2 Å compatible with the closed conformation of HPL (Fig. 8). When the lid opening was induced by the addition of bile salts (4 m*M* NaTDC) and colipase, a larger distance of 42 ± 2 Å between the two spin labels was estimated from DEER experiments (Fig. 8). These distances measured between spin labels grafted at positions 181 and 249 are in good agreement with those estimated from the known X-ray structures of HPL with the lid in the closed and open conformations (Fig. 8C).

4.3 HPL Lid Opening by Bile Salts in Solution

The effects of the bile salt concentration on EPR spectral shapes were studied by recording the EPR spectra of the spin-labeled HPL C181Y–D249C variant in solution, in the presence of colipase (Fig. 9A) (Belle et al., 2007). The shape of the EPR spectrum remained unchanged up to 0.5 m*M* NaTDC. Above 1 m*M* NaTDC, the proportion of the slow-motion component quickly increased with increasing bile salts concentration and reached 80% at 8 m*M* NaTDC (Fig. 9B). In the absence of colipase, the proportion of the slow-motion component spectrum was only 45%. Nevertheless, it indicates that bile salts alone can induce HPL lid opening, which is further enhanced by colipase. The major finding is however that the proportion of the slow-motion component of EPR spectrum, corresponding to the open conformation of the lid, can reach 60% below the CMC of bile salts

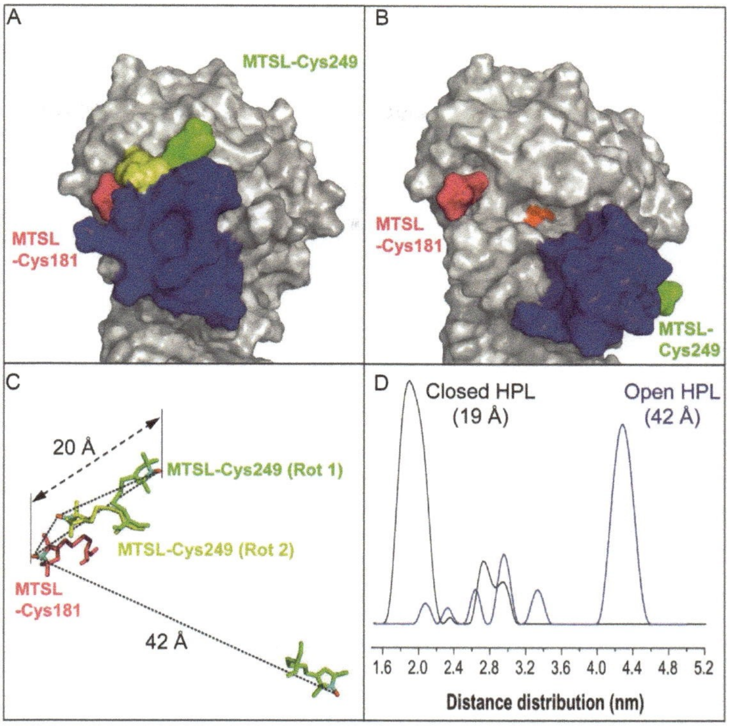

Fig. 8 Demonstration of HPL lid opening by bile salts using double electron-electron resonance (DEER) experiments. (A) Structural modeling of MTSL spin labels grafted to cysteine residues in HPL (C181 + C249). Surface representation of HPL with the lid (*blue* surface) in the closed conformation. (B) Surface representation of HPL with the lid (*blue* surface) in the open conformation and the active site serine (*red color*) accessible to solvent. The spin label at position 181 (MTSL-Cys181) is shown in *magenta* in both closed and open HPL. (C) The spin label at position 249 is shown in three different situations: two rotamers representative of the conformational subensembles observed by molecular dynamics when the lid is closed (MTSL-Cys249-Rot1, *dark green* and MTSL-Cys249-Rot2, *light green*), and the single conformation (*green*) modeled in the open structure of HPL. Images were generated using the PyMOL software program (http://www.pymol.org/). The distances between the oxygen atoms of spin labels as measured in silico are shown; (D) Interspin label distance distributions of bilabeled HPL variant (C181 + C249) either alone (*black line*) or in the presence of bile salts and colipase (*blue line*) as obtained by DEER experiments and processed by Tikhonov regularization methods using the DeerAnalysis software (Jeschke et al., 2006). When bilabeled HPL was alone in solution, a single-distance distribution of 19 ± 2 Å was found. In the presence of 4 m*M* NaTDC and colipase, the interspin distance distribution is shifted to 42 ± 2 Å indicating that residue 249 is pushed away from residue 181. The minor peaks between 25 and 35 Å are either artefacts generated by the mathematical processing of data or may correspond to minor intermediate conformations of the HPL lid. *Adapted from Ranaldi, S., Belle, V., Woudstra, M., Bourgeas, R., Guigliarelli, B., Roche, P., et al. (2010). Amplitude of pancreatic lipase lid opening in solution and identification of spin label conformational subensembles by combining continuous wave and pulsed EPR spectroscopy and molecular dynamics.* Biochemistry 49, *2140–2149.*

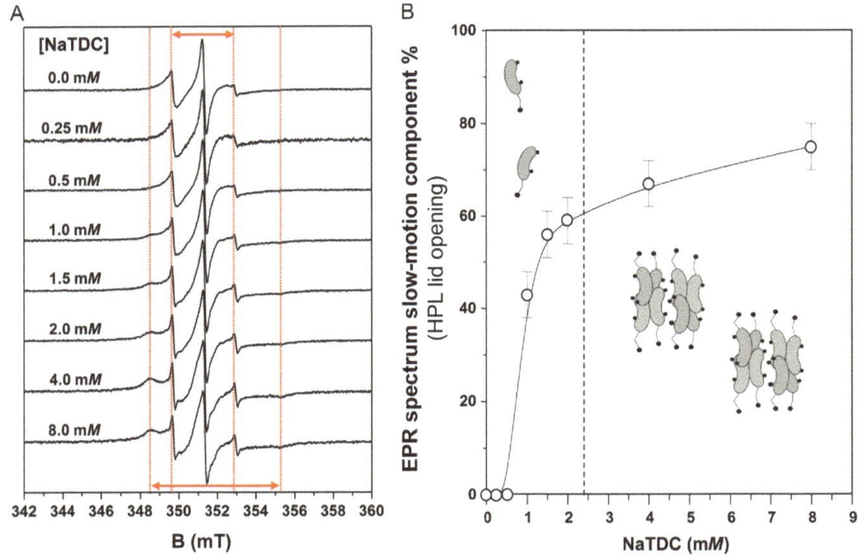

Fig. 9 Effects of bile salt concentration on the EPR spectrum of spin-labeled HPL. (A) Normalized room temperature EPR spectra of the spin-labeled D249C–C181Y HPL variant (80 μ*M*) as a function of NaTDC concentration, in the presence of colipase. EPR spectra were recorded at 9.9 GHz with a microwave power of 10 mW, with a magnetic field modulation frequency of 100 kHz and an amplitude of 0.1 mT. The *dotted vertical lines* show the outer line splitting between fast (closed HPL) and slow motion (open HPL) EPR spectrum components. (B) Proportion of the EPR spectrum slow-motion component as a function of NaTDC concentrations. The *dashed vertical line* represents the critical micellar concentration of NaTDC (2.4 m*M* in 0.15 *M* NaCl). *Adapted from Belle, V., Fournel, A., Woudstra, M., Ranaldi, S., Prieri, F., Thome, V., et al. (2007). Probing the opening of the pancreatic lipase lid using site-directed spin labeling and EPR spectroscopy.* Biochemistry, 46, *2205–2214.*

(2.4 m*M* for NaTDC; see vertical dotted line in Fig. 9B), which indicates that NaTDC monomers, and not micelles, are sufficient to induce HPL lid opening. These data are in good agreement with X-ray crystallography data showing surfactant monomers (BOG) interacting with the hydrophobic side of HPL lid (Fig. 5A and A′). When bile salt concentration is decreased, the fast-motion component of EPR spectrum is restored, indicating that HPL lid opening is a reversible process (Belle et al., 2007).

5. CONCLUSIONS

The use of surfactants allows stabilizing the open conformation of the lid in lipases and further structural studies on lipase activation and IRS. Most

X-ray structures of open lipases have thus been obtained from crystals formed in the presence of micellar concentrations of surfactants. SDSL–EPR spectroscopy reveals, however, that the opening of HPL lid is already largely induced by surfactants (bile salts) below the CMC, indicating that surfactant monomers are sufficient to stabilize the open conformation of the lid. These findings prove an excellent explanation for the fact that surfactant monomers are often observed in crystal structures of open conformations of lipases, in interaction with the lid. It is likely that the surfactant-dependent inhibition of lipases in solution mainly relies on the induction and stabilization of lid opening by surfactant monomers. Nevertheless, the use of surfactants at supramicellar concentrations results in a faster rate of lipase inhibition, with this being also true for lipases devoid of lid like *Fs*Cut and GPLRP2. In this latter case, the increase in the rate of lipase inhibition probably arises from the ability of micelles to allow a larger solubilization of poorly water-soluble inhibitors.

ACKNOWLEDGMENT

E.M.-D. is grateful to CONACYT for the financial support received during his PhD thesis research.

REFERENCES

Aloulou, A., Rodriguez, J. A., Fernandez, S., Van Oosterhout, D., Puccinelli, D., & Carriere, F. (2006). Exploring the specific features of interfacial enzymology based on lipase studies. *Biochimica et Biophysica Acta—Molecular and Cell Biology of Lipids*, *1761*, 995–1013.

Altenbach, C., Lopez, C. J., Hideg, K., & Hubbell, W. L. (2015). Exploring structure, dynamics, and topology of nitroxide spin-labeled proteins using continuous-wave electron paramagnetic resonance spectroscopy. *Methods in Enzymology*, *564*, 59–100.

Bakala N'Goma, J. C., Amara, S., Dridi, K., Jannin, V., & Carriere, F. (2012). Understanding the lipid-digestion processes in the GI tract before designing lipid-based drug-delivery systems. *Therapeutic Delivery*, *3*, 105–124.

Ball, A., Nielsen, R., Gelb, M. H., & Robinson, B. H. (1999). Interfacial membrane docking of cytosolic phospholipase A2 C2 domain using electrostatic potential-modulated spin relaxation magnetic resonance. *Proceedings of the National Academy of Sciences of the United States of America*, *96*, 6637–6642.

Barbe, S., Lafaquiere, V., Guieysse, D., Monsan, P., Remaud-Simeon, M., & Andre, I. (2009). Insights into lid movements of Burkholderia cepacia lipase inferred from molecular dynamics simulations. *Proteins: Structure, Function, and Genetics*, *77*, 509–523.

Bateman, A., & Sandford, R. (1999). The PLAT domain: A new piece in the PKD1 puzzle. *Current Biology*, *9*, R588–R590.

Belle, V., Fournel, A., Woudstra, M., Ranaldi, S., Prieri, F., Thome, V., et al. (2007). Probing the opening of the pancreatic lipase lid using site-directed spin labeling and EPR spectroscopy. *Biochemistry*, *46*, 2205–2214.

Benzonana, G., & Desnuelle, P. (1965). Kinetic study of the action of pancreatic lipase on emulsified triglycerides. Enzymology assay in heterogeneous medium. *Biochimica et Biophysica Acta*, *105*, 121–136.

Bezzine, S., Carrière, F., De Caro, J., Verger, R., & De Caro, A. (1998). Human pancreatic lipase: An exposed hydrophobic loop from the C-terminal domain may contribute to interfacial binding. *Biochemistry*, *37*, 11846–11855.

Bezzine, S., Ferrato, F., Ivanova, M. G., Lopez, V., Verger, R., & Carriere, F. (1999). Human pancreatic lipase: Colipase dependence and interfacial binding of lid domain mutants. *Biochemistry*, *38*, 5499–5510.

Bordes, F., Barbe, S., Escalier, P., Mourey, L., Andre, I., Marty, A., et al. (2010). Exploring the conformational states and rearrangements of Yarrowia lipolytica lipase. *Biophysical Journal*, *99*, 2225–2234.

Bourbon Freie, A., Ferrato, F., Carriere, F., & Lowe, M. E. (2006). Val407 and Ile408 in the β5' loop of pancreatic lipase mediate lipase-colipase interactions in the presence of bile salt micelles. *Journal of Biological Chemistry*, *281*, 7793–7800.

Bourbon-Freie, A., Dub, R. E., Xiao, X., & Lowe, M. E. (2009). Trp-107 and trp-253 account for the increased steady state fluorescence that accompanies the conformational change in human pancreatic triglyceride lipase induced by tetrahydrolipstatin and bile salt. *The Journal of Biological Chemistry*, *284*, 14157–14164.

Brady, L., Brzozowski, A. M., Derewenda, Z. S., Dodson, E., Dodson, G., Tolley, S., et al. (1990). A serine protease triad forms the catalytic centre of a triacylglycerol lipase. *Nature*, *343*, 767–770.

Brzozowski, A. (1993). Crystallization of a Humicola lanuginosa lipase-inhibitor complex with the use of polyethylene glycol monomethyl ether. *Acta Crystallographica*, *D49*, 352–354.

Brzozowski, A. M., Derewenda, U., Derewenda, Z. S., Dodson, G. G., Lawson, D. M., Turkenburg, J. P., et al. (1991). A model for interfacial activation in lipases from the structure of a fungal lipase-inhibitor complex. *Nature*, *351*, 491–494.

Brzozowski, A. M., Savage, H., Verma, C. S., Turkenburg, J. P., Lawson, D. M., Svendsen, A., et al. (2000). Structural origins of the interfacial activation in Thermomyces (Humicola) lanuginosa lipase. *Biochemistry*, *39*, 15071–15082.

Carrière, F., Thirstrup, K., Hjorth, S., & Boel, E. (1994). Cloning of the classical guinea pig pancreatic lipase and comparison with the lipase related protein 2. *FEBS Letters*, *338*, 63–68.

Carrière, F., Withers-Martinez, C., van Tilbeurgh, H., Roussel, A., Cambillau, C., & Verger, R. (1998). Structural basis for the substrate selectivity of pancreatic lipases and some related proteins. *Biochimica et Biophysica Acta*, *1376*, 417–432.

Chahinian, H., Bezzine, S., Ferrato, F., Ivanova, M. G., Perez, B., Lowe, M. E., et al. (2002). The beta 5' loop of the pancreatic lipase C2-like domain plays a critical role in the lipase-lipid interactions. *Biochemistry*, *41*, 13725–13735.

Chahinian, H., Sias, B., & Carriere, F. (2000). The C-terminal domain of pancreatic lipase: Functional and structural analogies with c2 domains. *Current Protein & Peptide Science*, *1*, 91–103.

Cudrey, C., van Tilbeurgh, H., Gargouri, Y., & Verger, R. (1993). Inactivation of pancreatic lipases by amphiphilic reagents 5-(dodecyldithio)-2-nitrobenzoic acid and tetrahydrolipstatin. Dependence upon partitioning between micellar and oil phases. *Biochemistry*, *32*, 13800–13808.

Delorme, V., Dhouib, R., Canaan, S., Fotiadu, F., Carriere, F., & Cavalier, J. F. (2011). Effects of surfactants on lipase structure, activity, and inhibition. *Pharmaceutical Research*, *28*, 1831–1842.

Derewenda, U., Brzozowski, A. M., Lawson, D. M., & Derewenda, Z. S. (1992). Catalysis at the interface: The anatomy of a conformational change in a triglyceride lipase. *Biochemistry, 31*, 1532–1541.

Derewenda, U., Swenson, L., Wei, Y. Y., Green, R., Kobos, P. M., Joerger, R., et al. (1994). Conformational lability of lipases observed in the absence of an oil-water interface. Crystallographic studies of enzymes from the fungi Humicola lanuginosa and Rhizopus delemar. *Journal of Lipid Research, 35*, 524–534.

Egloff, M.-P., Marguet, F., Buono, G., Verger, R., Cambillau, C., & van Tilbeurgh, H. (1995). The 2.46 Å resolution structure of the pancreatic lipase-colipase complex inhibited by a C_{11} alkyl phosphonate. *Biochemistry, 34*, 2751–2762.

Eydoux, C., Spinelli, S., Davis, T. L., Walker, J. R., Seitova, A., Dhe-Paganon, S., et al. (2008). Structure of human pancreatic lipase-related protein 2 with the lid in an open conformation. *Biochemistry, 47*, 9553–9564.

Gillmor, S. A., Villasenor, A., Fletterick, R., Sigal, E., Browner, M. F., Gillmor, S. A., et al. (1997). The structure of mammalian 15-lipoxygenase reveals similarity to the lipases and the determinants of substrate specificity. *Nature Structural Biology, 4*, 1003–1009.

Haiker, H., Lengsfeld, H., Hadvary, P., & Carrière, F. (2004). Rapid exchange of pancreatic lipase between triacylglycerol droplets. *Biochimica et Biophysica Acta, 1682*, 72–79.

Hermoso, J., Pignol, D., Kerfelec, B., Crenon, I., Chapus, C., & Fontecilla-Camps, J. C. (1996). Lipase activation by nonionic detergents. The crystal structure of the porcine lipase-colipase-tetraethylene glycol monooctyl ether complex. *Journal of Biological Chemistry, 271*, 18007–18016.

Hermoso, J., Pignol, D., Penel, S., Roth, M., Chapus, C., & Fontecilla-Camps, J. C. (1997). Neutron crystallographic evidence of lipase-colipase complex activation by a micelle. *The EMBO Journal, 16*, 5531–5536.

Hjorth, A., Carrière, F., Cudrey, C., Wöldike, H., Boel, E., Lawson, D. M., et al. (1993). A structural domain (the lid) found in pancreatic lipases is absent in the guinea pig (phospho)lipase. *Biochemistry, 32*, 4702–4707.

Hofmann, A. F. (1963). The function of bile salts in fat absorption. The solvent properties of dilute micellar solutions of conjugated bile salts. *The Biochemical Journal, 89*, 57–68.

Jensen, M. O., Jensen, T. R., Kjaer, K., Bjornholm, T., Mouritsen, O. G., & Peters, G. H. (2002). Orientation and conformation of a lipase at an interface studied by molecular dynamics simulations. *Biophysical Journal, 83*, 98–111.

Jeschke, G. (2002). Distance measurements in the nanometer range by pulsed EPR. *ChemPhysChem, 3*, 927–932.

Jeschke, G., Chechik, V., Godt, A., Zimmermann, H., Banham, J., Timmel, C. R., et al. (2006). DEER analysis 2006—A computational software package for analyzing pulsed ELDOR data. *Applied Magnetic Resonance, 30*, 473–498.

Lawson, D. M., Brzozowski, A. M., Rety, S., Verma, C., & Dodson, G. G. (1994). Probing the nature of substrate binding in *Humicola lanuginosa* lipase through X-ray crystallography and intuitive modelling. *Protein Engineering, 7*, 543–550.

Longhi, S., Belle, V., Fournel, A., Guigliarelli, B., & Carriere, F. (2011). Probing structural transitions in both structured and disordered proteins using site-directed spin-labeling EPR spectroscopy. *Journal of Peptide Science, 17*, 315–328.

Longhi, S., Mannesse, M., Verheij, H. M., De Haas, G. H., Egmond, M., Knoops-Mouthuy, E., et al. (1997). Crystal structure of cutinase covalently inhibited by a triglyceride analogue. *Protein Science, 6*, 275–286.

Lüthi-Peng, Q., & Winkler, F. K. (1992). Large spectral changes accompany the conformational transition of human pancreatic lipase induced by acylation with the inhibitor tetrahydrolipstatin. *European Journal of Biochemistry, 205*, 383–390.

Martinez, C., Nicolas, A., van Tilbeurgh, H., Egloff, M. P., Cudrey, C., Verger, R., et al. (1994). Cutinase, a lipolytic enzyme with a preformed oxyanion hole. *Biochemistry*, *33*, 83–89.

Mchaourab, H. S., Lietzow, M. A., Hideg, K., & Hubbell, W. L. (1996). Motion of spin-labeled side chains in T4 lysozyme. Correlation with protein structure and dynamics. *Biochemistry*, *35*, 7692–7704.

Miled, N., Roussel, A., Bussetta, C., Berti-Dupuis, L., Riviere, M., Buono, G., et al. (2003). Inhibition of dog and human gastric lipases by enantiomeric phosphonate inhibitors: A structure-activity study. *Biochemistry*, *42*, 11587–11593.

Moreau, H., Moulin, A., Gargouri, Y., Noël, J.-P., & Verger, R. (1991). Inactivation of gastric and pancreatic lipases by diethyl *p*-nitrophenyl phosphate. *Biochemistry*, *30*, 1037–1041.

Nalefski, E. A., & Falke, J. J. (1996). The C2 domain calcium-binding motif: Structural and functional diversity. *Protein Science*, *5*, 2375–2390.

Naylor, C. E., Eaton, J. T., Howells, A., Justin, N., Moss, D. S., Titball, R. W., et al. (1998). Structure of the key toxin in gas gangrene. *Nature Structural Biology*, *5*, 738–746.

Ollis, D. L., Cheah, E., Cygler, M., Dijkstra, B., Frolow, F., Franken, S. M., et al. (1992). The α/β hydrolase fold. *Protein Engineering*, *5*, 197–211.

Pannier, M., Veit, S., Godt, A., Jeschke, G., & Spiess, H. W. (2000). Dead-time free measurement of dipole-dipole interactions between electron spins. *Journal of Magnetic Resonance*, *142*, 331–340.

Perisic, O., Paterson, H. F., Mosedale, G., Lara-Gonzalez, S., & Williams, R. L. (1999). Mapping the phospholipid-binding surface and translocation determinants of the C2 domain from cytosolic phospholipase A2. *Journal of Biological Chemistry*, *274*, 14979–14987.

Pignol, D., Ayvazian, L., Kerfelec, B., Timmins, P., Crenon, I., Hermoso, J., et al. (2000). Critical role of micelles in pancreatic lipase activation revealed by small angle neutron scattering. *The Journal of Biological Chemistry*, *275*, 4220–4224.

Ranaldi, S., Belle, V., Woudstra, M., Bourgeas, R., Guigliarelli, B., Roche, P., et al. (2010). Amplitude of pancreatic lipase lid opening in solution and identification of spin label conformational subensembles by combining continuous wave and pulsed EPR spectroscopy and molecular dynamics. *Biochemistry*, *49*, 2140–2149.

Ranaldi, S., Belle, V., Woudstra, M., Rodriguez, J., Guigliarelli, B., Sturgis, J., et al. (2009). Lid opening and unfolding in human pancreatic lipase at low pH revealed by site-directed spin labeling EPR and FTIR spectroscopy. *Biochemistry*, *48*, 630–638.

Rehm, S., Trodler, P., & Pleiss, J. (2010). Solvent-induced lid opening in lipases: A molecular dynamics study. *Protein Science*, *19*, 2122–2130.

Rizo, J., & Südhof, T. C. (1998). C2-domains, structure and function of a universal Ca2+-binding domain. *Journal of Biological Chemistry*, *273*, 15879–15882.

Rouard, M., Sari, H., Nurit, S., Entressangles, B., & Desnuelle, P. (1978). Inhibition of pancreatic lipase by mixed micelles of diethyl *p*-nitrophenyl phosphate and the bile salts. *Biochimica et Biophysica Acta*, *530*, 227–235.

Roussel, A., Amara, S., Nyyssola, A., Mateos-Diaz, E., Blangy, S., Kontkanen, H., et al. (2014). A cutinase from Trichoderma reesei with a lid-covered active site and kinetic properties of true lipases. *Journal of Molecular Biology*, *426*, 3757–3772.

Roussel, A., Canaan, S., Egloff, M. P., Riviere, M., Dupuis, L., Verger, R., et al. (1999). Crystal structure of human gastric lipase and model of lysosomal acid lipase, two lipolytic enzymes of medical interest. *The Journal of Biological Chemistry*, *274*, 16995–17002.

Roussel, A., Miled, N., Berti-Dupuis, L., Riviere, M., Spinelli, S., Berna, P., et al. (2002). Crystal structure of the open form of dog gastric lipase in complex with a phosphonate inhibitor. *The Journal of Biological Chemistry*, *277*, 2266–2274.

Santini, S., Crowet, J. M., Thomas, A., Paquot, M., Vandenbol, M., Thonart, P., et al. (2009). Study of Thermomyces lanuginosa lipase in the presence of tributyrylglycerol and water. *Biophysical Journal*, *96*, 4814–4825.
Selvan, A., Seniya, C., Chandrasekaran, S. N., Siddharth, N., Anishetty, S., & Pennathur, G. (2010). Molecular dynamics simulations of human and dog gastric lipases: Insights into domain movements. *FEBS Letters*, *584*, 4599–4605.
Stein, R. A., Beth, A. H., & Hustedt, E. J. (2015). A straightforward approach to the analysis of double electron-electron resonance data. *Methods in Enzymology*, *563*, 531–567.
Tiss, A., Carriere, F., & Verger, R. (2001). Effects of gum arabic on lipase interfacial binding and activity. *Analytical Biochemistry*, *294*, 36–43.
van Tilbeurgh, H., Egloff, M.-P., Martinez, C., Rugani, N., Verger, R., & Cambillau, C. (1993). Interfacial activation of the lipase-procolipase complex by mixed micelles revealed by X-ray crystallography. *Nature*, *362*, 814–820.
van Tilbeurgh, H., Sarda, L., Verger, R., & Cambillau, C. (1992). Structure of the pancreatic lipase-procolipase complex. *Nature*, *359*, 159–162.
Verger, R. (1980). Enzyme kinetics of lipolysis. *Methods in Enzymology*, *64*, 340–392.
Winkler, F. K., d'Arcy, A., & Hunziker, W. (1990). Structure of human pancreatic lipase. *Nature*, *343*, 771–774.
Withers-Martinez, C., Carrière, F., Verger, R., Bourgeois, D., & Cambillau, C. (1996). A pancreatic lipase with a phospholipase A1 activity: Crystal structure of a chimeric pancreatic lipase-related protein 2 from guinea pig. *Structure*, *4*, 1363–1374.
Yang, Z., Bridges, M., Lerch, M. T., Altenbach, C., & Hubbell, W. L. (2015). Saturation recovery EPR and nitroxide spin labeling for exploring structure and dynamics in proteins. *Methods in Enzymology*, *564*, 3–27.
Yapoudjian, S., Ivanova, M. G., Brzozowski, A. M., Patkar, S. A., Vind, J., Svendsen, A., et al. (2002a). Binding of Thermomyces (Humicola) lanuginosa lipase to the mixed micelles of cis-parinaric acid/NaTDC. *European Journal of Biochemistry*, *269*, 1613–1621.
Yapoudjian, S., Ivanova, M., Douchet, I., Zenatti, A., Sentis, M., Marine, W., et al. (2002b). Surface fluorescence resonance energy transfer studies on interfacial adsorption of Thermomyces (humicola) lanuginosa lipase, using monomolecular films of cis-parinaric acid. *Biopolymers*, *65*, 121–128.

CHAPTER THIRTEEN

Measuring Phospholipase D Enzymatic Activity Through Biochemical and Imaging Methods

F. Philip, E.E. Ha, M.A. Seeliger, M.A. Frohman[1]
Center for Developmental Genetics, Stony Brook University School of Medicine, Stony Brook, NY, United States
[1]Corresponding author: e-mail address: michael.frohman@stonybrook.edu

Contents

Abstract

The phospholipase D (PLD) enzymatic superfamily regulates a wide range of cell biological and physiological pathways, including platelet activation, immune responses, cancer, and spermatogenesis. The three main enzymatic actions of the superfamily entail (i) hydrolyzing membrane phospholipids (phosphatidylcholine (PC) and cardiolipin) to generate choline and the second messenger signaling lipid phosphatidic acid (PA), (ii) using ethanol to transphosphatidylate PC to generate the long-lived metabolite phosphatidylethanol, and (iii) hydrolyzing RNA transcripts to generate piRNAs, the third form of endogenous RNAi. We discuss briefly previously published methods for *in vitro* and *in vivo* detection and imaging of PA, and focus on production, purification, and *in vitro* endonuclease activity analysis for human PLD6, a mitochondrial-tethered isoform with roles in fertility, cancer, and neuronal homeostasis.

Methods in Enzymology, Volume 583
ISSN 0076-6879
http://dx.doi.org/10.1016/bs.mie.2016.09.041

1. INTRODUCTION

The phospholipase D (PLD) superfamily of enzymes is defined by a catalytic site HxKxxxxD(6x)GG/SAN motif that mediates its trans-phosphalytic/hydrolytic actions on phosphodiester bonds in selected lipids and nucleic acid substrates (reviewed in Jenkins & Frohman, 2005). The canonical mammalian PLD1 (Hammond et al., 1995) and PLD2 (Colley et al., 1997) enzymes are best known for using water as the nucleophile to hydrolyze the abundant phospholipid phosphatidylcholine (PC) to generate the second messenger signaling lipid phosphatidic acid (PA).

Short-chained primary alcohols such as ethanol and 1-butanol can also serve as the nucleophile to generate phosphatidylethanol (PtdEtOH) and phosphatidylbutanol (PtdBut) (reviewed in McDermott, Wakelam, & Morris, 2004). Measuring PtdBut production in the presence of low amounts of 1-butanol has long been a favored method to quantify PLD1 and PLD2 cellular activity, since PtdBut is produced only by PLD, whereas PA is generated by several other families of enzymes, and since PtdBut is metabolically relatively inert, whereas PA is rapidly turned over (Morris, Frohman, & Engebrecht, 1997). Measurement of PtdEtOH by mass spectrometry of blood cells has recently become of great interest (Schrock, Thierauf, Wurst, Thon, & Weinmann, 2014), since the PtdEtOH generated subsequent to alcohol consumption can persist *in vivo* for up to a month (Gnann, Thierauf, Hagenbuch, Rohr, & Weinmann, 2014), whereas ethanol itself is much more rapidly metabolized. PtdEtOH is endogenously produced from ethanol generated by intestinal bacteria and has been proposed to be linked to the development of colon cancer in the context of heavy alcohol consumption (Pannequin et al., 2007).

PLD1- and PLD2 (PLD1/2)-generated PA has been associated with a variety of subcellular functions, such as intracellular trafficking, cytoskeletal dynamics, cell proliferation, and membrane remodeling (Jenkins & Frohman, 2005; McDermott et al., 2004). There are many proteins thought to be direct effectors of PA action, including mTOR (reviewed in Foster, Salloum, Menon, & Frias, 2014), PI4P5kinase (Honda et al., 1999), myosin phosphatase (Du & Frohman, 2009), sphingosine kinase 1 (Delon et al., 2004), kinesin (Manifava et al., 2001), lipin 1 (Huang et al., 2011), and Drp1 (Adachi et al., 2016). PA, as a cone-shaped, negatively charged lipid, has also been proposed to promote lipid bilayer negative curvature to lower the activation energy for membrane fusion and fission during exocytosis, endocytosis, and vesicular trafficking (Kooijman et al., 2005; Tanguy et al., 2016). PA can further be metabolized to generate the lipid signaling molecules diacylglycerol (DAG) and LysoPA, which regulate many additional pathways (Frohman, 2015). Physiologically, mice lacking PLD1 or PLD2 are viable but have defects in platelet activation (Elvers et al., 2010; Stegner et al., 2013), immune responses (Ali et al., 2013; Gobel et al., 2014; Schonberger et al., 2014), and cancer pathways (Chen et al., 2012).

PLD3 and 4 are transmembrane proteins that localize on the endoplasmic reticulum (ER) (Osisami, Ali, & Frohman, 2012; Yoshikawa et al., 2010). Catalytic activity has not been reported for either enzyme (Osisami et al., 2012; Yoshikawa et al., 2010). Mutations in PLD3 have been linked to Alzheimer's disease (Giri, Zhang, & Lu, 2016), whereas PLD4 mutations trigger autoimmune diseases such as systemic sclerosis (Terao et al., 2013). Mice lacking PLD5 have been generated, but no significant phenotypes were observed in a standardized phenotypic screen involving 25 tests, nor is it known whether PLD5 is catalytically active (http://www.mousephenotype.org/). PLD6 (also known as MitoPLD, and in *Drosophila*, Zucchini) is tail anchored into the outer surface of the mitochondrial membrane (Choi et al., 2006) and has been reported to have two catalytic functions, acting as a lipase to cleave the mitochondrial-localized lipid cardiolipin to generate PA (Baba et al., 2014; Choi et al., 2006; Huang et al., 2011; von Eyss et al., 2015; Zhang et al., 2016), and functioning as an endonuclease to cleave long RNAs during spermatogenesis to generate piRNAs (Ipsaro, Haase, Knott, Joshua-Tor, & Hannon, 2012; Nishimasu et al., 2012). piRNAs are small interfering RNAs that silence transposons to protect host germ lines during meiosis. PLD6's lipase function has been linked to mitochondrial fusion and fission (Adachi et al., 2016; Baba et al., 2014; Choi et al., 2006; Huang et al., 2011; von Eyss et al., 2015; Zhang et al., 2016), neuronal homeostasis (Zhang et al., 2016), and breast cancer progression (von Eyss et al., 2015).

PLD1 and PLD2 are regulated by signaling processes ensuing from G protein-coupled receptor and receptor tyrosine kinase activation events (Jenkins & Frohman, 2005; McDermott et al., 2004). Regulation of PLD6 is not currently understood, although it has been proposed that interaction with the protein Mitoguardin facilitates the lipase activity over the nuclease activity (Zhang et al., 2016).

Measurement of PLD activity has been key to advances in the field. Measurement approaches have involved *in vitro* and *in vivo* measurements of PA or PtdBut production, imaging of PA *in vivo* in cells, and analysis of oligonucleotide hydrolysis *in vitro* or piRNA production *in vivo*. PLD1 and PLD2 methods of analysis have been published and will be summarized here; biochemical, cellular, and organismal analysis of PLD6 lipase activity (Choi et al., 2006; Huang et al., 2011) and biochemical measurement of mouse and *Drosophila* PLD6 nuclease activity have been described as well (Ipsaro et al., 2012; Nishimasu et al., 2012). Here, we present a detailed method for purification and biochemical analysis of human PLD6 nuclease activity, which differs in detail from the published protocols for PLD6 analysis in other species.

2. METHODS

2.1 PLD1 and PLD2 Biochemical Measurement Approaches

There are several considerations in choosing a PLD1/2 assay approach (reviewed in Morris et al., 1997). PLD1/2 activity can be measured *in vitro* using exogenously provided substrates or in intact or broken cell systems that employ endogenous phospholipid substrates, and there are different options for what is quantified (choline, PA, or Ptd-alcohol). Some approaches involve using a radiolabeled PC substrate. In the *in vitro* assay procedure, the substrate is incorporated into mixed phospholipid vesicles (alternately, short-chained PCs (e.g., di-C10 PC) that are soluble in aqueous solution can be used). In the *in vivo* assay, PLD1/2 activity can be measured in intact cells by prelabeling the endogenous PC through feeding the cells with ^{3}H-palmitic acid. PLD1/2 activity is then determined by measuring conversion of the radiolabeled lipid substrate to PA, or Ptd-alcohol if ethanol or 1-butanol is included in the assay medium. PLD1/2 can also be assayed using nonradioactive approaches, which has become more popular in recent years. One method involves a coupled assay that measures choline; another employs mass spectrometry analysis to measure PtdEtOH.

2.1.1 In Vitro Measurement of PLD1/2 Activity Using Liposomes Containing ^{3}H-PC

This approach is ideal for assessing the effects of activators and inhibitors on PLD1/2 activity in isolation from cellular regulation. In the presence of the activators phosphatidylinositol 4,5-bisphosphate (PI4,5P_2), protein kinase C, and the small GTPases Rho and Arf, PLD1 activity increases 6000-fold over its basal activity. In contrast, PLD2 is activated only by PI4,5P_2 in this setting. Note that assay of cell lysates can be complicated, since varying amounts of the PLD1/2 activators will be present in stimulatory forms in the lysates, depending on the activation state of the cells at the moment of lysis. Methods for production and purification of PLD1, PLD2, and their activators, and the *in vitro* assay conditions using thin layer chromatography (TLC) to quantify PA and PtdBut have been published earlier in this series (Brown, Henage, Preininger, Xiang, & Exton, 2007; Frohman, Kanaho, Zhang, & Morris, 2000).

2.1.2 In Vivo *Measurement of PLD1/2 Activity Using ^{3}H-Palmitic Acid Cell Labeling*

This approach is ideal for assessing regulated activation of PLD1/2 in cells. In brief, cells are cultured with the labeled lipid for a day, during which time the palmitic acid is converted in part to ^{3}H-PC. PLD1/2 activation by signaling events accordingly generates ^{3}H-PA, or in the presence of primary alcohols, ^{3}H-Ptd-alcohol, which can again be quantified using TLC. Detailed methods for this assay have also previously been published in this series (Brown et al., 2007; Du, Morris, Sciorra, & Frohman, 2002).

2.1.3 In Vivo *Measurement of PLD1/2 Activity in Organisms*

These approaches are classically based on PLD1/2's transphosphatidylation activity. Introduction of ethanol into mice (and humans) results in the production of PtdEtOH, which can be detected immunologically in tissue sections (Pannequin et al., 2007) and quantitatively by mass spectrometry (Oliveira et al., 2010; Schrock et al., 2014; Scott et al., 2013). Sensitivity can be increased for LC–MS using labeled alcohol (Brown et al., 2007).

2.1.4 Measurement of PLD1/2 Activity Using a Coupled Assay

In this assay, PLD1/2 activity is monitored indirectly. PLD cleavage of PC generates choline (in addition to PA); the choline is then metabolized by choline oxidase to betaine and H_2O_2; finally, in the presence of horseradish peroxidase, the H_2O_2 reacts with 10-acetyl-3,7-dihydrophenoxazine to generate the highly fluorescent product resorufin, which is quantified in a plate reader. Commercially available as "Amplex Red" (Invitrogen), this approach has become widely used. The assay was initially targeted to *in vitro* analysis of purified PLD protein, since cellular factors that affect generation, turnover, or utilization of choline or H_2O_2 will alter the assay readout. However, we examined this approach in a side-by-side comparison with the traditional ^{3}H-palmitic acid cell labeling approach for platelets and found reasonable agreement between the assays (Elvers et al., 2010).

Nonetheless, this may not apply to all tissues, in particular neuronal tissues that metabolize choline rapidly, or cells under oxidative stress that are generating H_2O_2. Appropriate controls should be used to establish the validity of the coupled assay approach on a case-by-case basis.

2.1.5 Measurement of Cellular Generation of PA Using Fluorescent Sensors

A growing number of PA sensors fused to fluorescent proteins (e.g., EGFP) have been developed over the past 15 years, starting with the isolated PA-binding domains of the Raf1-kinase (Rizzo, Shome, Watkins, & Romero, 2000) and the yeast t-SNARE Spo20 (Nakanishi et al., 2006; Su et al., 2009), to modified ones that employ FRET (Ferraz-Nogueira, Diez-Guerra, & Llopis, 2014; Nishioka, Frohman, Matsuda, & Kiyokawa, 2010) or have increased sensitivity (Zhang et al., 2014). PA-binding domains are generally short (20–40 amino acid), nonstructured, polybasic sequences that do not exhibit stereospecificity for PA and seem to bind through charged interaction. Genetic analysis established specificity to PA for the yeast sensor Spo20 (Nakanishi et al., 2006), which became puzzling in connection to a subsequent report that showed that affinity of the PA-binding motif for membrane surfaces appeared responsive to other anionic lipids (i.e., "integrating" the lipid environment; Horchani, de Saint-Jean, Barelli, & Antonny, 2014). However, this was recently resolved with the finding that all of the known PA sensors are sensitive to membrane curvature, with negative curvature enhancing binding and positive curvature eliminating binding of the sensor to PA-containing membrane surfaces (Putta et al., 2016). It is now clear that the anionic lipids studied in (Horchani et al., 2014) affected membrane curvature. This finding should be taken into consideration in interpreting translocation of the sensor in the *in vivo* setting, and in particular, suggests that analysis of candidate PA-binding proteins to PA should be conducted using liposomes with physiological lipid composition and degrees of curvature, rather than the often employed lipid "dot blots."

2.1.6 General Comments on PLD1/2 Activity Assays

All of the approaches used to assay PLD1/2 activity require careful interpretation of the results, based on the use of appropriate controls. Measurement of PA is problematic, whether performed by TLC or by *in vivo* fluorescent sensor imaging, for several reasons. First, PA can be generated by other sources, specifically any of the many isoforms of DAG kinase, or LysoPA acetyltransferase enzymes (reviewed in Frohman, 2015). It has been compellingly demonstrated that pharmacological elimination of PLD1/2 activity (Ali et al., 2013; Su et al., 2009) or genetic ablation of PLD1/2 (Sato, Hongu, Sakamoto, Funakoshi, & Kanaho, 2013), in *some* settings, has little to no effect on PA generation at subcellular sites at which PLD1/2 normally generate PA during signaling events, due to a dominant or compensatory contribution of DAG kinase isoforms. Thus, in analysis of PA generation and localization, *it is critical to use pharmacological and/or genetic controls to enable reaching a conclusion that PLD1/2 are the source of the PA in that setting*. Second, PA is rapidly turned over, via the actions of PA phosphatases (Lipins) (Pascual & Carman, 2013) and PA-specific PLA enzymes (Baba et al., 2014). Third, radioactive labeling of lipid substrates may preferentially label membranes/membrane lipids in subcellular regions undergoing active remodeling. Ultimately, quantification or immunovisualization of Ptd-alcohol may be the most accurate means for assessment of PLD1/2 activity—but in all cases, every effort to employ the best possible set of negative controls will lead to the highest degree of *reproducibility* and *rigor* in the experimental outcomes. It should also be noted that although PLD1 and PLD2 become phosphorylated and undergo varied types of subcellular translocation during signaling and activation events, the phosphorylation and translocation, which can be monitored with immunostaining or other approaches, are in themselves are not accurate indicators of the degree of PLD1 or PLD2 activity.

2.2 Generation and Purification of Human PLD6

The PLD6 catalytic site is positively charged with a narrow groove that can accommodate a single-stranded but not double-stranded RNA or DNA oligonucleotide (Nishimasu et al., 2012). The conserved active site residues include the histidine and lysine of the HKD motif and are critical for PLD6 lipase (Choi et al., 2006) and nuclease (Nishimasu et al., 2012) activity. The N-terminus of PLD6 (amino acids 1–39) encodes a mitochondrial targeting sequence that anchors the enzyme into the mitochondrial outer membrane; this portion of the protein needs to be excluded from expression constructs to generate a soluble protein for biochemical assay (Choi et al., 2006).

2.2.1 Materials

pET-His6-GST-TEV (2-GT) bacterial expression vector (Addgene plasmid # 29707) or expression vector of choice.

2×YT media for growth of *Escherichia coli* BL21 (DE3) cells.
Buffer A: 50 m*M* Tris–HCl, pH 8, 300 m*M* NaCl, 3 m*M* β-ME, 20 m*M* imidazole, and protease inhibitors.
Buffer B: 50 m*M* Tris–HCl, pH 8, 300 m*M* NaCl, 3 m*M* β-ME, 500 m*M* imidazole, and protease inhibitors.
Buffer C: 50 m*M* Tris–HCl, pH 8, 300 m*M* NaCl, and 1 m*M* DTT.

2.2.2 Construct Preparation

A coding sequence for amino acids 36–252 of human PLD6 (Choi et al., 2006; available from multiple cDNA repositories or from the authors upon request) should be cloned in to the expression vector of choice. We chose to use pET-His6-GST-TEV (2-GT), cloning the PLD6 N-terminal-deleted fragment downstream of the His6-GST-TEV protease site leader sequence. The construct was transformed into *E. coli* BL21 (DE3) cells to enable inducible expression under control of the T7 promotor.

2.2.3 Protein Expression

1. A liter of 2×YT media containing ampicillin (100 μg/mL) is inoculated with bacteria harvested from a densely streaked 10-cm plate and incubated at 37°C at 220 rpm until an OD of 0.6 is attained (approximately 2–3 h).
2. The cells are then cooled to 20°C and protein expression induced using 0.5 m*M* IPTG (final concentration) and the incubation continued overnight at 20°C.
3. The following day, the cells are harvested by centrifugation at 4°C and the cell pellet homogenized twice at 15,000 psi in a cell homogenizer (Emulsiflex C3, Avestin) in 150 mL of Buffer A on ice.
4. The lysate is centrifuged at 18,000 × *g* at 4°C for 1 h to pellet insoluble proteins and cell debris and the supernatant recovered.
5. The supernatant is loaded onto an FPLC column containing Ni-NTA resin (5 mL HisTrap, GE Healthcare) previously equilibrated with Buffer A. After washing the column with Buffer A, the protein is eluted with an imidazole gradient over 15 column volumes from 20 to 500 m*M* using Buffers A and B at 4°C. Elution fractions are analyzed by SDS-PAGE and the ones that contain GST-PLD6 pooled.
6. The protein is then further purified through a Glutathione agarose affinity chromatography column (5 mL, GSTrap, GE Healthcare) equilibrated in Buffer C, washed with 10 column volumes of Buffer C, and eluted with 10 mL of Buffer C plus 20 m*M* reduced L-glutathione,

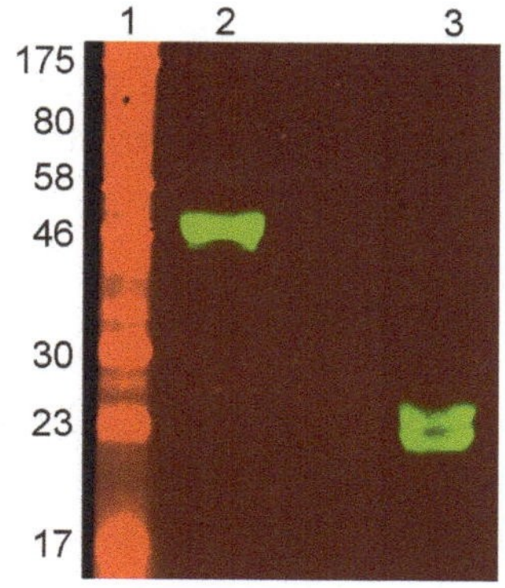

Fig. 1 PLD6 expression and purification. Lane 1, molecular markers; lane 2, a sample of the His6-GST-TEV-PLD6 protein after enrichment on an Ni-NTA column. Lane 3, PLD6 protein after cleavage at the TEV site, purification, and concentration. It is not known why the purified PLD6 protein sometimes runs as a doublet.

pH 7.2. Fractions are analyzed by SDS-PAGE, and the ones that contain PLD6 pooled. The concentration of GST-PLD6 is estimated by the Bradford assay (e.g., BioRad).

7. To cleave the GST tag, the protein is incubated with 0.1 mg of 6×His-tagged TEV protease per 2 mg of GST-PLD6 overnight at 4°C.
8. The cleaved tag and TEV protease are removed by adsorption to Ni-NTA. PLD6 remains in the flow through, for which the fractions are collected, analyzed by SDS-PAGE (Fig. 1), and quantified using Bradford reagent.
9. The protein is concentrated to 16 μ*M* and buffer exchanged to 20 m*M* Tris, pH 8.0, 150 m*M* NaCl, and 1 m*M* DTT through an Amicon filter before storing in aliquots at −80°C. A typical yield is 4 mg of PLD6 protein/liter of bacterial culture.

2.3 *In Vitro* PLD6 Assay

The nuclease activity of human PLD6, which is sequence independent, can be visualized using nonlabeled, radiolabeled, or fluorescent-labeled oligonucleotide substrates. The activity of the protein is inhibited by NaCl at concentrations >30 m*M*. The protein activity is not affected by EDTA or Mg^{2+}, consistent with prior descriptions of *Drosophila* and mouse PLD6 (Ipsaro et al., 2012; Nishimasu et al., 2012). Mouse PLD6 activity was also reported to be enhanced in the presence of the divalent cations Ca^{2+}, Mn^{2+}, and Zn^{2+}. For human PLD2, we have observed that Mn^{2+}, Zn^{2+}, Cu^{2+} have little effect on protein activity, but a concentration of Ca^{2+} of at least 2 m*M* is essential to observe activity. Whether this *in vitro* observation is physiologically relevant (i.e., human PLD6 may be activated by Ca^{2+} release from the ER and mitochondria into the cytoplasm) is unknown at present.

2.4 ^{32}P Assay

2.4.1 Materials

A DNA oligonucleotide, e.g., 5′-TTATCCACTTCCAATGTTATTA TTAGGTTTGGCTTTCGCTGGAGC-3′.

Buffer D: 20 m*M* Hepes, 2 m*M* $CaCl_2$, pH 7.0.

2.4.2 Protocol

1. Label the 5′ end of the ssDNA with γ^{32}P-ATP using T4 polynucleotide kinase.
2. Purify the labeled oligonucleotide using a microspin G25column (GE).
3. Incubate 200 n*M* oligonucleotide with 800 n*M* purified PLD6 for 1.5 h in Buffer D at 37°C.
4. Resolve the products on a 15% polyacrylamide urea gel (Fig. 2).

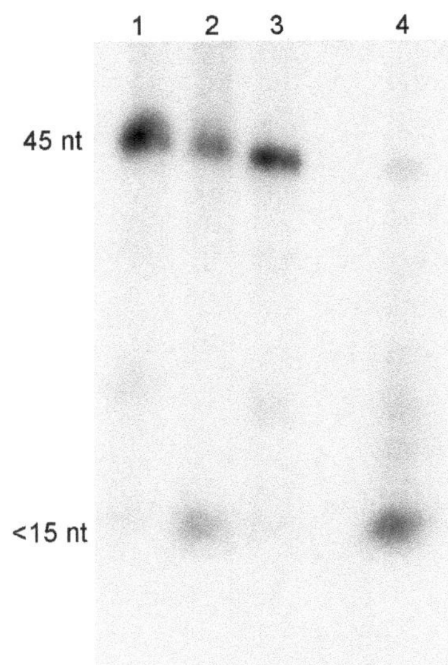

Fig. 2 Cleavage of a radioactive oligonucleotide substrate by PLD6 and analysis on a polyacrylamide urea gel. Lane 1, kinased oligonucleotide; lane 2, incubation with C3PO, a positive control single-stranded endoculease; lane 3, incubation with PLD6-H156N, a catalytically inactive mutant allele prepared in parallel to the wild-type protein; lane 4, incubation with wild-type PLD6.

3. FLUORESCENT-BASED PLD ACTIVITY ASSAYS

3.1 Polyacrylamide Gel-Based Assay

3.1.1 Materials

Substrate: A DNA oligonucleotide of length >40 nt, fluorescently labeled with an infrared (IR) dye with emission wavelength at 700 or 800 n*M* (IDT).

3.1.2 Protocol

1. Incubate 200 n*M* oligonucleotide with 800 n*M* purified PLD6 for 1.5 h in Buffer D at 37°C.
2. Resolve the products on a 15% polyacrylamide urea gel.
3. Image using an Odyssey CLx IR imager (Fig. 3).

Plate assay: A 96-well plate assay format for quantitative biochemical analyses of the protein activity can be used and is described as follows. This format is more sensitive than the gel-based activity assays.

3.1.3 Materials

A single-stranded DNA oligo labeled with biotin on the 5′ end and Alexa488 on the 3′ end (IDT).

Black well Neutravidin-coated plate (Life Technologies, catalog number 15217) pretreated with BSA to minimize nonspecific binding is used for the assay. Neutravidin is a deglycosylated form of Avidin with a near-neutral p*I* (6.3) that is responsible for its low nonspecific interactions.

Round-bottom black 96-well plate (Eppendorf, catalog number 951040102).

0.02 mg/mL BSA (Pierce, catalog number 23209).

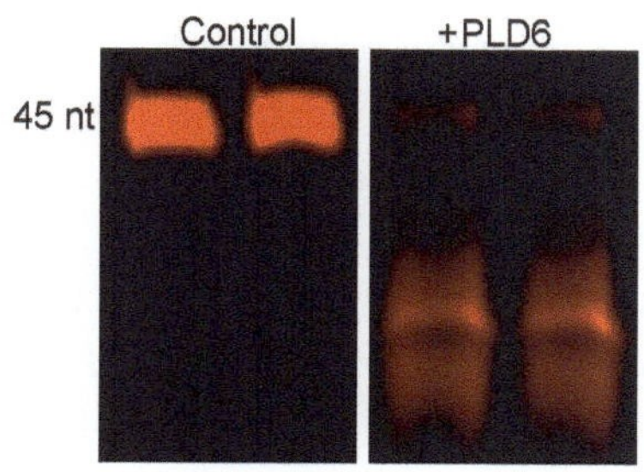

Fig. 3 Cleavage of a fluorescent oligonucleotide substrate by PLD6 and analysis on a polyacrylamide urea gel. Assay performed in duplicate. The control incubation lacked PLD6 protein.

3.1.4 Protocol

1. Wash the wells of the Neutravidin-coated plate twice with 200 μL of Buffer D.
2. Add 100 μL of 30 n*M* Biotin-ssDNA-Alexa488 to the Neutravidin-coated plate and incubate for 1.5 h at RT with gentle shaking.
3. Measure the initial fluorescence of the Biotin-ssDNA-Alexa488 and the empty wells (background) on a plate reader (Molecular Devices, Model—FilterMax F5).
4. Remove unbound substrate and wash wells with 200 μL of Buffer D twice for 20 min of gentle shaking each time.
5. Add 100 μL of 120 n*M* PLD6 from a 16 μ*M* stock (133.3-fold dilution) to the well and incubate for 1.5 h at 37°C.
6. Pretreat the wells of the round-bottom black 96-well plate with 200 μL of 0.02 mg/mL BSA for 30 min at RT.
7. Transfer supernatant (containing the cleaved substrate) to the round-bottom black 96-well plate.
8. Measure fluorescence of both plates to quantify released substrate (Fig. 4A) and the amount remaining bound to the Neutravidin plate (Fig. 4B).

3.1.5 Notes

(1) Since the fluorescence of the Biotin-ssDNA-Alexa488 substrate decays over time and is sensitive to temperature variations and Neutravidin coating, the substrate is also added to two additional wells in the Neutravidin and round-bottom plate at the beginning of the assay, and the fluorescence quantified at the beginning and end of the assay. We found that Neutravidin substantially quenches the Alexa488 fluorescence emission. Hence, the intensity values obtained for the cleaved fractions are larger than that originally bound to the wells. The intensity values obtained are background subtracted and corrected for the changes in the fluorescence that occur over time.

(2) Optimal conditions for PLD6 activity: [NaCl] in the assay buffer should be kept to a minimum since [NaCl] >30 m*M* inhibits PLD6 activity. The protein is stored in 1 m*M* DTT/20 m*M* Tris (pH 8.0) containing 150 m*M* NaCl and is diluted using NaCl-free assay buffer for use, resulting in final assay conditions of 20 m*M* Hepes, pH 7, 1.1 m*M* NaCl, and 2 m*M* $CaCl_2$. Ca^{2+} is required for activity of human PLD6; pH 7.0 is the optimal pH for the assay.

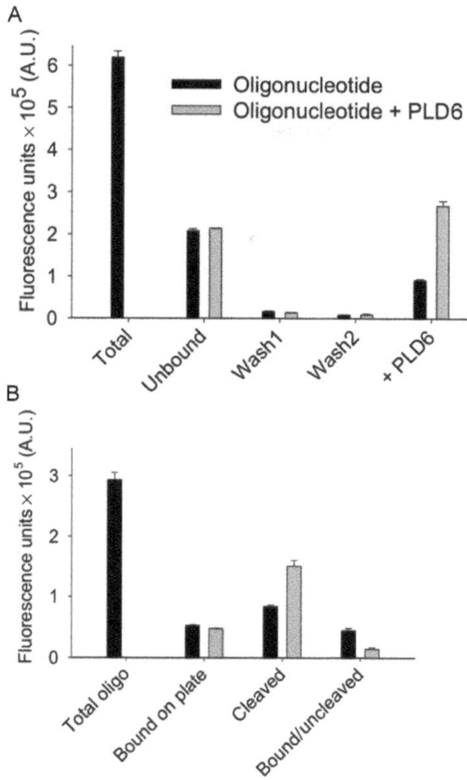

Fig. 4 A plate assay for PLD6 activity. (A) The five columns show the total fluorescence added to the wells, the amount of fluorescence remaining unbound at 1 h at RT; the amount released in two washes; and the amount released in the assay, in the presence or absence of PLD6 protein. All of the measurements in (A) were made on soluble fluorescence transferred to a round-bottom black well plate. (B) Fluorescence measurements of bound and released fluorescence in the Neutravidin plate. The fluorescence values increase in the presence of PLD6 (lane 3, "cleaved") since cleavage of the oligo substrate and release into the medium dequenches the fluorescence. Correspondingly, once the assay buffer is removed (lane 4), there is a weaker signal in the wells that contained PLD6, since less substrate remains on the walls of the wells. Experiment preformed in triplicate.

(3) This assay requires transfer of the cleaved substrate from the Neutravidin plate to a second plate for fluorescence quantification. However, since Neutravidin quenches the fluorescence of Alexa488 on the bound, uncleaved substrate, measurement of the fluorescence of the total substrate (cleaved and uncleaved) in the Neutravidin plate after the enzymatic reaction yields an output almost as large as measuring just the cleaved substrate in a second plate after transfer (Fig. 4B).

4. SUMMARY

We describe here a variety of approaches for assaying PLD6 nuclease activity *in vitro*, which suffice for experimental purposes and medium-throughput screens. PLD6 lipase activity (Choi et al., 2006), which can be readily assessed in cells using TLC to follow conversion of cardiolipin into PA (Choi et al., 2006) or fluorescent PA sensors to visualize PA formation on the mitochondrial surface (Baba et al., 2014; Choi et al., 2006; Huang et al., 2011; von Eyss et al., 2015; Zhang et al., 2016), has been much more difficult to reproducibly demonstrate *in vitro*; possibly the inclusion of the protein Mitoguardin (Zhang et al., 2016) in combination with an appropriate membrane surface will assist in development of an assay to measure this catalytic function.

ACKNOWLEDGMENTS

This work was supported by NIH Grants GM100109 and GM084251 to M.A.F. We thank Y. Altshuller for assistance with cloning of the human PLD6 expression construct, Weibing Zhang with assistance with the PLD6 biochemical purification, and Matthew Burak with assistance with the radiolabeled PLD6 assay.

The authors declare no potential conflicts of interest.

REFERENCES

Adachi, Y., Itoh, K., Cerveny, K., Suzuki, T., Macdonald, P., Frohman, M., ... Sesaki, H. (2016). Coincident phosphatidic acid interaction restrains Drp1 in mitochondrial division. *Molecular Cell*, *63*, 1034–1043.

Ali, W. H., Chen, Q., Delgiorno, K. E., Su, W., Hall, J. C., Hongu, T., ... Frohman, M. A. (2013). Deficiencies of the lipid-signaling enzymes phospholipase D1 and D2 alter cytoskeletal organization, macrophage phagocytosis, and cytokine-stimulated neutrophil recruitment. *PLoS One*, *8*(1), e55325. http://dx.doi.org/10.1371/journal.pone.0055325.

Baba, T., Kashiwagi, Y., Arimitsu, N., Kogure, T., Edo, A., Maruyama, T., ... Tani, K. (2014). Phosphatidic acid (PA)-preferring phospholipase A1 regulates mitochondrial dynamics. *The Journal of Biological Chemistry*, *289*(16), 11497–11511. http://dx.doi.org/10.1074/jbc.M113.531921.

Brown, H. A., Henage, L. G., Preininger, A. M., Xiang, Y., & Exton, J. H. (2007). Biochemical analysis of phospholipase D. *Methods in Enzymology*, *434*, 49–87. http://dx.doi.org/10.1016/s0076-6879(07)34004-4.

Chen, Q., Hongu, T., Sato, T., Zhang, Y., Ali, W., Cavallo, J. A., ... Frohman, M. A. (2012). Key roles for the lipid signaling enzyme phospholipase d1 in the tumor microenvironment during tumor angiogenesis and metastasis. *Science Signaling*, *5*(249), ra79. http://dx.doi.org/10.1126/scisignal.2003257.

Choi, S. Y., Huang, P., Jenkins, G. M., Chan, D. C., Schiller, J., & Frohman, M. A. (2006). A common lipid links Mfn-mediated mitochondrial fusion and SNARE-regulated exocytosis. *Nature Cell Biology*, *8*(11), 1255–1262. http://dx.doi.org/10.1038/ncb1487.

Colley, W. C., Sung, T. C., Roll, R., Jenco, J., Hammond, S. M., Altshuller, Y., ... Frohman, M. A. (1997). Phospholipase D2, a distinct phospholipase D isoform with novel regulatory properties that provokes cytoskeletal reorganization. *Current Biology*, 7(3), 191–201.

Delon, C., Manifava, M., Wood, E., Thompson, D., Krugmann, S., Pyne, S., & Ktistakis, N. T. (2004). Sphingosine kinase 1 is an intracellular effector of phosphatidic acid. *The Journal of Biological Chemistry, 279*(43), 44763–44774. http://dx.doi.org/10.1074/jbc.M405771200.

Du, G., & Frohman, M. A. (2009). A lipid-signaled myosin phosphatase surge disperses cortical contractile force early in cell spreading. *Molecular Biology of the Cell, 20*(1), 200–208. http://dx.doi.org/10.1091/mbc.E08-06-0555.

Du, G., Morris, A. J., Sciorra, V. A., & Frohman, M. A. (2002). G-protein-coupled receptor regulation of phospholipase D. *Methods in Enzymology, 345*, 265–274.

Elvers, M., Stegner, D., Hagedorn, I., Kleinschnitz, C., Braun, A., Kuijpers, M. E., … Nieswandt, B. (2010). Impaired alpha(IIb)beta(3) integrin activation and shear-dependent thrombus formation in mice lacking phospholipase D1. *Science Signaling, 3*(103), ra1. http://dx.doi.org/10.1126/scisignal.2000551.

Ferraz-Nogueira, J. P., Diez-Guerra, F. J., & Llopis, J. (2014). Visualization of phosphatidic acid fluctuations in the plasma membrane of living cells. *PLoS One, 9*(7), e102526. http://dx.doi.org/10.1371/journal.pone.0102526.

Foster, D. A., Salloum, D., Menon, D., & Frias, M. A. (2014). Phospholipase D and the maintenance of phosphatidic acid levels for regulation of mammalian target of rapamycin (mTOR). *The Journal of Biological Chemistry, 289*(33), 22583–22588. http://dx.doi.org/10.1074/jbc.R114.566091.

Frohman, M. A. (2015). The phospholipase D superfamily as therapeutic targets. *Trends in Pharmacological Sciences, 36*(3), 137–144. http://dx.doi.org/10.1016/j.tips.2015.01.001.

Frohman, M. A., Kanaho, Y., Zhang, Y., & Morris, A. J. (2000). Regulation of phospholipase D1 activity by Rho GTPases. *Methods in Enzymology, 325*, 177–189.

Giri, M., Zhang, M., & Lu, Y. (2016). Genes associated with Alzheimer's disease: An overview and current status. *Clinical Interventions in Aging, 11*, 665–681. http://dx.doi.org/10.2147/cia.s105769.

Gnann, H., Thierauf, A., Hagenbuch, F., Rohr, B., & Weinmann, W. (2014). Time dependence of elimination of different PEth homologues in alcoholics in comparison with social drinkers. *Alcoholism, Clinical and Experimental Research, 38*(2), 322–326. http://dx.doi.org/10.1111/acer.12277.

Gobel, K., Schuhmann, M. K., Pankratz, S., Stegner, D., Herrmann, A. M., Braun, A., … Meuth, S. G. (2014). Phospholipase D1 mediates lymphocyte adhesion and migration in experimental autoimmune encephalomyelitis. *European Journal of Immunology, 44*(8), 2295–2305. http://dx.doi.org/10.1002/eji.201344107.

Hammond, S. M., Altshuller, Y. M., Sung, T. C., Rudge, S. A., Rose, K., Engebrecht, J., … Frohman, M. A. (1995). Human ADP-ribosylation factor-activated phosphatidylcholine-specific phospholipase D defines a new and highly conserved gene family. *The Journal of Biological Chemistry, 270*(50), 29640–29643.

Honda, A., Nogami, M., Yokozeki, T., Yamazaki, M., Nakamura, H., Watanabe, H., … Kanaho, Y. (1999). Phosphatidylinositol 4-phosphate 5-kinase alpha is a downstream effector of the small G protein ARF6 in membrane ruffle formation. *Cell, 99*(5), 521–532.

Horchani, H., de Saint-Jean, M., Barelli, H., & Antonny, B. (2014). Interaction of the Spo20 membrane-sensor motif with phosphatidic acid and other anionic lipids, and influence of the membrane environment. *PLoS One, 9*(11), e113484. http://dx.doi.org/10.1371/journal.pone.0113484.

Huang, H., Gao, Q., Peng, X. X., Choi, S.-Y., Sarma, K., Ren, H., … Frohman, M. A. (2011). piRNA-associated germline nuage formation and spermatogenesis require MitoPLD pro-fusogenic mitochondrial-surface lipid signaling. *Developmental Cell, 20*, 376–387.

Ipsaro, J. J., Haase, A. D., Knott, S. R., Joshua-Tor, L., & Hannon, G. J. (2012). The structural biochemistry of Zucchini implicates it as a nuclease in piRNA biogenesis. *Nature*, *491*(7423), 279–283. http://dx.doi.org/10.1038/nature11502.

Jenkins, G. M., & Frohman, M. A. (2005). Phospholipase D: A lipid centric review. *Cellular and Molecular Life Sciences*, *62*(19–20), 2305–2316. http://dx.doi.org/10.1007/s00018-005-5195-z.

Kooijman, E. E., Chupin, V., Fuller, N. L., Kozlov, M. M., de Kruijff, B., Burger, K. N., & Rand, P. R. (2005). Spontaneous curvature of phosphatidic acid and lysophosphatidic acid. *Biochemistry*, *44*(6), 2097–2102. http://dx.doi.org/10.1021/bi0478502.

Manifava, M., Thuring, J. W., Lim, Z. Y., Packman, L., Holmes, A. B., & Ktistakis, N. T. (2001). Differential binding of traffic-related proteins to phosphatidic acid- or phosphatidylinositol (4,5)-bisphosphate-coupled affinity reagents. *The Journal of Biological Chemistry*, *276*(12), 8987–8994. http://dx.doi.org/10.1074/jbc.M010308200.

McDermott, M., Wakelam, M. J., & Morris, A. J. (2004). Phospholipase D. *Biochemistry and Cell Biology*, *82*(1), 225–253. http://dx.doi.org/10.1139/o03-079.

Morris, A. J., Frohman, M. A., & Engebrecht, J. (1997). Measurement of phospholipase D activity. *Analytical Biochemistry*, *252*(1), 1–9. http://dx.doi.org/10.1006/abio.1997.2299.

Nakanishi, H., Morishita, M., Schwartz, C. L., Coluccio, A., Engebrecht, J., & Neiman, A. M. (2006). Phospholipase D and the SNARE Sso1p are necessary for vesicle fusion during sporulation in yeast. *Journal of Cell Science*, *119*(Pt. 7), 1406–1415. http://dx.doi.org/10.1242/jcs.02841.

Nishimasu, H., Ishizu, H., Saito, K., Fukuhara, S., Kamatani, M. K., Bonnefond, L., … Nureki, O. (2012). Structure and function of Zucchini endoribonuclease in piRNA biogenesis. *Nature*, *491*(7423), 284–287. http://dx.doi.org/10.1038/nature11509.

Nishioka, T., Frohman, M. A., Matsuda, M., & Kiyokawa, E. (2010). Heterogeneity of phosphatidic acid levels and distribution at the plasma membrane in living cells as visualized by a Foster resonance energy transfer (FRET) biosensor. *The Journal of Biological Chemistry*, *285*(46), 35979–35987. http://dx.doi.org/10.1074/jbc.M110.153007.

Oliveira, T. G., Chan, R. B., Tian, H., Laredo, M., Shui, G., Staniszewski, A., … Di Paolo, G. (2010). Phospholipase d2 ablation ameliorates Alzheimer's disease-linked synaptic dysfunction and cognitive deficits. *The Journal of Neuroscience*, *30*(49), 16419–16428. http://dx.doi.org/10.1523/jneurosci.3317-10.2010.

Osisami, M., Ali, W., & Frohman, M. A. (2012). A role for phospholipase D3 in myotube formation. *PLoS One*, 7(3), e33341. http://dx.doi.org/10.1371/journal.pone.0033341.

Pannequin, J., Delaunay, N., Darido, C., Maurice, T., Crespy, P., Frohman, M. A., … Hollande, F. (2007). Phosphatidylethanol accumulation promotes intestinal hyperplasia by inducing ZONAB-mediated cell density increase in response to chronic ethanol exposure. *Molecular Cancer Research*, *5*(11), 1147–1157. http://dx.doi.org/10.1158/1541-7786.mcr-07-0198.

Pascual, F., & Carman, G. M. (2013). Phosphatidate phosphatase, a key regulator of lipid homeostasis. *Biochimica et Biophysica Acta*, *1831*(3), 514–522. http://dx.doi.org/10.1016/j.bbalip.2012.08.006.

Putta, P., Rankenberg, J., Korver, R. A., van Wijk, R., Munnik, T., Testerink, C., & Kooijman, E. E. (2016). Phosphatidic acid binding proteins display differential binding as a function of membrane curvature stress and chemical properties. *Biochimica et Biophysica Acta*, *1858*(11), 2709–2716. http://dx.doi.org/10.1016/j.bbamem.2016.07.014.

Rizzo, M. A., Shome, K., Watkins, S. C., & Romero, G. (2000). The recruitment of Raf-1 to membranes is mediated by direct interaction with phosphatidic acid and is independent of association with Ras. *The Journal of Biological Chemistry*, *275*(31), 23911–23918. http://dx.doi.org/10.1074/jbc.M001553200.

Sato, T., Hongu, T., Sakamoto, M., Funakoshi, Y., & Kanaho, Y. (2013). Molecular mechanisms of N-formyl-methionyl-leucyl-phenylalanine-induced superoxide generation and degranulation in mouse neutrophils: Phospholipase D is dispensable. *Molecular and Cellular Biology*, *33*(1), 136–145. http://dx.doi.org/10.1128/mcb.00869-12.

Schonberger, T., Jurgens, T., Muller, J., Armbruster, N., Niermann, C., Gorressen, S., … Elvers, M. (2014). Pivotal role of phospholipase D1 in tumor necrosis factor-alpha-mediated inflammation and scar formation after myocardial ischemia and reperfusion in mice. *The American Journal of Pathology*, *184*(9), 2450–2464. http://dx.doi.org/10.1016/j.ajpath.2014.06.005.

Schrock, A., Thierauf, A., Wurst, F. M., Thon, N., & Weinmann, W. (2014). Progress in monitoring alcohol consumption and alcohol abuse by phosphatidylethanol. *Bioanalysis*, *6*(17), 2285–2294. http://dx.doi.org/10.4155/bio.14.195.

Scott, S. A., Xiang, Y., Mathews, T. P., Cho, H. P., Myers, D. S., Armstrong, M. D., … Brown, H. A. (2013). Regulation of phospholipase D activity and phosphatidic acid production after purinergic (P2Y6) receptor stimulation. *The Journal of Biological Chemistry*, *288*(28), 20477–20487. http://dx.doi.org/10.1074/jbc.M113.451708.

Stegner, D., Thielmann, I., Kraft, P., Frohman, M. A., Stoll, G., & Nieswandt, B. (2013). Pharmacological inhibition of phospholipase D protects mice from occlusive thrombus formation and ischemic stroke—Brief report. *Arteriosclerosis, Thrombosis, and Vascular Biology*, *33*(9), 2212–2217. http://dx.doi.org/10.1161/atvbaha.113.302030.

Su, W., Yeku, O., Olepu, S., Genna, A., Park, J. S., Ren, H., … Frohman, M. A. (2009). 5-Fluoro-2-indolyl des-chlorohalopemide (FIPI), a phospholipase D pharmacological inhibitor that alters cell spreading and inhibits chemotaxis. *Molecular Pharmacology*, *75*(3), 437–446. http://dx.doi.org/10.1124/mol.108.053298.

Tanguy, E., Carmon, O., Wang, Q., Jeandel, L., Chasserot-Golaz, S., Montero-Hadjadje, M., & Vitale, N. (2016). Lipids implicated in the journey of a secretory granule: From biogenesis to fusion. *Journal of Neurochemistry*, *137*(6), 904–912. http://dx.doi.org/10.1111/jnc.13577.

Terao, C., Ohmura, K., Kawaguchi, Y., Nishimoto, T., Kawasaki, A., Takehara, K., … Mimori, T. (2013). PLD4 as a novel susceptibility gene for systemic sclerosis in a Japanese population. *Arthritis and Rheumatism*, *65*(2), 472–480. http://dx.doi.org/10.1002/art.37777.

von Eyss, B., Jaenicke, L. A., Kortlever, R. M., Royla, N., Wiese, K. E., Letschert, S., … Eilers, M. (2015). A MYC-driven change in mitochondrial dynamics limits YAP/TAZ function in mammary epithelial cells and breast cancer. *Cancer Cell*, *28*(6), 743–757. http://dx.doi.org/10.1016/j.ccell.2015.10.013.

Yoshikawa, F., Banno, Y., Otani, Y., Yamaguchi, Y., Nagakura-Takagi, Y., Morita, N., … Furuichi, T. (2010). Phospholipase D family member 4, a transmembrane glycoprotein with no phospholipase D activity, expression in spleen and early postnatal microglia. *PLoS One*, *5*(11), e13932. http://dx.doi.org/10.1371/journal.pone.0013932.

Zhang, Y., Liu, X., Bai, J., Tian, X., Zhao, X., Liu, W., … Tong, C. (2016). Mitoguardin regulates mitochondrial fusion through MitoPLD and is required for neuronal homeostasis. *Molecular Cell*, *61*(1), 111–124. http://dx.doi.org/10.1016/j.molcel.2015.11.017.

Zhang, F., Wang, Z., Lu, M., Yonekubo, Y., Liang, X., Zhang, Y., … Du, G. (2014). Temporal production of the signaling lipid phosphatidic acid by phospholipase D2 determines the output of extracellular signal-regulated kinase signaling in cancer cells. *Molecular and Cellular Biology*, *34*(1), 84–95. http://dx.doi.org/10.1128/mcb.00987-13.

CHAPTER FOURTEEN

PLD Protein–Protein Interactions With Signaling Molecules and Modulation by PA

J. Gomez-Cambronero*^,[1], A.J. Morris†, K.M. Henkels*

*Wright State University, Boonshoft School of Medicine, Dayton, OH, United States
†The Gill Heart Institute, College of Medicine, Lexington Veterans Affairs Medical Center, University of Kentucky, Lexington, KY, United States
[1]Corresponding author: e-mail address: julian.cambronero@wright.edu

Contents

Abstract

We describe methods for studying phospholipase D (PLD) interactions with signaling proteins and modulation of these interactions by the PLD reaction product, phosphatidic acid (PA). PLD is fundamental to the physiological maintenance of cellular/intracellular membranes, protein trafficking, cytoskeletal dynamics, membrane remodeling, cell proliferation, meiotic division and sporulation. PA is an acidic phospholipid involved in the biosynthesis of many other lipids that affects the enzymatic activities of many different signaling proteins via protein–lipid interactions or as a substrate. The involvement of PLD as an effector of protein–protein interactions and downstream signaling via PA-mediated processes has led to the investigation of PA-binding domains in target protein partners. We present here data and protocols detailing the interaction between PLD2–Rac2 interaction and modulation of this interaction by PA. We describe

Methods in Enzymology, Volume 583
ISSN 0076-6879
http://dx.doi.org/10.1016/bs.mie.2016.09.042

biochemical techniques to measure interactions between PLD, PA, and the small GTPase Rac2, which are associated in the cell. We found two maxima concentrations of PA that contributed to association or dissociation of Rac2 with PLD2, as well as the PLD2 lipase and guanine nucleotide exchange factor (GEF) activities. Fluctuations in the Rac2–PLD2 protein–protein binding interaction facilitate shuttling of Rac2 and/or PLD2 within the cell dependent on local cellular PA concentration. Fluorescence resonance emission transfer stoichiometry for PLD2 and Rac2 binding yielded a 3:1 ratio of Rac2:PLD2. Detection of PA in mammalian cells with a new biosensor showed colocalization in and around the nucleus. We also described methods for quantitation of PA in biological materials by HPLC electrospray ionization tandem mass spectrometry.

ABBREVIATIONS

DOPA 1,2-dioleoyl-*sn*-glycero-3-phosphate
FRET fluorescence resonance emission transfer
GEF guanine nucleotide exchange factor
PA phosphatidic acid
PABD PA-binding domain
PBD Pak1-binding domain
PLD phospholipase D
PLD1 phospholipase D1
PLD2 phospholipase D2

1. INTRODUCTION

Phospholipase D (PLD) converts phosphatidylcholine (PC) to phosphatidic acid (PA) and choline and is important for physiological maintenance of cellular/intracellular membranes, intracellular protein trafficking, cytoskeletal dynamics, membrane remodeling and cell proliferation, and meiotic division and sporulation (Hammond et al., 1997; Powner & Wakelam, 2002; Wang, Xu, & Zheng, 1994). The initial research into PLD began with plants in the 1940s and progressed to mammals in the 1970s (Hanahan & Chaikoff, 1947a, 1947b; Saito & Kanfer, 1973). Research in the 1990s led to identification of PLD genes in plants and then in yeast and mammals in a very short period of time (Colley et al., 1997; Hammond et al., 1995; Kodaki & Yamashita, 1997; Rose, Rudge, Frohman, Morris, & Engebrecht, 1995; Wang, Dyer, & Zheng, 1993; Wang et al., 1994). Since then, PLD's involvement in lipid signaling via its main reaction product, PA, has become a topic of mounting interest to a wide variety of researchers.

PA is an acidic phospholipid involved in the biosynthesis of many other lipids that affects either directly via protein–lipid interactions or as a substrate the enzymatic activities of many different signaling proteins involved in cytoskeletal dynamics, vesicle trafficking and cell proliferation, spreading, and survival (Hermansson, Hokynar, & Somerharju, 2011; Jang, Lee, Hwang, & Ryu, 2012; Stace & Ktistakis, 2006). Interaction of protein-binding partners with PA requires a conserved polybasic amino acid domain that functions as a membrane-localization sequence, such as lipin1β, p70 S6 kinase (S6K), and aurora kinase A (AURKA) (Henkels, Mallets, Dennis, & Gomez-Cambronero, 2015; Mahankali, Henkels, Speranza, & Gomez-Cambronero, 2015; Ren et al., 2010).

PA induces production of TNFα and IL-1β in macrophages (Lee et al., 2007). IL-1β primes IL-8-activated human neutrophils for elastase release and PLD activity (Brandolini et al., 1996). IL-8 activates PLD2, while PLD-generated PA mediates IL-8 expression (Cummings et al., 2002). Stimulation with IL-8 causes a strong polarizing effect and selective accumulation of PLD2 bright caps (horseshoe formations) at the edge of the cell (Lehman et al., 2006). Additionally, the PLD–PA complex regulates fibroblast dorsal ruffle development by controlling DOCK1 cellular localization (Sanematsu et al., 2013).

Upstream signaling from tyrosine and serine/threonine kinases mediates PLD's interaction with other target proteins via direct protein–protein binding contact that influences PLD catalysis and downstream events, such as cell migration and metastatic invasion (Gomez-Cambronero, 2014). As PA is the catalytic product of the reaction of PLD with phospholipids, the involvement of PLD as an effector of protein–protein interactions and downstream signaling via PA-mediated processes has led to the investigation of PA-binding domains (PABD) on target protein partners.

2. PLD2 AND RAC2 PROTEIN–PROTEIN INTERACTION AND THE EFFECT OF PA

At least one large GTPase (dynamin) and many different small GTPases (i.e., Arf1, RalA, RhoA, Rac1, Rac2, Cdc42, Ras, and Sar1p) interact with and activate PLD via their Switch I domains, which interact with the C-terminus of PLD (Bae, Min, Fleming, & Exton, 1998; Henkels, Mahankali, & Gomez-Cambronero, 2013; Kim et al., 1998; Lee, Kim, Ghim, Suh, & Ryu, 2015; Lee et al., 2006; Pathre et al., 2003; Peng, Henkels, Mahankali, Dinauer, & Gomez-Cambronero, 2011; Peng, Henkels, Mahankali, Marchal, et al., 2011; Yamazaki et al., 1999). Likewise, PLD can regulate small GTPases via an N-terminal PA-mediated mechanism or via the C-terminal PLD-guanine nucleotide exchange factor (GEF) activity (Chae et al., 2010; Jeon et al., 2011; Mahankali, Peng, Henkels, Dinauer, & Gomez-Cambronero, 2011).

PLD2 interacts with Rac2 via the PLD2 amino acids F107, F129, L166, R172 and L173, which is independent of PLD's lipase activity (Mahankali, Henkels, Alter, & Gomez-Cambronero, 2012). This site is also important to PLD's interaction with dynamin via R128, which further confirms a GTPase mediating a switch-on mechanism via a PLD interaction and its role as an effector (Lee et al., 2015). Once cells are stimulated, both catalytic activities of PLD2 (lipase and GEF) are upregulated. PA is quickly produced from the PLD2 lipase activity, which then positively regulates the PLD2-GEF function and activates it. GTP-bound Rac2 inhibits PLD2 lipase activity via increased PA accumulation, which upon further accumulation also interferes with PLD2-GEF binding to Rac2-GTP as substrate (Mahankali, Henkels, & Gomez-Cambronero, 2013).

3. PROTOCOL

3.1 Overview

There are numerous biochemical techniques currently being used to measure interactions between PLD, PA and the small GTPases, which range from immunoreactivity interactions as a measure of protein–protein binding to GEF activity assays in the presence of relevant protein cofactors and/or phospholipids to fluorescence resonance emission transfer (FRET) analysis via PABD on target proteins using newly developed PA sensors. Recent studies have focused on developing methods to directly measure changes in PA levels in cells and tissues, which can be used to monitor PLD activity and even determine the effectiveness of small-molecule PLD inhibitors.

3.2 *In Vivo* Rac2–PLD2 Interaction by Coimmunoprecipitation and PLD/Rac Ratio Calculations

Both PLD isoforms have many different signaling partners that bind to each other and facilitate enzymatic activities necessary for a multitude of signaling pathways required for the cell to function properly. We first suspected that the PLD2 isoform interacted with the small GTPase Rac2, as PLD and Rac were both activated with a profound effect on the kinetics of cell adhesion (Peng, Henkels, Mahankali, Dinauer, et al., 2011). We were able to determine that these two proteins are in close proximity in cells where they regulate each other via two PLD2 cdc42/Rac-interactive binding (CRIB) domains that PLD2 uses to bind to Rac2 and modulate the biological activities of this small GTPase via other interactions with β-actin, Arp3, and β-catenin (Peng, Henkels, Mahankali, Dinauer, et al., 2011; Peng, Henkels, Mahankali, Marchal, et al., 2011; Speranza, Mahankali, & Gomez-Cambronero, 2013; Speranza, Mahankali, Henkels, & Gomez-Cambronero, 2014). We used coimmunoprecipitation (co-IP) as a tool to discern this novel interaction, whereby one protein of interest in this proposed dimer was immunoprecipitated with antibodies specific to only it, and the second protein of interest was indirectly pulled down as a result of this interaction. Additionally, the converse experimental design yielded similar protein–protein interaction results, which suggests there was no antibody epitope interference under the conditions and reagents we used.

3.2.1 Material: Rac2–PLD2 co-IP

- pcDNA3.1-HA-PLD2-WT and pRK5-myc-Rac2-WT plasmids (from Dr. Cambronero's lab) (Wright State University, Dayton, OH, USA)
- COS-7 cells for transfection (cat. # CRL-1651) (ATCC, Manassas, VA, USA)
- Transit 20/20 transfection reagent (cat. # MIR 5400) (Mirus Bio, Madison, WI, USA)
- 1,2-Dioleoyl-*sn*-glycero-3-phosphate, sodium salt (DOPA) (cat. # 840875) (Avanti Polar Lipids, Alabaster, AL, USA)
- Liposome Buffer (0.5% BSA in PBS, pH 7.2) (50 mg BSA into 10 mL of 1 × PBS, pH 7.2, make certain to check the pH)
- Rabbit α-myc and Rabbit α-HA IgG antibodies for co-IP (cat. # 2272 and 3724, respectively) (Cell Signaling, Danver, MA, USA)
- Mouse α-myc and Mouse α-HA IgG antibodies and goat antimouse HRP antibodies for western blotting (cat. # 2276, 2367, and 7076, respectively) (Cell Signaling, Danvers, MA, USA)
- Protein G Agarose FastFlow (cat. # 16–156) (EMD Millipore, Temecula, CA, USA)
- Mini-SDS-PAGE gels (cat. # 456–1094) (BioRad Laboratories, Hercules, CA, USA), Immobilon-P PVDF transfer membrane for western blotting (cat. # IPVH00010) (EMD Millipore, Temecula, CA, USA)
- SDS-PAGE Stopping Solution (1.04 g Tris base, 15 mL glycerol, 50 mg Bromophenol Blue, 9 g SDS in a final volume of 100 mL made up with good-quality water, pH adjusted to 6.7)
- Amersham ECL reagent (cat. # RPN2106) (GE Healthcare, Pittsburgh, PA, USA)
- Autoradiograph film (cat. # 100 NIF) (MidSci, St. Louis, MO, USA)

3.2.2 Protocol

1. To characterize the protein–protein interactions between the small GTPase Rac2 and PLD2, we performed co-IP experiments from COS-7 cell lysates that were mock-transfected or that overexpressed both Rac2 and PLD2 proteins via transient transfection of 2 μg pRK5-myc-Rac2-WT and/or 2 μg pcDNA3.1-HA-PLD2-WT plasmid DNAs using 5–8 μL Transit 20/20 transfection reagent per each transfection reaction.
2. Additionally, 48 h posttransfection COS-7 cells were incubated in the presence of increasing concentrations of DOPA in Liposome Buffer for 20 min to determine the effect of exogenous addition of the product of the PLD reaction (PA), according to Hatton, Lintz, Mahankali, Henkels, and Gomez-Cambronero (2015) and Mahankali, Farkaly, Bedi, Hostetler, and Gomez-Cambronero (2015). To prepare 1 m*M* DOPA stock solution, weigh out 1 mg of DOPA and add to 1.4 mL of Liposome Buffer. Sonicate twice on ice for 5 s each sonication, if bubbles form that is acceptable. Keep DOPA stock on ice to prepare final concentrations. Do not store overnight but prepare fresh daily as needed. Prepare intermediate stocks in relevant cell culture media as needed to achieve the concentration you require. Final concentrations used in our lab range from 1 n*M* to 1 μ*M* with 300 n*M* being the norm. Final volumes of your DOPA in cell culture media should be in the range of 1–2 mL added to cells in 35 mm well or 1-well of a 6-well plate. (*Note*: other phospholipids from Avanti Polar Lipids can be used in place of DOPA if you prefer, such as 1,2-dioleoyl-*sn*-glycero-3-phosphocholine (DOPC) (cat. # 850675), L-α-phosphatidylinositol (PI) (cat. # 840042), L-α-phosphatidylinositol (DOPS) (cat. # 830035), 1-oleoyl-2-hydroxy-*sn*-glycero-3-phosphate (Lyso-PA) (cat. # 857130), or phosphatidylinositol-4,5-bisphosphate (PIP_2) (cat. # 840046).
3. Following DOPA treatment, cells were either scraped from the well or trypsinized and then sedimented to acquire all of the cells from the cell culture vessel(s). Cells were lysed in 1.5-mL Eppendorf tubes using 50–150 μL Special Lysis Buffer (5 m*M* HEPES, pH 7.4, 100 μ*M* sodium orthovanadate, and 0.1 Triton X-100) + 1 mg/mL leupeptin and aprotinin protease inhibitors proteins.
4. Subsequent equal concentrations of protein from lysates (50–500 ng) were coimmunoprecipitated for 4 h at 4°C using either 1 μg Rabbit

anti-HA or 1 μg Rabbit anti-myc antibodies specific for the HA-tagged PLD2 and the myc-tagged Rac2 proteins conjugated to 20 μL Protein G agarose beads. To accomplish this step, for each reaction that you will need, first wash 20 μL Protein G agarose beads by sedimenting in 1.5 mL screw cap Eppendorf tubes at 14,000 × *g* for 1 min at 4°C. Next, aspirate the supernatant to waste using a yellow pipet tip attached to gentle vacuum suction and without disturbing the sedimented beads. Then, add 1 μg of the relevant antibody with 20 μL Protein G agarose in 100 μL Special Lysis Buffer for each reaction that you require. If you require more reactions, then just scale up accordingly. Allow the antibody–Protein G agarose complex to incubate for 1 h at 4°C with gentle rotisserie mixing, then sediment at 14,000 × *g* for 1 min at 4°C. Next, aspirate supernatant to waste as listed previously in this step.

5. Measure 1 mL of the SDS-PAGE Stopping Buffer into a 1.5-mL Eppendorf tube and add 60 μL of β-mercaptoethanol to it. Mix thoroughly. Then, resuspend pellets from Step #4 in 40 μL SDS-PAGE Loading Buffer-containing β-mercaptoethanol and heat denature at 95°C for 7 min. Then, briefly vortex and finally sediment at 14,000 × *g* for 1 min at 4°C.
6. Load only the remaining supernatants after heat-denaturing into their own individual lanes on mini-SDS-PAGE gels (i.e., 1 mm thick, 10-well 4–20% gels). Equal concentrations of protein samples (100–500 ng) were analyzed by SDS-PAGE for 1 h at 100–200 V and subsequent western blot analyses to confirm the presence of overexpressed Rac2 and PLD2 proteins in the cell lysates and level of protein–protein interaction.
7. Primary antibodies specific for Rac2 or PLD2 (or their respective tags) and then HRP-conjugated secondary antibodies were incubated sequentially with PVDFs and products visualized using ECL reagents and autoradiograph film.

As shown in Fig. 1A, Rac2 and PLD2 are both associated in the cell, as both were coimmunoprecipitated with antibodies specific for the other protein. First, HA-tagged PLD2 was immunoprecipitated with Rabbit α-HA antibodies, and when western blots were probed using Mouse α-myc antibodies, the presence of Rac2 was detected (bottom panel), as was PLD2 (top panel). Conversely, myc-tagged Rac2 was immunoprecipitated with Rabbit α-myc antibodies, and when western blots were probed with Mouse α-HA antibodies, PLD2 was detected (Fig. 1B, bottom panel),

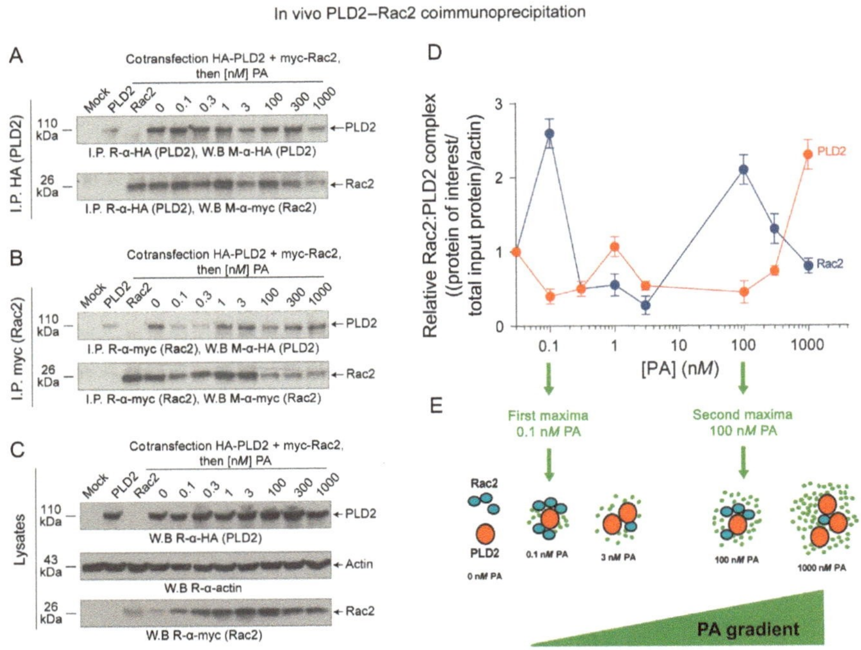

Fig. 1 Demonstration of a molecular association between Rac2 and PLD2. Coimmunoprecipitation (co-IP) of COS-7 cells overexpressing PLD2 and Rac2. Cells were transfected with HA-PLD2 and myc-Rac2 plasmids for 2 days, and PA was added directly to cells on the day of the experiment and then lysates were prepared. (A) Samples were used for co-IP with anti-HA antibodies (for PLD2), and western blots were probed with anti-myc antibodies to detect Rac2 (*bottom*) and with anti-HA antibodies to detect PLD2 (*top*). (B) Samples were used for co-IP with anti-myc antibodies (for Rac2), and western blots were probed with anti-HA antibodies to detect PLD2 (*top*) and with anti-myc antibodies to detect Rac2 (*bottom*). (C) Equal protein loading controls for both input PLD2, input Rac2, and total actin per lane prior to I.P. (D) Quantification of both Rac2 (*blue*) and PLD2 (*red*) binding from co-IP experiments shown in Fig. 1A–C in terms of protein of interest divided by total input protein of interest divided by actin. (E) Schematic representing effects of PA concentration on Rac2–PLD2-binding interaction. Note two maxima concentrations of PA that contribute to association or dissociation of Rac2 with PLD2 (0.1 and 100 n*M* PA). At maxima PA, PLD2 is optimally bound by Rac2 and not free to dissociate to function elsewhere in the cell independent of the small GTPase. Experiments were performed a total of three separate times ($n=3$).

as was Rac2 (Fig. 1B, top panel). To confirm the presence of overexpressed/input proteins, we performed SDS-PAGE and western blot analyses of myc-Rac2 or HA-PLD2 proteins that were overexpressed in COS-7 cells and equal protein loading controls are shown in Fig. 1C, as well as actin.

PLD2 and Rac2 interacted in cells (i.e., cells ectopically transfected with PLD2 + Rac2 plasmids) (Fig. 1A and B), and in both cases, the presence of exogenous PA resulted in biphasic fluctuations in the Rac2–PLD2 protein–protein interaction (Fig. 1D). We found two maxima concentrations of PA that contributed to association or dissociation of Rac2 with PLD2 (0.1 and 100 n*M* PA). At maxima PA, PLD2 is optimally bound by Rac2 and not free to dissociate to functions elsewhere independent of the small GTPase (Fig. 1E). The strength of the Rac2–PLD2 interaction (protein–protein), which was measured using co-IP, is a function of PA availability in cells or cell media at these two maxima PA concentrations of protein–protein binding. Fluctuations in the Rac2–PLD2 protein–protein-binding interaction like this would allow shuttling of Rac2 and/or PLD2 back and forth in the cell depending on the local concentration of PA in the cell.

3.3 Rac2 Activity (GTP Loading)

There are a number of cellular signaling pathways that contribute to the transport of PLD and the product of its enzymatic reaction PA in the cell. One mechanism that could shuttle PA around and throughout the cell is that of receptor trafficking via small GTPases that occurs in cells *in vivo* as a result of growth factor-dependent stimulation. As we have shown that PA affects the binding affinity of the Rac2–PLD2 protein–protein interaction, we concentrated this next study on the effect of PA on the Rac2 GTP-binding activity, which is important for vesicular trafficking inside the cell. The tools that we used to accurately measure Rac2 activity include Pak1-binding domain (PBD) pull-down assays using Rac2 binding to Pak1 that is immunoprecipitated using Pak1 antibodies and subsequent western blot detection using Rac2 antibodies and one half of the GTP–GDP exchange reaction using PLD2 as GEF for Rac2 via the $[^{35}S]GTP\gamma S$-binding reaction.

3.3.1 Material: Rac2 GTP-Binding Assays

- Purified, recombinant HA-tagged Rac2 and purified, myc-tagged PLD2 from overexpression of either the pBac-C1-6xHN-HA-Rac2-WT or the pBac-C1-6xHN-myc-PLD2-WT plasmids (Dr. Cambronero's lab) in Sf21 insect cells (Clontech)
- Rac2/cdc42 assay reagent (PAK1 PBD beads) (cat. # 14325) (EMD Millipore, Temecula, CA, USA)
- Rabbit α-HA IgG antibody to detect HA-tagged Rac2 (cat. # 3724) (Cell Signaling, Danvers, MA, USA)

- Mini-SDS-PAGE gels (cat. # 456–1094) (BioRad Laboratories, Hercules, CA, USA), Immobilon-P PVDF transfer membrane for western blotting (cat. # IPVH00010) (EMD Millipore, Temecula, CA, USA)
- SDS-PAGE Stopping Solution (1.04 g Tris base, 15 mL glycerol, 50 mg Bromophenol Blue, 9 g SDS in a final volume of 100 mL made up with good-quality water, pH adjusted to 6.7)
- Amersham ECL reagent (cat. # RPN2106) (GE Healthcare, Pittsburgh, PA, USA)
- Autoradiograph film (cat. # 100 NIF) (MidSci, St. Louis, MO, USA)
- COS-7 cells for transfection (cat. # CRL-1651) (ATCC, Manassas, VA, USA)
- Transit 20/20 transfection reagent (cat. # MIR 5400) (Mirus Bio, Madison, WI, USA)
- 1,2-Dioleoyl-*sn*-glycero-3-phosphate, sodium salt (DOPA) (cat. # 840875) (Avanti Polar Lipids, Alabaster, AL, USA)
- Liposome Buffer (0.5% BSA in PBS, pH 7.2) (50 mg BSA into 10 mL of 1 × PBS, pH 7.2, make certain to check the pH)
- RK5-myc-Rac2-WT, pcDNA3.1-HA-PLD2WT, and pcDNA3.1-HA-PLD2-K758R plasmids for overexpression in cells (Dr. Cambronero's lab)
- [^{35}S]-guanosine 5′-(γ-thio)triphosphate (cat. # NEG030X250UC) (Perkin-Elmer, Waltham, MA, USA)
- Whatman Protran BA85 nitrocellulose filter paper (cat. # 10401261), liquid scintillation vials, and Scintiverse BD liquid scintillation cocktail (cat. # SX18-4) (Fisher Scientific, Hanover Park, IL, USA)

3.3.2 Protocol

1. We first analyzed the GTP binding of purified, recombinant Rac2 in response to increasing PA *in vitro* using the PBD pull-down assay. Approximately, 19 pmol of purified Rac2 protein was incubated with increasing concentrations of DOPA (0–3 n*M*) for 20 min on ice in a 20-μL reaction.
2. Next, 5 μL of Rac2/cdc42 assay reagent (PAK-1-PBD agarose beads) was added to each sample and incubated at 4°C for 30 min in the presence of Magnesium Lysis Buffer (25 m*M* HEPES, 150 m*M* NaCl, 1% Igepal CA-630, 10 m*M* $MgCl_2$, 1 m*M* EDTA, and 2% glycerol).
3. Measure 1 mL of the SDS-PAGE Stopping Buffer into a 1.5-mL Eppendorf tube and add 60 μL of β-mercaptoethanol to it. Mix thoroughly. Then, the PA-containing, Rac2-PAK-1-PBD agarose bead

samples from Step #2 were combined with 40 μL SDS-PAGE Loading Buffer-containing β-mercaptoethanol and heat denatured at 95°C for 7 min. Then, briefly vortex and finally sediment at 14,000 × *g* for 1 min at 4°C.

4. Load only the remaining supernatants after heat-denaturing into their own individual lanes on mini-SDS-PAGE gels (i.e., 1 mm thick, 10-well 4–20% gels). Equal concentrations of protein samples (100–500 ng) were analyzed by SDS-PAGE for 1 h at 100–200 V and subsequent western blot analyses using α-HA antibodies to confirm the presence of recombinant, GTP-bound Rac2.
5. Primary antibodies specific for Rac2 (or its tag) and then HRP-conjugated secondary antibodies were sequentially incubated with PVDFs and products visualized using ECL reagents and autoradiograph film.

As shown in Fig. 2A, we found a biphasic effect of PA on Rac2 GTP loading with two maxima binding events at 0.1 and 100 n*M* PA, the effect of which is also graphically depicted in Fig. 2B. These data suggest that in the absence of PLD2, the product of the PLD reaction, PA, is sufficient to mediate a biphasic effect on Rac2 activity and that the PA lipid mediates Rac2 GTP binding.

3.3.3 Alternative Protocol

1. This pattern of PA-mediated fluctuation in Rac2 activity was also evident using an alternative method.
2. To examine if PLD2 could exchange GDP for [^{35}S]GTPγS on Rac2, 19 pmol PLD2-WT and 19 pmol Rac2 were incubated with 8 μ*M* GDP, 6 m*M* $MgCl_2$ (20 μL volume) for 10 min at room temperature.
3. GDP-bound Rac2 was added to 25 μL 100 μ*M* AMP-PNP, 1 m*M* $MgCl_2$, and 1 μ*M* [^{35}S]GTPγS in the absence or presence of PA (75 μL volume).
4. PA (0–3 n*M*) was added directly to the resulting cell lysates for 20 min at 37°C.
5. Aliquots (30 μL) were taken after the 20 min incubation period or for a different set of time (i.e., 0, 15, or 30 min) and diluted in 30 μL of STOP Buffer (10 m*M* HEPES, pH 7.4, 100 m*M* NaCl, and 5 m*M* $MgCl_2$). Samples were spotted onto Whatman BA85 nitrocellulose filters (2.53 cm^2), washed 3 × with ice-cold STOP Buffer, air-dried, and then placed into 7 mL scintillation vials containing 4–5 mL Scintiverse BD liquid scintillation cocktail.

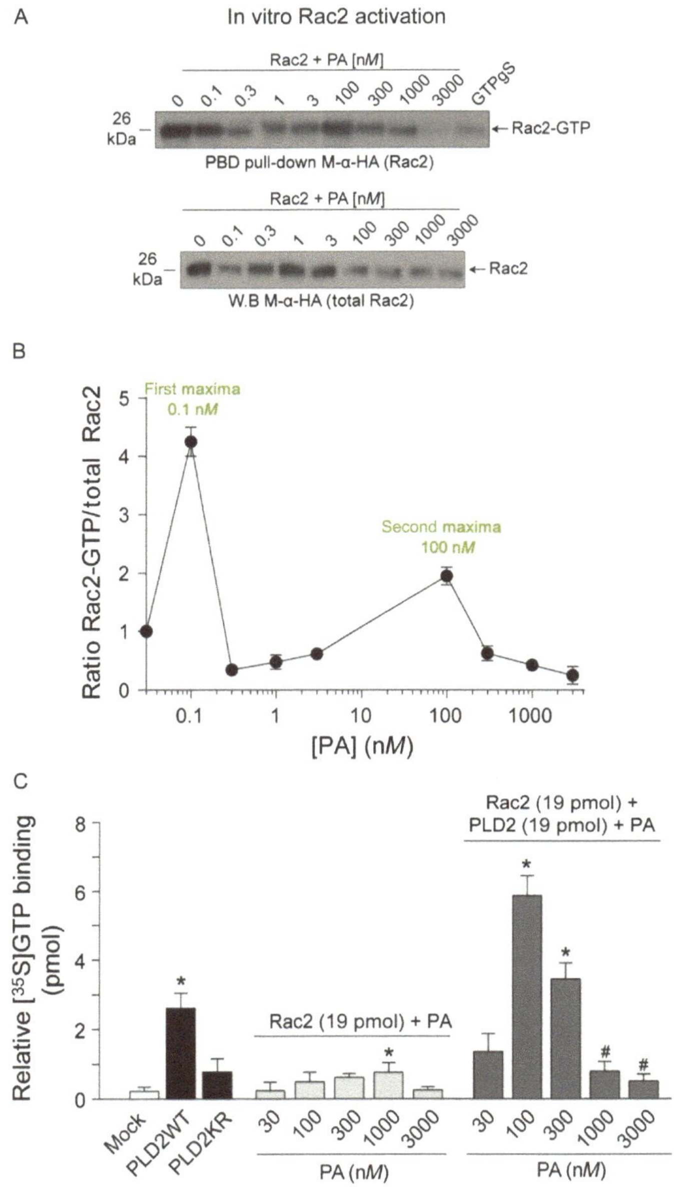

Fig. 2 Phosphatidic acid biphasically affects Rac2 GTP-binding activity. (A) *In vitro* Rac2 activation (PBD pull-down) assay using recombinant HA-Rac2 in the presence of increasing concentrations of PA (0–3 n*M*), as a readout of increased Rac2 GTPase activity due to PA. (B) Graphical representation of the ratio of Rac2-GTP divided by total Rac2 shown in (A). (C) 300 n*M* PA significantly increased PLD2-GEF-mediated Rac2 GTP loading. The (*) *symbols* denote statistically significant ($P < 0.05$) ANOVA increases between samples and controls. The (#) *symbols* denote statistically significant ($P < 0.05$) ANOVA decreases between samples and controls. Experiments were performed a total of three separate times ($n = 3$).

6. The relative amount of [^{35}S]GTPγS-bound Rac2 on the dried membranes was measured using liquid scintillation counting and expressed as picomoles bound, as in (Mahankali et al., 2011). Samples were counted in a liquid scintillation counter for 3 min intervals using an appropriate [^{35}S] protocol.
7. Purified, recombinant Rac2, PLD2-WT, and PLD2-K758R proteins were used as positive and negative controls, accordingly.

As shown in Fig. 2C, PLD2 and Rac2 interacted in cells that were ectopically transfected with PLD2 + Rac2 plasmids. Additionally, a fluctuation in Rac2 GTP loading via PLD2-GEF activity was evident in the presence of PA when PLD2 was cooverexpressed compared to Rac2 only overexpressing cells. Maximal Rac2 GTP binding occurred at 100 n*M* PA, while high concentrations of PA (>300 n*M*) resulted in inhibition of Rac2 activity. These data are consistent with that shown in Fig. 1 that used co-IP to measure the effect of PA on the Rac2–PLD2-binding interaction. PA-mediated fluctuations in the Rac2 GTP-binding activity like this could also contribute to shuttling of Rac2 via a lipid-binding mechanism, potentially as a result of PLD interaction.

3.4 FRET Stoichiometry for PLD2 and Rac2 Binding

FRET is a useful tool to investigate the spatial and molecular proximity of the proteins contained within a protein complex (Best et al., 2007; Gambin & Deniz, 2010; Kong, Polte, Alsberg, & Mooney, 2005). FRET can provide contrast between colocalized proteins of interest within a protein–protein complex better than that of conventional optical microscopy. Three criteria must be met for successful FRET: there must be a close proximity between the donor and acceptor molecules (<100 Å), the absorption spectrum of the acceptor molecule needs to have overlap with the emission spectra of the donor molecule, and the transition dipole orientations between the donor/acceptor transition need to be virtually parallel (Iqbal et al., 2008; VanBeek, Zwier, Shorb, & Krueger, 2007). Since we know that Rac2 and PLD2 can be coimmunoprecipitated and that fluorescence microscopy of stimulated leukocytes revealed an arc of Rac2 at the leading edge of pseudopodia, while PLD2 was physically posterior to the wave of Rac2 during chemotaxis (Peng, Henkels, Mahankali, Dinauer, et al., 2011; Peng, Henkels, Mahankali, Marchal, et al., 2011; Speranza et al., 2013, 2014), we hypothesized that PLD2 and its binding partner Rac2 would be suitable subjects for further FRET analysis.

3.4.1 Material: FRET Between Rac2 and PLD2

- pCerulean-C1-HA-Rac2-WT, mCit-C1-mycPLD2-WT, and mCit-N1-HA-PLD2-WT plasmids (from Dr. Cambronero's lab)
- COS-7 cells for transfection (cat. # CRL-1651) (ATCC, Manassas, VA, USA)
- FuGENE 6 transfection reagent (cat. # E2691) (Promega, Madison, WI, USA)

3.4.2 Protocol

1. To further confirm PLD2 interaction with Rac2 in the cell, the highly sensitive FRET technique was used.
2. We used mCitrine-C1-mycPLD2W-WT that contained a YFP tag as the FRET acceptor and pCerulean-C1-HA-Rac2-WT that contained a CFP tag as the FRET donor (Fig. 3A), similar to (Peng, Henkels, Mahankali, Dinauer, et al., 2011).
3. Pilot experiments were performed to determine relative levels of expression of the fluorescently tagged overexpressed Rac2 and PLD2 as determined by immunofluorescence microscopy.
4. COS-7 cells were mock-transfected or transfected with 0.5 μg CFP-Rac2 or 2 μg mCit-PLD2 plasmids in the presence of 5–8 μL Mirus Transit 20/20 transfection reagent and 48 h posttransfection, media was aspirated away from cells that were then fixed onto glass coverslips using 1 mL 4% paraformaldehyde for 10 min.
5. Cells were subsequently stained with subsequent incubations using 1:200 phalloidin-TRITC in PBS for 20 min and 1:2000 DAPI in PBS for 5 min to visualize polymerized actin and nuclei as cellular reference markers. Each of these incubations was followed by three washes with 1 mL PBS.

Resulting images shown in Fig. 3B are from merging relevant images of interest with companion TRITC- and DAPI-stained images, which indicate overexpression of FRET-tagged Rac2 and PLD2 in cells. The next logical step was to overexpress both proteins simultaneously using cotransfections prior to FRET analysis. As shown in Fig. 3C, there was considerable colocalization of both CFP-tagged Rac2 + mCit-tagged PLD2 to the nucleus and vesicular locations, which suggests potential protein–protein interaction between PLD2 and Rac2. Vesicles could be used for both the delivery and retrieval of PLD2 to the plasma membrane during agonist stimulation once the small GTPase Rac2 has been delivered to the plasma membrane.

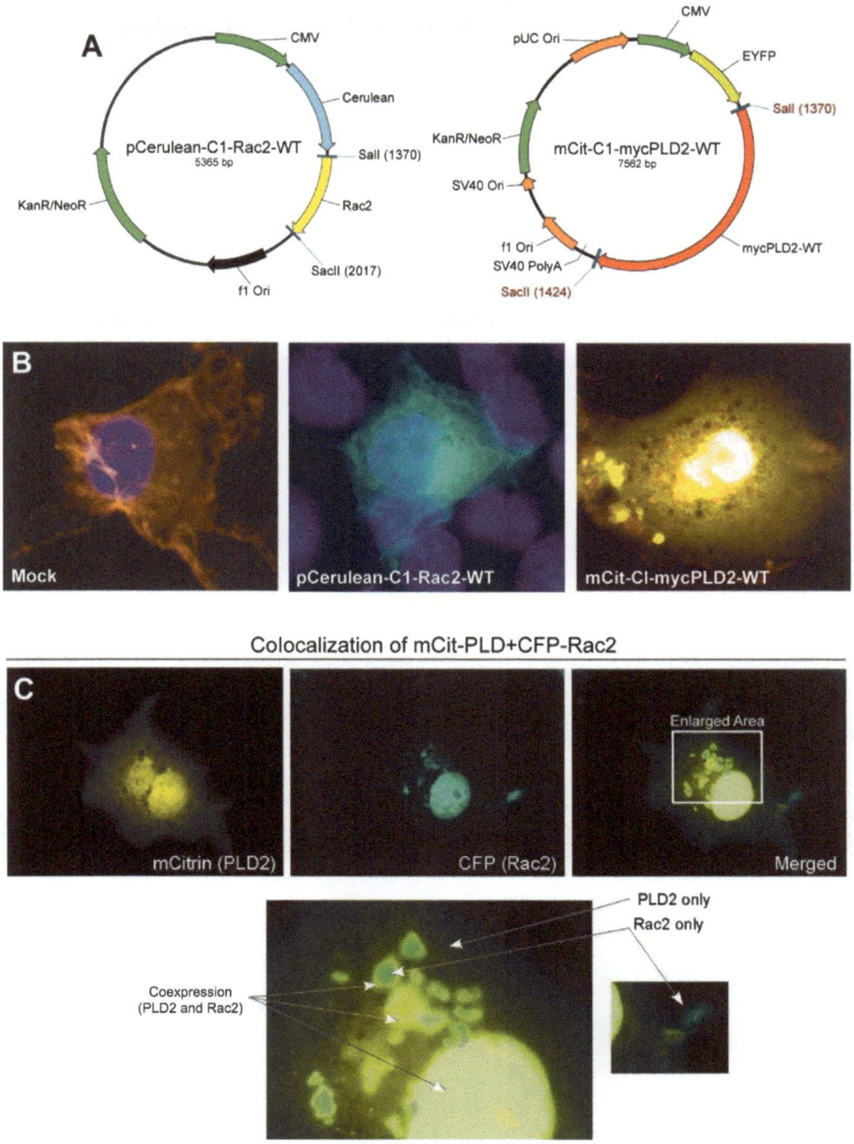

Fig. 3 Overexpressed fluorescently tagged PLD2 and Rac2 colocalize in cells. (A) Vector maps of fluorescently tagged Rac2 and PLD2 plasmids used for FRET. (B) Immunofluorescence microscopy of mock-treated or overexpressed CFP-Rac2 and mCit-PLD2 proteins in COS-7 cells, respectively. Polymerized β-actin was visualized using phalloidin-TRITC staining for mock- and PLD2-transfected cells only, and nuclei were stained in all samples with DAPI. Resulting images shown are from merging relevant image of interest with companion TRITC- and/or DAPI-stained images. (C) Immunofluorescence microscopy of COS-7 cotransfected with CFP-tagged Rac2+mCit-tagged PLD2, as either independent images or merged image. *Lower panels*, enlarged area of merged image. Shown in this figure are images representative of $n=3$ independent experiments with visualization of five similar fields from each condition.

To establish the existence of FRET between PLD2 + Rac2, COS-7 cells were cotransfected with equimolar concentrations of PLD2-mCitrine and Rac2-mCFP using FuGENE 6 reagent for 48 h (the cells were serum-starved for the last 16 h) so that, therefore, neither protein was in excess following overexpression. When ready for microscopy, cells were left unstimulated. Fluorescence microscopy was performed with appropriate excitation and emission filter sets (Chroma Corporation) for CFP, YFP, and FRET (CFP excitation and YFP emission). Images were collected on a 14-bit Cascade HQII digital camera and were processed as described in Hoppe et al. to determine the FRET stoichiometry of the interaction between PLD2 and Rac2 (Hoppe, Christensen, & Swanson, 2002).

Fluorescence imaging showed FRET microscopy localized to the nucleus in the absence of any stimulation in the following images shown in Fig. 4A: I_a (PLD2-mCitrine), I_d (Rac2-mCFP), and I_f (FRET) images, as labeled on the left. Calculated FRET images are E_a, E_d, and R_m, as labeled on the right. The mean average FRET efficiency was predominantly in the

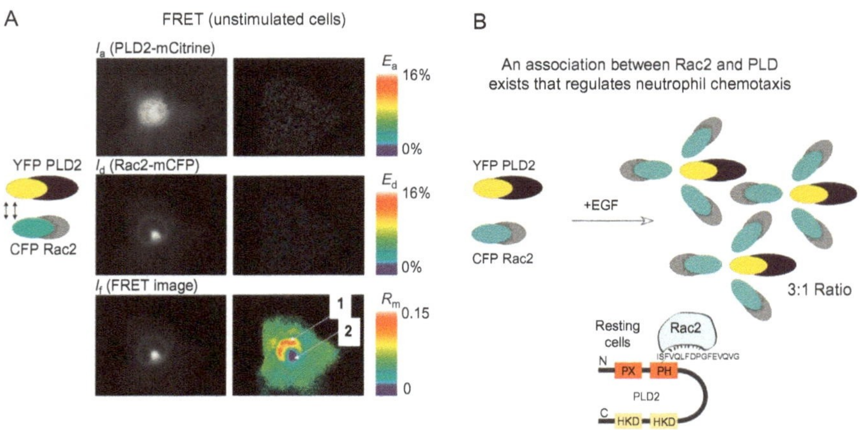

Fig. 4 Demonstration of FRET between fluorescent-tagged PLD2 and Rac2. (A) COS-7 cells were transfected with mCit-PLD2 (FRET acceptor) and CFP-Rac2 (FRET donor) for 48 h. When ready for FRET, cells were left unstimulated. Microscopic images shown FRET in the nucleus: I_a (mCit-PLD2), I_d (CFP-Rac2), and I_f (FRET) images. *Color panels* on right show predicted FRET images: E_a, E_d, and R_m, magnitude of E_a, E_d, or the ratio of the acceptor to the donor molecules (R_m), respectively. (B) Model of PLD2–Rac2 stoichiometry during neutrophil chemotaxis. Rac2 has a dual effect on PLD2 that controls the onset and termination of leukocyte chemotaxis, which is mediated by the accumulation of excess Rac2 GTPase compared to PLD2 in the cells.

nucleus (*E*avg) as determined using Metamorph and Matlab software and which reached a maximum of 0.15. For comparison, the linked CFP-Citrine positive control used for calibration of the microscope yielded an *E*avg of 0.37. Control experiments cooverexpressing mCitrine-PLD2 and free-CFP showed no FRET. The existence of FRET using mCit-PLD2 and CFP-Rac2 overexpression under equimolar conditions shows that PLD2 and Rac2 are in close enough proximity in the cell's nucleus to form a stable heterodimeric complex (<7 nm distance).

Data shown here and in other studies suggest that cytoplasmic localization of Rac2 + PLD2 to vesicles is both dependent and independent of cell stimulation. Rac2 interacts with and binds to PA-containing vesicles in neutrophil cytosol, which suggests that PLD2 mediates this process (Faugaret, Chouinard, Harbour, El azreq, & Bourgoin, 2010). Further, Rac2 has a dual effect on PLD2 such that a positive-feedback loop controls the onset of leukocyte chemotaxis, while Rac2-led PLD2 inhibition mediates the termination of leukocyte chemotaxis, which is mediated by the accumulation of excess Rac2 GTPase compared to PLD2 in the cells (Fig. 4B; Peng, Henkels, Mahankali, Marchal, et al., 2011).

3.5 Detection of PA in the Cells With a Biosensor

The use of a biosensor as a tool to detect cellular PA has been extensively characterized from sporulation-specific protein 20 (Spo20p), which is a yeast protein required for the fusion of exocytic vesicles with the plasma membrane during yeast sporulation through its interactions with the soluble *N*-ethylmaleimide sensitive-factor attachment protein receptor (SNARE) complex (Nakanishi, de los Santos, & Neiman, 2004; Zeniou-Meyer et al., 2007; Zhang et al., 2014). Spo20p is a dual function protein that contains both an inhibitory region that sequesters the protein in the nucleus and a positive regulatory region that binds to phospholipids in the cell membrane (Nakanishi et al., 2004). This positive regulatory region comprises a 40-amino acid region (51–91) containing an amphipathic helix with hydrophobic and positive charges that binds to cell membrane phospholipids and preferentially to PA, which has been termed the PABD. Wild type and mutant enhanced green fluorescent protein (EGFP)-tagged Spo20PABD plasmids have been used to visualize PA localization in mammalian cells in previous studies by many different labs. Some of the difficulties other researchers have encountered while using this EGFP-tagged PABD sensor include lack of response using moderate stimulation or ligands specific to membrane-bound receptors. EGFP-tagged proteins have been widely utilized to monitor cellular processes, which is fully dependent on EGFP being a completely passive reporter protein.

Evidence suggests that two populations of EGFP have the potential to exist in cells, one being cytosolic and monomeric and the other being secretory, nuclear, and oligomeric (Chen, Muller, Ruan, & Gratton, 2002; Jain, Joyce, Molinete, Halban, & Gorr, 2001). The use of EGFP-tagged vectors has evolved to the next generation of monomeric pAcGFP plasmids to circumvent the issue of EGFP oligomerization (Gurskaya et al., 2003; Matz et al., 1999). Therefore, the Spo20 PABD sequence comprising amino acids 51–91 (MDNCSGSRRRDRLHVKLKSLRNKIHKQLHP NCRFDDATKTSDDKC) was subcloned into the pAcGFP-C1 and pAcGFP-N1 vectors using *Bgl*II and *Eco*RI or *Xho*I and *Bam*HI, respectively (Mutagenex, Suwanee, GA, USA) (Fig. 5A and B), which we used as our biosensor for cellular PA localization following transient transfections (Fig. 5C).

3.5.1 Material: PLD and the PA Sensor Colocalize in Cells

- pAcGFP-C1-PABD-WT, pcDNA3.1-mycPLD2-WT, and pcDNA3.1-mycPLD2-K758R *plasmids* (Cambronero's lab)
- 5-Fluoro-2-indolyl des-chlorohalopemide (FIPI) (cat. # 13563) (Cayman Chemical, Ann Arbor, MI, USA)
- Rabbit α-PLD2 IgG (H-133) antibody and goat antirabbit FITC IgG antibody (cat. # sc-25513 and sc-2012, respectively) (Santa Cruz Biotechnology, Santa Cruz, CA, USA)
- 4′,6-Diamidino-2-phenylindole (DAPI) (cat. # D1306) (Invitrogen, Carlsbad, CA, USA)
- Human epidermal growth factor (EGF) (cat. # 300-110P) (Gemini Bio-Products, West Sacramento, CA, USA)
- COS-7 cells (cat. # CRL-1651) (ATCC, Manassas, VA, USA)
- Transit 20/20 transfection reagent (cat. # MIR 5400) (Mirus Bio, Madison, WI, USA)
- Starvation Media (DMEM + 0.1% BSA, sterile filtered)
- Glass coverslips and microscope slides
- Vectashield immunofluorescence mounting media (cat. # H-1000) (Vector Labs, Burlingame, CA, USA)

We next studied intracellular localization of endogenous and overexpressed PLD2 compared to that of the overexpressed PA sensor. Our rationale is that

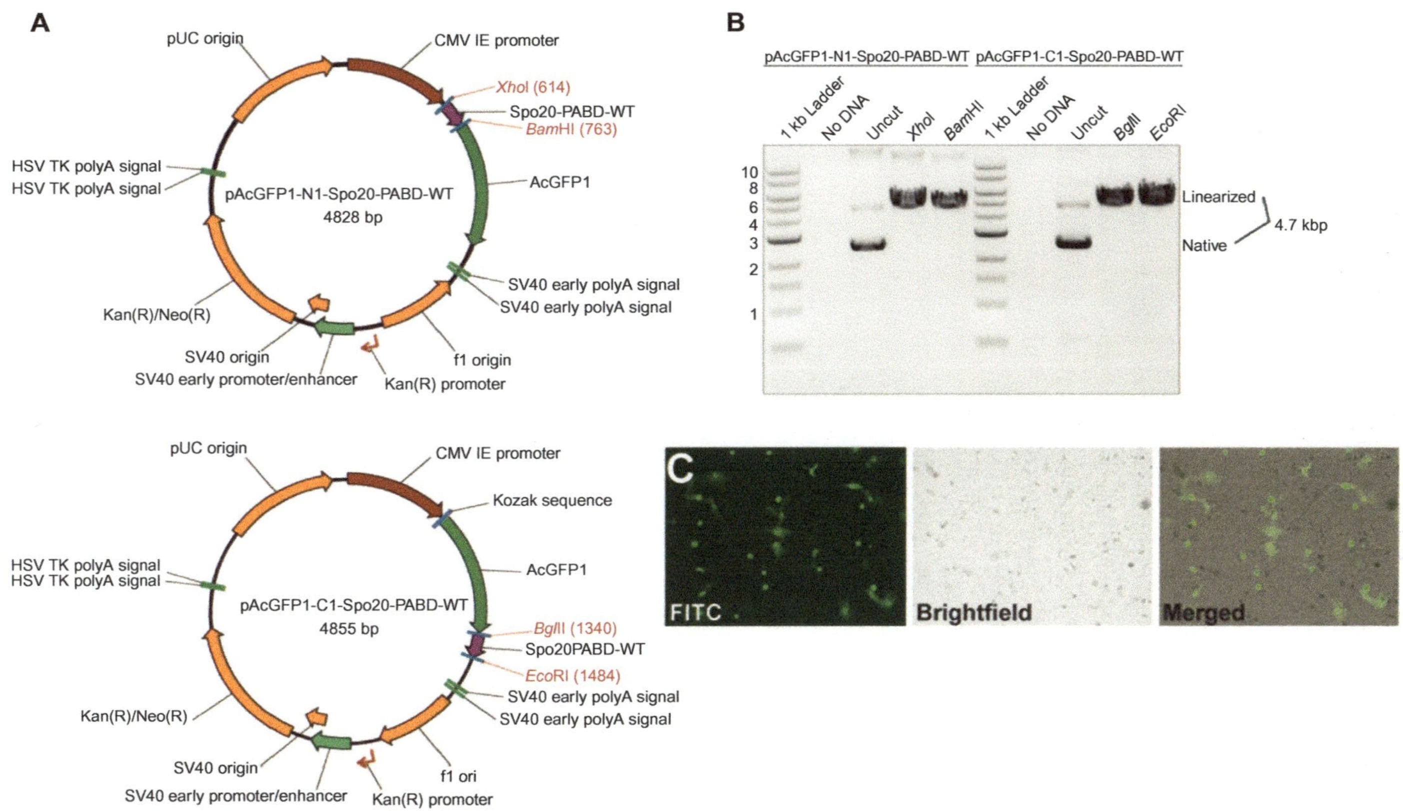

Fig. 5 Generation of a new PABD sensor into a monomeric GFP vector. (A) Vector maps of N-terminal and C-terminal Spo20-PABD AcGFP-tagged plasmids used for colocalization with PLD protein in cells. (B) Restriction enzyme digestion of cloning scheme: Spo20 PABD was cloned into pAcGFP-N1 vector using *Xho*I and *Bam*HI, while *Bgl*II and *Eco*RI were used for cloning into the pAcGFP-C1 vector. (C) Immunofluorescent and brightfield microscopic images (20 ×), indicating a positive result for overexpression of pAcGFP-C1-Spo20-PABD-WT into COS-7 cells 48 h posttransfection of 1 μg of plasmid DNA.

if the PABD of the PA sensor actually binds to PA then we would expect a similar localization pattern as that of PLD2. For this to occur, we confirmed the cellular localization pattern of both PLD2 and the PA sensor using mammalian cells.

3.5.2 Protocol

1. COS-7 cells were plated onto glass coverslips contained in 35 mm wells or 6-well plates and then once adhered were transfected with 2 μg of the GFP-tagged Spo20-PABD sensor or the myc-tagged PLD2 (either WT or lipase-inactive K758R) plasmids using 5–8 μL Transit 20/20 transfection reagent per each transfection reaction.
2. Forty-eight hour posttransfection cells were serum-starved for 2 h using 1 mL Starvation Media per each well. At this point, cells were next untreated or treated with the addition of a final concentration of 300 n*M* FIPI added directly to the 1 mL Starvation Media already on the cells for 20 min and then unstimulated or stimulated with a final concentration of 10 n*M* human EGF for 15 additional minutes. (*Note*: to prepare the FIPI and EGF stock solutions, initially prepare each stock solution so that it is ~1000 × more concentrated than the final required concentration. Therefore, prepare 300 μ*M* FIPI and a 10 μ*M* EGF stock solutions. Also, check the manufacturer's recommendations for initial solubilization for each compound. Once correctly solubilized, each compound can then be diluted accordingly into the correct/relevant buffer.)
3. Cells were fixed with 1 mL of 4% paraformaldehyde, permeabilized with 0.1% Triton X-100 in PBS, pH 7.4, blocked for 4 h in blocking buffer (5% BSA in TBS-T), stained with α-PLD2-TRITC antibodies in blocking buffer (5% BSA in TBS-T) for 1 h to overnight to visualize either overexpressed PLD2 or endogenous PLD2, washed 3 × with 1 mL PBS for 5 min each wash and then with 1 mL of 1:2000 DAPI in PBS for 5 min to visualize nuclei, and then lastly followed by another set of washes 3 × with 1 mL PBS for 5 min each wash. Coverslips were washed once briefly with water and then allowed to air-dry. Once dried, coverslips are then mounted onto glass microscope slides using Vectashield mounting media. Allow slides to set overnight and then paint the edge of the coverslip with clear nail polish to seal the slides. Once the nail polish is completely dry, it is safe to proceed to microscopy.

4. The PA sensor was visualized using its GFP tag and therefore a FITC filter. PLD2 was visualized using the TRITC filter, while DAPI was visualized using the UV filter.
5. Merged images of overlaid FITC and DAPI images or TRITC and DAPI images are shown.

Immunofluorescence microscopic images of COS-7 cells indicate that overexpression of either the Spo20-derived PABD sensor or PLD2-WT resulted in nuclear/perinuclear accumulation of the PA sensor and PLD2-WT in the absence of EGF stimulation (control) (Fig. 6A, top panels), which was abrogated following EGF stimulation for 7–15 min (Fig. 6A, bottom panels). Additionally, preincubation of cells with either a PLD-specific inhibitor (300 n*M* FIPI) or overexpression of the lipase-inactive PLD2-K758R resulted in reduced nuclear localization of PLD independent of EGF stimulation. Following EGF stimulation, both the PABD sensor and PLD2-WT moved to cellular locations outside of the nucleus/perinucleus (Fig. 6A, lower panels).

This change in PA sensor localization was also evidenced as a change in the ratio of nuclear vs cytoplasmic staining of the PA sensor, which moved from Golgi/perinuclear to plasma membrane localization following EGF stimulation (Fig. 6B). Cooverexpression of both the PABD sensor and PLD2-WT resulted in colocalization in and around the nucleus in the absence of EGF (Fig. 6C). Collectively, these data suggest colocalization and trafficking of the PA sensor with PLD2 occurs in cells as a result of EGFR stimulation, which is dependent on catalytically active PLD2 and therefore the product of its reaction, PA.

3.6 PA Measurement in Tissues and After PLD Inhibitors: Lipid Extraction From Cells and Tissues

PA is a relatively minor cellular phospholipid. PA can be generated by sequential acylation of glycerol 3-phosphate which is a key step in the de novo synthesis of some glycerophospholipids and triglycerides (Athenstaedt & Daum, 1999). PA can also be formed by phosphorylation of diacylglycerol, and indeed PA and DG can be interconverted in cells through the actions of the relevant kinases and phosphatases (Merida, Avila-Flores, & Merino, 2008). As discussed elsewhere, PLD generates PA by hydrolysis of PC. PLD has the unique property of catalyzing transphosphatidylation reactions in which primary alcohols replace water as the nucleophile in the hydrolysis step of the PLD reaction. PLD-catalyzed transphosphatidylation generates phosphatidylalcohols which are not

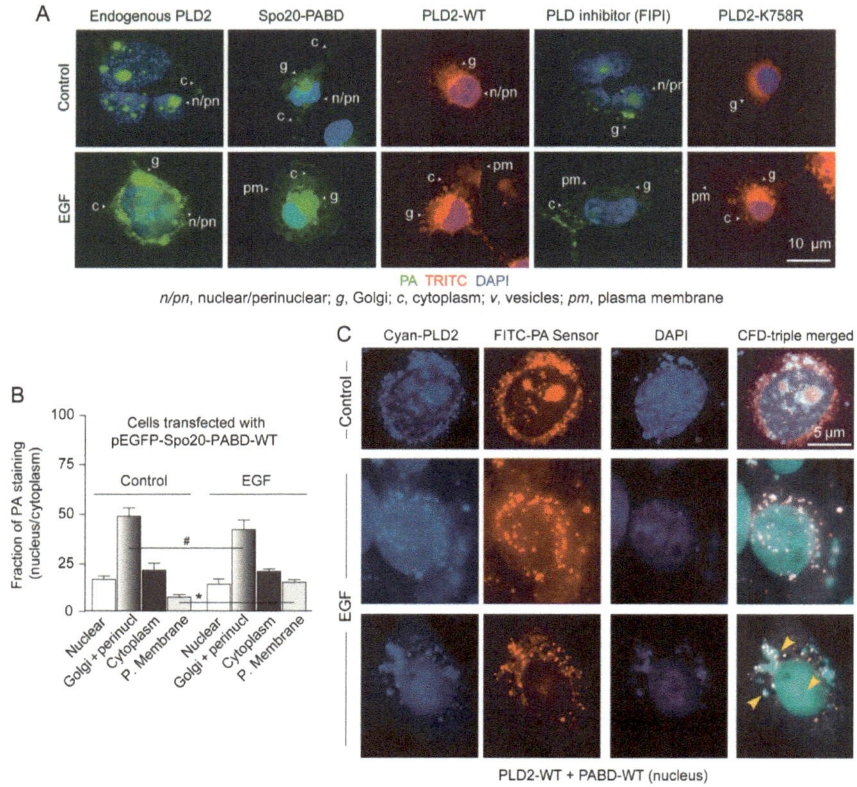

Fig. 6 PLD2 and PA colocalize to the nucleus. (A) COS-7 cells were fixed and then stained with α-PLD2-FITC antibodies to visualize endogenous PLD2 (in the absence or presence of the PLD-specific inhibitor FIPI), with α-PLD2-TRTC antibodies to visualize overexpressed PLD2 (-WT and -K758R) and/or with DAPI to visualize nuclei. The PA sensor was visualized using its GFP tag. Merged images of overlaid FITC and DAPI images or TRITC and DAPI images are shown. Abbreviations: *n/pn*, nuclear/perinuclear; *g*, Golgi; *c*, cytoplasm; *v*, vesicles; *pm*, plasma membrane. Scale bar = 10 μm. (B) Quantification of cellular immunofluorescence PA staining initially shifted from the Golgi and perinucleus (resting state) after 15 min incubation with 10 n*M* EGF to the plasma membrane using ImageJ software. The (*) *symbols* above *bars* denote statistically significant ($P < 0.05$) ANOVA increases between samples and controls. The (#) *symbols* above *bars* denote statistically significant ($P < 0.05$) ANOVA decreases between samples and controls. (C) Initial nuclear colocalization of both overexpressed PLD2-WT and overexpressed PA sensor in the absence of EGF is then colocalized away from the nucleus and out into the cytoplasm. Shown in this figure are images representative of $n = 3$ independent experiments with visualization of five similar fields in each condition. Scale bar = 5 μm.

naturally occurring in most cells and tissues and can be separated from other phospholipids by thin-layer chromatography (Liscovitch, Czarny, Fiucci, & Tang, 2000). Phosphatidylalcohols also appear to be more metabolically stable than PA, so their levels offer a robust way to report PLD activity in cells and tissues. However, the relative efficiency of the PLD-catalyzed hydrolysis and transphosphatidylation reactions in cells and tissues is difficult to determine and the relatively high concentrations of alcohols needed to achieve effective transphosphatidylation in cells and tissues are sufficiently high that nonspecific effects on cellular signaling and other biological processes remain a concern, as does the possibility that accumulation of phosphatidylalcohols has biological effects.

Accordingly, we have developed methods to directly measure changes in PA levels in cells and tissues that can be used to monitor PLD activity or test the efficacy of small-molecule PLD inhibitors (McDermott, Sigal, Crump, & Morris, 2006; Ren et al., 2010). These methods employ HPLC-coupled electrospray ionization tandem mass spectrometry on triple quadrupole instruments. Because PA is a family of lipids containing different fatty acids at the sn1 and sn2 groups of the glycerol moiety, these methods necessarily detect PA molecular species on the basis of the mass of their molecular ions, which is indicative of the total number of carbon atoms and degree of saturation of these fatty acid substituents. These methods can be modified to monitor fatty acid-derived anions generated by collisional dissociation of the molecular ion species. However, the methods described later monitor a common product ion species derived from the glycerol phosphate moiety which ionizes effectively ion negative ionization mode.

3.6.1 Material

- 8 mL Borosilicate glass tubes with Teflon cap
- PBS
- MeOH, $CHCl_3$, 0.1 *M* HCl
- C17 lysophosphatidic acid (LPA) (cat. # 857127), C-17 sphingosine 1-phosphate (S1P) (cat. # 860641) (both from Avanti Polar Lipids, Alabaster, AL, USA)

3.6.2 Protocol: Lipid Extraction From Cells and Tissues (Method Is Scaled for 35 mm Diameter Dishes/6-Well Plates)

1. Transfer cells to ice, remove culture medium, wash in ice-cold PBS, remove all liquid from cells.

2. Add 1 mL ice-cold MeOH to cells, scrape, transfer to borosilicate glass tube.
3. Add another 1 mL ice-cold MeOH to cells, scrape, transfer to same borosilicate glass tube.
4. Add 1 mL $CHCl_3$ and 0.5 mL 0.1 *M* HCl to tube.
5. Add internal standard (5–50 pmol C17 LPA, C-17 S1P as appropriate).
6. Vortex for 5 min (works best if you have a multitube vortexer).
7. Add 1 mL $CHCl_3$ and 1.3 mL 0.1 *M* HCl to separate phases.
8. Vortex for 5 min.
9. Centrifuge at $>3000 \times g$ for 10 min.
10. Transfer lower phase to 4 mL screw cap vial using a Pasteur pipette. Take care to avoid the protein interface and upper phase. Leave behind ~50 μL if needed (do not worry about being more accurate than this, as we have an internal standard to correct for recovery).
11. Evaporate to dryness under N_2 using N-evaporation in fume hood.
12. Resuspend in 1 mL 1:4 $CHCl_3$:MeOH.
13. If desired, remove 0.1–0.2 mL to a borosilicate 12 × 17 tube for phosphate determination.
14. Dry remaining sample under N_2, resuspend in 0.1 mL MeOH, vortex, and allow to air-dry for ~10 min.
15. Transfer to autosampler vial with glass or polypropylene insert for LC–MS analysis.

3.6.3 Protocol: Measurement of PA by HPLC ESI MS/MS

Extracted lipids are analyzed by reverse phase HPLC-coupled electrospray ionization tandem mass spectrometry. Studies reported from our laboratory have used ABSciex 4000 and 6500 series Q-Trap (hybrid triple quadrupole/linear ion trap) instruments operated in selected ion monitoring mode. These methods could be adapted to work with any gradient HPLC system coupled to a triple quadrupole mass spectrometer although in that case the instrument settings detailed later would have to be optimized for the selected system. The HPLC conditions and mass spectrometer settings used in our laboratory with an ABSciex 4000 Q-Trap system and a Shimadzu binary HPLC system are listed later.

LC conditions:

1. Column: Agilent Zorbax Eclipse XDB C8 column, 5 μm, 4.6 × 150 mm.
2. Solvent A: 75:25 = methanol/water, 0.5% formic acid + 0.1% ammonium formate.

Table 1 Gradient for LC Conditions

Time (min)	%B
0	0
2	0
3	100
11	100
12	100

3. Solvent B: 80 (99/1 = methanol/water, 0.5% formic acid + 0.1% ammonium formate): 20 (chloroform).
4. Flow rate: 0.5 mL/min; injection volume: 10 μL; oven temperature: 30°C; equilibration time: 3 min.
5. The gradient that should be used is shown in Table 1.

MS parameters (negative ESI):

1. Declustering potential (DP): −120.0 V; collision energy (CE): −56.0 V; collision cell exit potential (CXP): −13.0 V; CAD gas: medium; curtain gas: 20; ion spray voltage: −4500 V; temperature: 550.0°C; gas 1: 40.0 psi; gas 2: 40.0 psi.
2. The precursor and product ion masses monitored for the indicated PA species are shown in Table 2.

The data generated are integrated to determine chromatographic peak areas for the indicated ion pairs. In the method described earlier, we included C17 LPA as an internal standard hich was quantitated separately by HPLC MS/MS to calculate recovery of lipids throughout the procedure. The individual PA species were then quantitated by correcting integrated peak areas from the extracted ion chromatograms and use of external calibrations generated using authentic PA standards obtained from commercial sources that were independently quantitated by phosphorous determination after wet digestion in perchloric acid.

4. CONCLUSIONS AND OUTLOOK

Numerous biochemical approaches exist to discern PLD and PLD-mediated PA interactions with downstream signaling pathways in the cells, which we have defined herein to assist other researchers interested in phospholipases and phospholipid metabolism. Recent studies have focused on developing methods to directly measure changes in PA levels in cells and

Table 2 MS Parameters (Negative ESI) for Precursor and Product Ion Masses Monitored for the Indicated PA Species

		Q1 (Da)	Q3 (Da)	Time (ms)	DP (V)	CE (V)	CXP (V)
34-1	16:0–18:1(9z)	690.52	153	150	−120	−56	−13
43-6	21:0–22:6	806.58	153	150	−120	−56	−13
38-4	18:0–20:4	740.53	153	150	−120	−56	−13
34-1	16:0–18:1(11z)	673.49	153	150	−120	−56	−13
24-0	12:0–12:0	535.35	153	150	−120	−56	−13
28-0	14:0–14:0	591.41	153	150	−120	−56	−13
30-0	15:0–15:0	619.44	153	150	−120	−56	−13
32-0	16:0–16:0	647.47	153	150	−120	−56	−13
36-0	18:0–18:0	703.54	153	150	−120	−56	−13
36-4	18:2–18:2	695.47	153	150	−120	−56	−13
36-6	18:3–18:3	691.44	153	150	−120	−56	−13
32-1	14:0–18:1	645.46	153	150	−120	−56	−13
20-0	10:0–10:0	479.29	153	150	−120	−56	−13
34-0	16:0–18:2	671.47	153	150	−120	−56	−13
38-4	18:0–20:4	723.5	153	150	−120	−56	−1
36-4	16:0–20:4	699.5	153	150	−120	−56	−1
36-2	18:1–18:1	759.6	153	150	−120	−56	−1
40-0	18:0–22:0	701.52	153	150	−120	−56	−1
36-1	18:0–18:1	719.47	153	150	−120	−56	−1
38-6	16:0–22:6	747.5	153	150	−120	−56	−1
40-6	18:0–22:6	423.226	152.9	150	−95	−30	−7

tissues, which can be used to monitor PLD activity and even determine the effectiveness of small-molecule PLD inhibitors. Current and ongoing research interest in this field incorporating the use of PLD1 and PLD2-deficient mice and/or PLD-specific small-molecule inhibitors has identified possible roles for these two PLD isoforms in the development of cardiovascular disease and tumorigenesis, brain development, lipid metabolism, and leukocyte functions. PLD1 and PLD2 knockout mice are more prone to be overweight, maintain a higher percent body fat, and develop diabetes as a function of age as a result of increased appetite and caloric intake compared to control mice (Trujillo Viera, El-Merahbi, Nieswandt, Stegner, & Sumara, 2016). Additionally, genetic and pharmacologic inhibition of PLD in mice results in protection against formation of stationary blood clots leading to improved stroke outcome (Elvers et al., 2010; Stegner et al., 2013). Tumor vascularization of melanoma or lung carcinoma was abolished in the tumor microenvironment of PLD1$^{-/-}$ mice (Chen et al., 2012), while use of PLD2-specific small-molecule inhibitors contributed to suppression of primary tumor growth and secondary metastases (Henkels, Boivin, Dudley, Berberich, & Gomez-Cambronero, 2013). Newborn mice lacking the two most common PLD isoforms suffer from diminished brain growth and astroglial proliferation, while adult mice deficient in PLD have impaired cognitive function, possibly due to significantly decreased brain acetylcholine levels (Burkhardt, Beyer, & Klein, 2015; Burkhardt et al., 2014). Lymphocytes from PLD1-deficient mice are less prone to adhere and migrate *in vitro* in mouse models of blood–brain barrier and cell migration (Gobel et al., 2014). Overall, future interest in PLD signaling research will undoubtedly focus on improving drug delivery and therapeutic effectiveness in many different models of human disease.

ACKNOWLEDGMENTS

Dr. Joel Swanson (University of Michigan School of Medicine) originally supplied the mCitrine plasmids used for subcloning of PLD2 and Rac2 for FRET analysis and also performed the FRET analyses. We would like to thank Denise Kramer for her help generating the PA sensor plasmids.

Funding: The following grants to Dr. Cambronero (J.G-C.) have supported this work: HL05665314 from the National Institutes of Health (NIH), 13GRNT17230097 from the American Heart Association (AHA), and 229102 20053 from the WSU Boonshoft School of Medicine (BSOM). Research in Dr. Morris's laboratory is supported by grants from the NIH, Department of Veterans Affairs, and Department of Defense.

REFERENCES

Athenstaedt, K., & Daum, G. (1999). Phosphatidic acid, a key intermediate in lipid metabolism. *European Journal of Biochemistry*, *266*, 1–16.

Bae, C. D., Min, D. S., Fleming, I. N., & Exton, J. H. (1998). Determination of interaction sites on the small G protein RhoA for phospholipase D. *The Journal of Biological Chemistry*, *273*, 11596–11604.

Best, R. B., Merchant, K. A., Gopich, I. V., Schuler, B., Bax, A., & Eaton, W. A. (2007). Effect of flexibility and cis residues in single-molecule FRET studies of polyproline. *Proceedings of the National Academy of Sciences of the United States of America*, *104*, 18964–18969.

Brandolini, L., Bertini, R., Bizzarri, C., Sergi, R., Caselli, G., Zhou, D., et al. (1996). IL-1 beta primes IL-8-activated human neutrophils for elastase release, phospholipase D activity, and calcium flux. *Journal of Leukocyte Biology, 59*, 427–434.

Burkhardt, U., Beyer, S., & Klein, J. (2015). Role of phospholipases D1 and 2 in astroglial proliferation: Effects of specific inhibitors and genetic deletion. *European Journal of Pharmacology, 761*, 398–404.

Burkhardt, U., Stegner, D., Hattingen, E., Beyer, S., Nieswandt, B., & Klein, J. (2014). Impaired brain development and reduced cognitive function in phospholipase D-deficient mice. *Neuroscience Letters, 572*, 48–52.

Chae, Y. C., Kim, K. L., Ha, S. H., Kim, J., Suh, P. G., & Ryu, S. H. (2010). Protein kinase Cdelta-mediated phosphorylation of phospholipase D controls integrin-mediated cell spreading. *Molecular and Cellular Biology, 30*, 5086–5098.

Chen, Q., Hongu, T., Sato, T., Zhang, Y., Ali, W., Cavallo, J. A., et al. (2012). Key roles for the lipid signaling enzyme phospholipase d1 in the tumor microenvironment during tumor angiogenesis and metastasis. *Science Signaling, 5*, ra79.

Chen, Y., Muller, J. D., Ruan, Q., & Gratton, E. (2002). Molecular brightness characterization of EGFP in vivo by fluorescence fluctuation spectroscopy. *Biophysical Journal, 82*, 133–144.

Colley, W. C., Sung, T. C., Roll, R., Jenco, J., Hammond, S. M., Altshuller, Y., et al. (1997). Phospholipase D2, a distinct phospholipase D isoform with novel regulatory properties that provokes cytoskeletal reorganization. *Current Biology, 7*, 191–201.

Cummings, R. J., Parinandi, N. L., Zaiman, A., Wang, L., Usatyuk, P. V., Garcia, J. G., et al. (2002). Phospholipase D activation by sphingosine 1-phosphate regulates interleukin-8 secretion in human bronchial epithelial cells. *The Journal of Biological Chemistry, 277*, 30227–30235.

Elvers, M., Stegner, D., Hagedorn, I., Kleinschnitz, C., Braun, A., Kuijpers, M. E., et al. (2010). Impaired alpha(IIb)beta(3) integrin activation and shear-dependent thrombus formation in mice lacking phospholipase D1. *Science Signaling, 3*, ra1.

Faugaret, D., Chouinard, F. C., Harbour, D., El azreq, M. A., & Bourgoin, S. G. (2010). An essential role for phospholipase D in the recruitment of vesicle amine transport protein-1 to membranes in human neutrophils. *Biochemical Pharmacology, 81*, 144–156.

Gambin, Y., & Deniz, A. A. (2010). Multicolor single-molecule FRET to explore protein folding and binding. *Molecular BioSystems, 6*, 1540–1547.

Gobel, K., Schuhmann, M. K., Pankratz, S., Stegner, D., Herrmann, A. M., Braun, A., et al. (2014). Phospholipase D1 mediates lymphocyte adhesion and migration in experimental autoimmune encephalomyelitis. *European Journal of Immunology, 44*, 2295–2305.

Gomez-Cambronero, J. (2014). Phosphatidic acid, phospholipase D and tumorigenesis. *Advances in Biological Regulation, 54*, 197–206.

Gurskaya, N. G., Fradkov, A. F., Pounkova, N. I., Staroverov, D. B., Bulina, M. E., Yanushevich, Y. G., et al. (2003). A colourless green fluorescent protein homologue from the non-fluorescent hydromedusa Aequorea coerulescens and its fluorescent mutants. *The Biochemical Journal, 373*, 403–408.

Hammond, S. M., Altshuller, Y. M., Sung, T. C., Rudge, S. A., Rose, K., Engebrecht, J., et al. (1995). Human ADP-ribosylation factor-activated phosphatidylcholine-specific phospholipase D defines a new and highly conserved gene family. *The Journal of Biological Chemistry, 270*, 29640–29643.

Hammond, S. M., Jenco, J. M., Nakashima, S., Cadwallader, K., Gu, Q., Cook, S., et al. (1997). Characterization of two alternately spliced forms of phospholipase D1. Activation of the purified enzymes by phosphatidylinositol 4,5-bisphosphate, ADP-ribosylation factor, and Rho family monomeric GTP-binding proteins and protein kinase C-alpha. *The Journal of Biological Chemistry, 272*, 3860–3868.

Hanahan, D. J., & Chaikoff, I. L. (1947a). A new phospholipide-splitting enzyme specific for the ester linkage between the nitrogenous base and the phosphoric acid grouping. *The Journal of Biological Chemistry*, *169*, 699–705.

Hanahan, D. J., & Chaikoff, I. L. (1947b). The phosphorous-containing lipids of the carrot. *The Journal of Biological Chemistry*, *168*, 233–240.

Hatton, N., Lintz, E., Mahankali, M., Henkels, K. M., & Gomez-Cambronero, J. (2015). Phosphatidic acid increases epidermal growth factor receptor expression by stabilizing mRNA decay and by inhibiting lysosomal and proteasomal degradation of the internalized receptor. *Molecular and Cellular Biology*, *35*, 3131–3144.

Henkels, K. M., Boivin, G. P., Dudley, E. S., Berberich, S. J., & Gomez-Cambronero, J. (2013). Phospholipase D (PLD) drives cell invasion, tumor growth and metastasis in a human breast cancer xenograph model. *Oncogene*, *32*, 5551–5562.

Henkels, K. M., Mahankali, M., & Gomez-Cambronero, J. (2013). Increased cell growth due to a new lipase-GEF (Phospholipase D2) fastly acting on Ras. *Cellular Signalling*, *25*, 198–205.

Henkels, K. M., Mallets, E. R., Dennis, P. B., & Gomez-Cambronero, J. (2015). S6K is a morphogenic protein with a mechanism involving Filamin-A phosphorylation and phosphatidic acid binding. *FASEB Journal*, *29*, 1299–1313.

Hermansson, M., Hokynar, K., & Somerharju, P. (2011). Mechanisms of glycerophospholipid homeostasis in mammalian cells. *Progress in Lipid Research*, *50*, 240–257.

Hoppe, A., Christensen, K., & Swanson, J. A. (2002). Fluorescence resonance energy transfer-based stoichiometry in living cells. *Biophysical Journal*, *83*, 3652–3664.

Iqbal, A., Arslan, S., Okumus, B., Wilson, T. J., Giraud, G., Norman, D. G., et al. (2008). Orientation dependence in fluorescent energy transfer between Cy3 and Cy5 terminally attached to double-stranded nucleic acids. *Proceedings of the National Academy of Sciences of the United States of America*, *105*, 11176–11181.

Jain, R. K., Joyce, P. B., Molinete, M., Halban, P. A., & Gorr, S. U. (2001). Oligomerization of green fluorescent protein in the secretory pathway of endocrine cells. *The Biochemical Journal*, *360*, 645–649.

Jang, J. H., Lee, C. S., Hwang, D., & Ryu, S. H. (2012). Understanding of the roles of phospholipase D and phosphatidic acid through their binding partners. *Progress in Lipid Research*, *51*, 71–81.

Jeon, H., Kwak, D., Noh, J., Lee, M. N., Lee, C. S., Suh, P. G., et al. (2011). Phospholipase D2 induces stress fiber formation through mediating nucleotide exchange for RhoA. *Cellular Signalling*, *23*, 1320–1326.

Kim, J. H., Lee, S. D., Han, J. M., Lee, T. G., Kim, Y., Park, J. B., et al. (1998). Activation of phospholipase D1 by direct interaction with ADP-ribosylation factor 1 and RalA. *FEBS Letters*, *430*, 231–235.

Kodaki, T., & Yamashita, S. (1997). Cloning, expression, and characterization of a novel phospholipase D complementary DNA from rat brain. *The Journal of Biological Chemistry*, *272*, 11408–11413.

Kong, H. J., Polte, T. R., Alsberg, E., & Mooney, D. J. (2005). FRET measurements of cell-traction forces and nano-scale clustering of adhesion ligands varied by substrate stiffness. *Proceedings of the National Academy of Sciences of the United States of America*, *102*, 4300–4305.

Lee, C. S., Kim, J. M., Ghim, J., Suh, P. G., & Ryu, S. H. (2015). GTP-dependent interaction between phospholipase D and dynamin modulates fibronectin-induced cell spreading. *Cellular Signalling*, *27*, 2363–2370.

Lee, C. S., Kim, I. S., Park, J. B., Lee, M. N., Lee, H. Y., Suh, P. G., et al. (2006). The phox homology domain of phospholipase D activates dynamin GTPase activity and accelerates EGFR endocytosis. *Nature Cell Biology*, *8*, 477–484.

Lee, J. G., Lee, S. H., Park, D. W., Bae, Y. S., Yun, S. S., Kim, J. R., et al. (2007). Phosphatidic acid as a regulator of matrix metalloproteinase-9 expression via the TNF-alpha signaling pathway. *FEBS Letters, 581,* 787–793.

Lehman, N., Di Fulvio, M., McCray, N., Campos, I., Tabatabaian, F., & Gomez-Cambronero, J. (2006). Phagocyte cell migration is mediated by phospholipases PLD1 and PLD2. *Blood, 108,* 3564–3572.

Liscovitch, M., Czarny, M., Fiucci, G., & Tang, X. (2000). Phospholipase D: Molecular and cell biology of a novel gene family. *The Biochemical Journal, 345*(Pt. 3), 401–415.

Mahankali, M., Farkaly, T., Bedi, S., Hostetler, H. A., & Gomez-Cambronero, J. (2015). Phosphatidic acid (PA) can displace PPARalpha/LXRalpha binding to the EGFR promoter causing its transrepression in luminal cancer cells. *Scientific Reports, 5,* 15379.

Mahankali, M., Henkels, K. M., Alter, G., & Gomez-Cambronero, J. (2012). Identification of the catalytic site of phospholipase D2 (PLD2) newly described guanine nucleotide exchange factor activity. *The Journal of Biological Chemistry, 287,* 41417–41431.

Mahankali, M., Henkels, K. M., & Gomez-Cambronero, J. (2013). A GEF-to-phospholipase molecular switch caused by phosphatidic acid, Rac and JAK tyrosine kinase that explains leukocyte cell migration. *Journal of Cell Science, 126,* 1416–1428.

Mahankali, M., Henkels, K. M., Speranza, F., & Gomez-Cambronero, J. (2015). A non-mitotic role for Aurora kinase A as a direct activator of cell migration upon interaction with PLD, FAK and Src. *Journal of Cell Science, 128,* 516–526.

Mahankali, M., Peng, H. J., Henkels, K. M., Dinauer, M. C., & Gomez-Cambronero, J. (2011). Phospholipase D2 (PLD2) is a guanine nucleotide exchange factor (GEF) for the GTPase Rac2. *Proceedings of the National Academy of Sciences of the United States of America, 108,* 19617–19622.

Matz, M. V., Fradkov, A. F., Labas, Y. A., Savitsky, A. P., Zaraisky, A. G., Markelov, M. L., et al. (1999). Fluorescent proteins from nonbioluminescent Anthozoa species. *Nature Biotechnology, 17,* 969–973.

McDermott, M. I., Sigal, Y. J., Crump, J. S., & Morris, A. J. (2006). Enzymatic analysis of lipid phosphate phosphatases. *Methods, 39,* 169–179.

Merida, I., Avila-Flores, A., & Merino, E. (2008). Diacylglycerol kinases: At the hub of cell signalling. *The Biochemical Journal, 409,* 1–18.

Nakanishi, H., de los Santos, P., & Neiman, A. M. (2004). Positive and negative regulation of a SNARE protein by control of intracellular localization. *Molecular Biology of the Cell, 15,* 1802–1815.

Pathre, P., Shome, K., Blumental-Perry, A., Bielli, A., Haney, C. J., Alber, S., et al. (2003). Activation of phospholipase D by the small GTPase Sar1p is required to support COPII assembly and ER export. *The EMBO Journal, 22,* 4059–4069.

Peng, H. J., Henkels, K. M., Mahankali, M., Dinauer, M. C., & Gomez-Cambronero, J. (2011a). Evidence for two CRIB domains in phospholipase D2 (PLD2) that the enzyme uses to specifically bind to the small GTPase Rac2. *The Journal of Biological Chemistry, 286,* 16308–16320.

Peng, H. J., Henkels, K. M., Mahankali, M., Marchal, C., Bubulya, P., Dinauer, M. C., et al. (2011b). The dual effect of Rac2 on phospholipase D2 regulation that explains both the onset and termination of chemotaxis. *Molecular and Cellular Biology, 31,* 2227–2240.

Powner, D. J., & Wakelam, M. J. (2002). The regulation of phospholipase D by inositol phospholipids and small GTPases. *FEBS Letters, 531,* 62–64.

Ren, H., Federico, L., Huang, H., Sunkara, M., Drennan, T., Frohman, M. A., et al. (2010). A phosphatidic acid binding/nuclear localization motif determines lipin1 function in lipid metabolism and adipogenesis. *Molecular Biology of the Cell, 21,* 3171–3181.

Rose, K., Rudge, S. A., Frohman, M. A., Morris, A. J., & Engebrecht, J. (1995). Phospholipase D signaling is essential for meiosis. *Proceedings of the National Academy of Sciences of the United States of America, 92,* 12151–12155.

Saito, M., & Kanfer, J. (1973). Solubilization and properties of a membrane-bound enzyme from rat brain catalyzing a base-exchange reaction. *Biochemical and Biophysical Research Communications*, *53*, 391–398.

Sanematsu, F., Nishikimi, A., Watanabe, M., Hongu, T., Tanaka, Y., Kanaho, Y., et al. (2013). Phosphatidic acid-dependent recruitment and function of the Rac activator DOCK1 during dorsal ruffle formation. *The Journal of Biological Chemistry*, *288*, 8092–8100.

Speranza, F. J., Mahankali, M., & Gomez-Cambronero, J. (2013). Macrophage migration arrest due to a winning balance of Rac2/Sp1 repression over beta-catenin-induced PLD expression. *Journal of Leukocyte Biology*, *94*, 953–962.

Speranza, F., Mahankali, M., Henkels, K. M., & Gomez-Cambronero, J. (2014). The molecular basis of leukocyte adhesion involving phosphatidic acid and phospholipase D. *The Journal of Biological Chemistry*, *289*, 28885–28897.

Stace, C. L., & Ktistakis, N. T. (2006). Phosphatidic acid- and phosphatidylserine-binding proteins. *Biochimica et Biophysica Acta*, *1761*, 913–926.

Stegner, D., Thielmann, I., Kraft, P., Frohman, M. A., Stoll, G., & Nieswandt, B. (2013). Pharmacological inhibition of phospholipase D protects mice from occlusive thrombus formation and ischemic stroke—Brief report. *Arteriosclerosis, Thrombosis, and Vascular Biology*, *33*, 2212–2217.

Trujillo Viera, J., El-Merahbi, R., Nieswandt, B., Stegner, D., & Sumara, G. (2016). Phospholipases D1 and D2 suppress appetite and protect against overweight. *PLoS One*, *11*, e0157607.

VanBeek, D. B., Zwier, M. C., Shorb, J. M., & Krueger, B. P. (2007). Fretting about FRET: Correlation between kappa and R. *Biophysical Journal*, *92*, 4168–4178.

Wang, X., Dyer, J. H., & Zheng, L. (1993). Purification and immunological analysis of phospholipase D from castor bean endosperm. *Archives of Biochemistry and Biophysics*, *306*, 486–494.

Wang, X., Xu, L., & Zheng, L. (1994). Cloning and expression of phosphatidylcholine-hydrolyzing phospholipase D from Ricinus communis L. *The Journal of Biological Chemistry*, *269*, 20312–20317.

Yamazaki, M., Zhang, Y., Watanabe, H., Yokozeki, T., Ohno, S., Kaibuchi, K., et al. (1999). Interaction of the small G protein RhoA with the C terminus of human phospholipase D1. *The Journal of Biological Chemistry*, *274*, 6035–6038.

Zeniou-Meyer, M., Zabari, N., Ashery, U., Chasserot-Golaz, S., Haeberle, A. M., Demais, V., et al. (2007). Phospholipase D1 production of phosphatidic acid at the plasma membrane promotes exocytosis of large dense-core granules at a late stage. *The Journal of Biological Chemistry*, *282*, 21746–21757.

Zhang, F., Wang, Z., Lu, M., Yonekubo, Y., Liang, X., Zhang, Y., et al. (2014). Temporal production of the signaling lipid phosphatidic acid by phospholipase D2 determines the output of extracellular signal-regulated kinase signaling in cancer cells. *Molecular and Cellular Biology*, *34*, 84–95.

CHAPTER FIFTEEN

Analysis of Phosphatidic Acid Binding and Regulation of PIPKI *In Vitro* and in Intact Cells

L.W.R. Tay, Z. Wang, G. Du[1]

University of Texas Health Science Center at Houston, Houston, TX, United States

[1]Corresponding author: e-mail address: guangwei.du@uth.tmc.edu

Contents

Abstract

Phosphatidylinositol 4,5-bisphosphate [PI(4,5)P2] is a lipid second messenger that regulates a wide array of essential cellular events, such as signal transduction, vesicle trafficking, actin cytoskeleton dynamics, adhesion, and motility. To control the spatiotemporal production of PI(4,5)P2, the activity of type 1 phosphotidylinositol-4-phosphate-5-kinases (PIPKIs) is tightly regulated by small GTPases and another signaling lipid, phosphatidic acid (PA). It is of interest that PI(4,5)P2 is also a critical cofactor for the activation of the PA-generating enzyme, phospholipase D (PLD). It has been proposed that the reciprocal stimulation of PLD and PIPKI enzymes enables a rapid feedforward stimulation loop for the localized and acute generation of signaling lipids that are critical for the regulation of actin cytoskeletal reorganization and membrane trafficking. Here, we outline the methods for the expression and purification of PIPKIγ from bacteria,

Methods in Enzymology, Volume 583
ISSN 0076-6879
http://dx.doi.org/10.1016/bs.mie.2016.09.043

determination of direct PA binding, and activation of PIPKIγ using *in vitro* liposomes assays, and examination of actin cytoskeletal reorganization promoted by the PA–PIPKIγ signaling in intact cells using fluorescent microscopy.

1. INTRODUCTION

Phosphoinositides are important lipid second messengers involved in a variety of cellular processes (Martin, 1998; Viaud et al., 2016). The seven phosphoinositides are defined by the phosphorylation of the common inositol ring head group at different sites and combinations. Phosphoinositides regulate the localization and/or activity of their effector proteins through specifically binding to protein domains, such as PH and FYVE domain (Di Paolo & De Camilli, 2006; Martin, 1998; Viaud et al., 2016). Upon stimulation, the activation or inhibition of phosphoinositide metabolizing enzymes results in the generation of certain phosphoinositides at specific subcellular locations, which in turn recruit and activate their interacting signaling proteins at these sites (Di Paolo & De Camilli, 2006; Martin, 1998; Viaud et al., 2016). One of the most studied phosphoinositides is phosphatidylinositol 4,5-bisphosphate [PI(4,5)P2]. PI(4,5)P2 has been known to participate in the regulation of many essential cellular events, such as cell proliferation, death, metabolism, vesicle trafficking, intracellular calcium release, actin cytoskeleton reorganization, adhesion, and migration (Sun, Thapa, Hedman, & Anderson, 2013; Tan, Thapa, Choi, & Anderson, 2015). To perform the diverse functions of PI(4,5)P2, enzymes generating PI(4,5)P2 and its effector proteins must be tightly regulated spatially and temporally (Sun et al., 2013; Tan et al., 2015; van den Bout & Divecha, 2009). There are two families of phosphatidylinositol (PIP) kinase: type I (PIPKI) and type II (PIPKII), which generate PI(4,5)P2 from two biochemically different pools of substrate. PIPKI and PIPKII synthesize PI(4,5)P2 from phosphatidylinositol-4-bisphosphate (PI4P) and phosphatidylinositol-5-bisphosphate (PI5P), respectively (Doughman, Firestone, & Anderson, 2003). The three family members of PIPKI: PIPKIα, PIPKIβ, and PIPKIγ, as well as their splice variants, are responsible for the majority of cellular PI(4,5)P2 production (Doughman et al., 2003; Sun et al., 2013; Tan et al., 2015).

Compared to PI(4,5)P2, phosphatidic acid (PA) is a less understood signaling lipid. PA has been implicated in many important cellular processes similar to those regulated by PI(4,5)P2, such as cell proliferation, signaling transduction, vesicle trafficking, and cytoskeleton dynamics (Jang, Lee, Hwang, & Ryu, 2012; Stace & Ktistakis, 2006; Zhang & Du, 2009). Signaling PA is mainly generated by two families of enzymes, phospholipase D (PLD) and diacylglycerol kinase (DGK), each of which utilizes different phospholipid substrates. The two PLD family members, PLD1 and PLD2, hydrolyze the most abundant membrane phospholipid, phosphatidylcholine (PC), to generate PA (Frohman, 2015; Jenkins & Frohman, 2005). The 10 DGK family members generate PA through the phosphorylation of diacylglycerol (Shulga, Topham, & Epand, 2011). PA directly regulates a variety of signaling proteins (Jang et al., 2012; Stace & Ktistakis, 2006), among which are PIPKIs that generate PI(4,5)P2. PA specifically promotes the activity of all three members of the PIPKI family (Ishihara et al., 1998; Jenkins, Fisette, & Anderson, 1994). Among the three PIPKIs, PIPKIγ is the most robustly stimulated by PA (Ishihara et al., 1998). Conversely, the activity of both PLD1 and PLD2 is also strongly dependent on PI(4,5)P2 interaction (Du et al., 2000; Sciorra et al., 1999). It has been proposed that the reciprocal stimulation of PLD (PA) and PIPKI [PI(4,5)P2] enables a positive feedforward regulation loop that is critical for rapid generation of signaling lipids at specific subcellular locations upon stimulations (Cockcroft, 2001; Roach et al., 2012).

This chapter highlights several molecular and cellular biology techniques for studying PA regulation of PIPKIs, including expression and purification of recombinant PIPKIγ from bacteria, examination of direct binding between PA and PIPKIγ using a liposome pulldown assay, measurement of PA-stimulated PIPKIγ using liposomes, and examination of actin reorganization in PIPKIγ overexpressing cells. Using these techniques, we identified a PA-binding site located in the flat membrane interacting surface of PIPKI and the role of PLD-generated PA in PIPKIγ-promoted formation of actin comets and foci (Roach et al., 2012).

2. BACTERIAL EXPRESSION AND PURIFICATION OF PIPKIγ

PIPKIs expressed and purified from bacteria using affinity purification with an affinity epitope tag, such as 6×His, retain its enzymatic and regulatory properties (Loijens & Anderson, 1996; Roach et al., 2012). Both N-terminal- and C-terminal-tagged PIPKIs have been reported to be enzymatically active (Ishihara et al., 1998; Loijens & Anderson, 1996; Roach et al., 2012); therefore, the position of tag does not seem to alter the enzymatic activity. However, we found that PIPKIγ proteins expressed in bacteria appeared as multiple bands on sodium dodecyl sulfate polyacrylamide gel electrophoresis (SDS-PAGE), most likely due to early termination of protein translation. Expression of PIPKIγ in Rosetta cells, BL21 derivatives designed to enhance the expression of eukaryotic proteins that contain codons rarely used in *Escherichia coli*, at lower temperature (25°C), reduced, but not completely eliminated, the truncated proteins. To solve the problem associated with PIPKIγ truncation, we cloned PIPKIγ cDNA into a bacteria expression vector, pET24, which contains an N-terminal T7 tag and a C-terminal 6×His tag (Roach et al., 2012). Purification of PIPKIγ using the C-terminal 6×His tag led to a homogenous full-length PIPKIγ protein product (Roach et al., 2012). A detailed methodology is described later:

1. Transform pET24-PIPKIγ wild-type or mutant plasmid into Rosetta *E. coli* cells (Millipore, Billerica, MA). Grow the cells on Lysogeny broth (LB) agar plates containing 50 μg/mL kanamycin and 25 μg/mL chloramphenicol in a 37°C bacteria incubator overnight.
2. The next evening, pick a colony about 2 mm in diameter from the plate and inoculate in LB containing a final concentration of 50 μg/mL kanamycin and 25 μg/mL chloramphenicol. Shake the culture overnight in a bacterial shaker at 37°C, 250 rpm.
3. The following morning, transfer 8 mL of the cultured bacteria into 400 mL of fresh LB containing kanamycin and chloramphenicol. Continue shaking at 37°C, 250 rpm, until the OD_{600} reaches 0.6. This takes approximately 3 h.
4. Add 1 *M* isopropyl β-D-1-thiogalactopyranoside into the media to make a final concentration of 1 m*M*. Incubate culture in a bacterial shaker for 4 h at 25°C, 250 rpm.
5. Spin cells at 5000 rpm for 10 min and fully resuspend the pellet in 20 mL of cold lysis buffer [50 m*M* NaH_2PO_4 (pH 8.0), 300 m*M* NaCl, 10 m*M* imidazole. Add fresh EDTA-free protease inhibitor (Roche Diagnostics, Indianapolis, IN) before use]. From this point forward, ensure that all samples remain on ice.
6. Break bacteria cell wall and membrane using a French press. Allow sample to run through three cycles in the French press to ensure complete lysis.
7. Centrifuge samples at 18,000 rpm at 4°C for 30 min. Save supernatant for 6×His-affinity purification.
8. In a fresh 1.5-mL microfuge tube, pipette 400 μL of Ni-NTA agarose (Cat # 30210, Qiagen, Valencia, CA).

9. Centrifuge the resin for 1 min at $700 \times g$ and carefully discard the supernatant.
10. Resuspend pellet with 1.2 mL of lysis buffer.
11. Centrifuge for 1 min at $700 \times g$ and carefully remove the supernatant.
12. Transfer the bead slurry to 20 mL of cell lysate collected in step 7. Shake the sample on a head-over-end rotator for 1 h at 4°C.
13. Load the mixture of beads and lysate into a column. Wash the beads with 5 mL of wash buffer [50 m*M* NaH_2PO_4 (pH 8.0), 300 m*M* NaCl, 20 m*M* imidazole. Add fresh EDTA-free protease inhibitor before use].
14. Repeat the wash process twice.
15. Elute bound 6×His-tagged proteins by adding 250 μL of elution buffer [50 m*M* NaH_2PO_4 (pH 8.0), 300 m*M* NaCl, 250 m*M* imidazole] into the column and cap of both the top and bottom ends of the column. Allow the column to stand for 10 min. Uncap the column and collect the flow through.
16. Repeat the elution process twice. Collect each elution faction in separated tubes.
17. Dialyzed eluted proteins against 20 m*M* Tris (pH 7.6), 200 m*M* NaCl, 5 m*M* β-mercaptoethanol, followed by 20 m*M* Tris (pH 7.6), 100 m*M* NaCl, and 5 m*M* β-mercaptoethanol, to exchange the buffer and remove the imidazole used in purification and elution.
18. To check for protein purity, mix 8 μL of eluted proteins with 2 μL of 5 × SDS loading buffer [250 m*M* Tris–HCl (pH 6.8), 10% SDS, 50% glycerol, 5% β-mercaptoethanol, 0.05% bromophenol blue], and run a SDS-PAGE. Proteins can be visualized by Coomassie Blue staining. The recombinant PIPKIγ proteins are usually more than 95% pure on Coomassie Blue-stained SDS-PAGE.
19. Measure protein concentrations using Coomassie Plus protein assay (Thermo Fisher, Waltham, MA). Keep purified proteins in refrigerator and use within 1 week. For long-term storage, proteins may be kept at –80°C.

3. DETERMINATION OF THE DIRECT BINDING BETWEEN PA AND PIPKIγ USING A LIPOSOME PULLDOWN ASSAY

Many PA-binding proteins have been identified and belong to different classes of proteins, such as kinases, phosphatases, nucleotide-binding proteins, and phospholipases (Jang et al., 2012; Stace & Ktistakis, 2006;

Zhang & Du, 2009). Unlike phosphoinositides, there is no defined protein domain or motif for PA binding. Instead, PA-binding proteins often bind to the negatively charged head group of PA through their positively charged amino acid residues, including lysine, arginine, and histidine (Jang et al., 2012; Stace & Ktistakis, 2006). Despite the fact that PA strongly stimulates the activity of PIPKIs (Ishihara et al., 1998; Jenkins et al., 1994; Loijens & Anderson, 1996), the specific residues that mediate the interaction between PA and PIPKIs remained unknown for a long time.

Lipid–protein interaction can be studied using several methods, e.g., protein–lipid overlay assay on nitrocellulose membrane (or equivalent) or microtiter plates, surface plasmon resonance, fluorescence resonance energy transfer (FRET)-based liposome binding assays, liposome flotation, and liposome pulldown (Rusten & Stenmark, 2006; Zhao & Lappalainen, 2012). Liposome-based assays are generally preferred in investigating lipid–protein interaction as liposomes better mimic the membrane environments in the cell. Here, we describe a sucrose-loaded liposome pulldown assay as a convenient tool to analyze the direct binding between PA and PIPKIγ. This experiment requires an analytical bench top ultracentrifuge and a liposome extruder. With the use of this assay, we identified that amino acid residues, K97, R100, H126, and H127, are critical for the direct binding of PIPKIγ to PA (Roach et al., 2012). The same protocol can be also used to examine protein binding of other acidic phospholipids.

3.1 Liposome Preparation

1. Mix POPC (1-palmitoyl-2-oleoyl-*sn*-glycero-3-phosphocholine, Cat # 850457C, Avanti Polar Lipids, Alabaster, AL), DOPE (1,2-dioleoyl-*sn*-glycero-3-phosphoethanolamine, Cat # 850725C, Avanti Polar Lipids), and lipids of interest at a mole ratio of 6:3:1. The total amount of lipids is 8 μmol. Add additional chloroform to a volume of 1 mL. In addition to PA liposomes [POPC/DOPE/POPA (1-palmitoyl-2-oleoyl-*sn*-glycero-3-phosphate, Cat # 840857C, Avanti Polar Lipids)], PS liposomes [POPC/DOPE/POPS(1-palmitoyl-2-oleoyl-*sn*-glycero-3-phospho-L-serine, Cat # 840034C, Avanti Polar Lipids)] and PI(4)P liposomes [POPC/DOPE/PI(4)P(brain L-α-phosphatidylinositol-4-phosphate, Cat # 840045, Avanti Polar Lipids)] can be also used as nonspecific charge and substrate binding controls, respectively.

2. Add the lipid mixture into a 100-mL round-bottom boiling flask. Attach the flask to a rotary vacuum evaporator that is connected to a gas pressure regulator. Adjust the level of the flask to ensure that the lipid mixture is below the water level in a preheated 37°C water bath.

   Note: Start with a very clean flask. Rinse flask with chloroform and allow it to air dry in the chemical hood between different liposomes. Heating is critical for some phospholipids as the temperature drop during organic solvent evaporation may lead to the precipitation of some phospholipids and the formation of uneven lipid films.
3. Rotate the flask at 120 rpm. Ensure that the gas pressure regulator is closed.
4. Turn the vacuum on and slowly turn on the valve of gas pressure regulator to raise the pressure. Keep the flask rotating until the organic solvents completely evaporate. The lipid mixture should form a thin and uniform film on the side of the flask.
5. Once fully evaporated, raise the level of flask to keep it out of the water bath. Stop rotation, but keep the vacuum on for 2 h to remove residual organic solvents.
6. Add 1 mL of inside buffer [176 m*M* sucrose in 20 m*M* Tris–HCl (pH 7.6)] to resuspend the lipids. Shake and vortex the lipid mixture to make sure that no lipid film is left on the wall of the flask.
7. Transfer lipids to a 5-mL cryotube and perform 10 cycles of freeze–thaw using liquid nitrogen and a 37°C water bath.
8. To make liposomes of 100 nm in size, prewet filter paper (25 mm) and polycarbonate membranes (Nuclepore Track-Etch membrane, pore size 100 nm, Cat # 110605, GE Healthcare Bio-Sciences, Pittsburgh, PA) with outside buffer. Assemble the Lipex Extruder (Northern Lipids INC, Burnaby, BC, Canada) by stacking one filter paper and then two polycarbonate membranes with the shiny-side facing up (a miniextruder from Avanti Polar Lipids can also be used for liposome extrusion). Connect the Lipex Extruder to a nitrogen gas tank with a regulator.
9. Slowly turn on nitrogen gas. Ensure that there is no leak from membranes when nitrogen gas pressure is applied. Extrude the liposomes 10 times with nitrogen gas at a pressure of 260 psi.
10. Add 4 mL outside (binding) buffer [20 m*M* Tris–HCl (pH 7.6), 100 m*M* NaCl, and 1 m*M* EDTA] to the liposomes. Transfer the liposomes into ultracentrifugation tubes and centrifuge liposomes at 100,000 × *g* for 1 h at room temperature.

Note: It might be desirable to perform sucrose-loaded liposome pulldown assays in binding buffers with different salt concentrations. It is important to keep equal osmotic strength for the inside and outside (binding) buffers. The different combinations of inside and outside buffers for the equal osmotic strength have been described in a previous publication (Buser & McLaughlin, 1998).

11. Resuspend the pellet with 1 mL outside (binding) buffer.
12. Liposomes can be stored at 4°C for up to 1 week. If needed, dilute liposomes with outside (binding) buffer for desired concentrations.

3.2 Liposome Pulldown

1. *Dilute proteins* of interest with outside (binding) buffer plus 1 m*M* DTT (DTT is added fresh before the experiment) to a total volume of 200 μL in a clean 1.5-mL ultracentrifuge tube (for PIPKIγ, we used 0.6 μg of purified protein). Centrifuge to remove aggregated proteins at 100,000 × *g* for 30 min at 4°C.
2. Carefully transfer the supernatant to another clean ultracentrifuge tube. Add liposomes at desired concentrations (e.g., 1000, 200, 40, 10, and 0 μ*M*) into the tube containing the target protein. Incubate the liposome and protein mixtures at 4°C for 30 min with rotation.
3. Centrifuge samples at 100,000 × *g* at 4°C for 30 min.
4. Carefully remove all supernatant and resuspend pellet with 1 × SDS loading buffer. Run the samples in SDS-PAGE and check for bound proteins by Western blot using an antibody against epitope tags, such as T7 or 6×His.

4. MEASUREMENT OF PA-STIMULATED PIPKIγ ACTIVITY USING LIPOSOMES

PI(4,5)P2 production generated from PIPKI activity can be measured either in intact cells or *in vitro* using purified proteins. By quantifying the phosphorylation of PI4P to PI(4,5)P2, the use of an *in vitro* activity assay allows the study of the direct regulation of PIPKI activity by certain lipids or proteins. The activity of PIPKI can be measured *in vitro* using either micelles (Ishihara et al., 1998; Loijens & Anderson, 1996; Parker, Loijens, & Anderson, 1998) or liposomes (Roach et al., 2012). Here, we describe the use of liposomes to measure the activity of PIPKIγ in the presence or absence of PA. The described kinase activity assay involves the use of radioactive materials. Investigators are required to follow the radiation safety guidelines in their institution regarding proper handling, storage, disposal, and/or clean-up of radioactive materials. To reduce radioactive exposure, wear the appropriate personal protective equipment and use proper shielding. In addition, it is critical to use filtered pipette tips to reduce contamination of samples and pipettes.

4.1 PIPKI Liposome Kinase Assay

1. Prepare liposomes using the same method for liposome pulldown described in Section 3. Replace both inside and outside buffers with 1 × PIPKI kinase reaction buffer [50 m*M* Tris–HCl (pH 7.5), 100 m*M* NaCl, 10 m*M* $MgCl_2$, 1 m*M* EGTA]. The lipids in liposomes with or without PA are PC:PA:PI4P (60:20:20) and PC:PI4P (80:20), respectively.
2. Place each sample to be assayed into a 1.5-mL Eppendorf tube. We use 0.5–20 ng of purified PIPKIγ proteins diluted in 10 μL 1 × PIPKI kinase reaction buffer [500 m*M* Tris–HCl (pH 7.5), 1 *M* NaCl, 100 m*M* $MgCl_2$, 10 m*M* EGTA].
3. Prepare the reaction mixture as below in 40 μL of 1 × PIPKI kinase reaction buffer. It is suggested to set up one or half more reaction to have enough reaction mixture for the kinase assay. Save at least 1 μL of the reaction mix for quantifying PI(4,5)P2 production later (see step 6 in Section 4.3).

Stock Solution	Volume/ Tube	Final Concentration in 50 μL	Total Volume for *n* Samples
10 × PIPKI kinase reaction buffer	5 μL	1 ×	5 μL × *n*
10 m*M* ATP	0.25 μL	50 μ*M*	0.25 μL × *n*
[^{32}P]-γ-ATP (10 μCi/μL) (Perkin Elmer, Waltham, MA)	0.5 μL	5 μCi	0.5 μL × *n*
4 m*M* liposomes	5 μL	400 μ*M*	5 μL × *n*
ddH_2O	29.75 μL	–	29.75 μL × *n*

4. Start the reaction by adding 40 μL of the reaction mixture to each sample in step 2. Incubate the reaction mixture at room temperature for 10 min.

5. To stop the reaction, add 100 μL of 1N HCl. Immediately after stopping the reaction, add 200 μL of chloroform:methanol (1:1) and vortex.
6. Centrifuge at maximum speed in a microcentrifuge for 1 min. Remove and discard the upper aqueous phase in a liquid radioactive waste container.
7. Wash with 80 μL of methanol:1N HCl (1:1) and vortex.
8. Centrifuge at maximum speed in a microcentrifuge for 1 min. Remove and discard the upper aqueous phase in a liquid radioactive waste container. Save the bottom organic phase. Discard radioactive waste accordingly.

4.2 Separation of PI(4,5)P2 Lipids by Thin Layer Chromatography

1. Carefully mark the site for sample spotting 2.5 cm from the bottom edge of the potassium oxalate pretreated silica gel H TLC plate (Analtech, Newark, DE). Leave at least 1 cm interval between each samples.
2. Bake the TLC plate in an oven at 100°C for 10 min and allow it to cool to room temperature.
3. Prepare a chromatography tank by lining it with filter paper and adding 209 mL of chloroform:methanol:15N NH_4OH:dH_2O (90:90:7:22). Allow it to equilibrate for at least 30 min.
4. Spot lipid samples using 50-μL glass capillary pipets.
5. Place the TLC plate loaded with samples into the chromatography tank with chloroform:methanol:15N NH_4OH:dH_2O (90:90:7:22), allow it to develop until the solvent front reaches the top of the TLC plate. This step takes about 2 h.
6. Remove the TLC plate from the chromatography tank and allow it to air dry for 10 min.

4.3 Quantification of PI(4,5)P2 Production

1. Phospholipids can be visualized with iodine vapor. Similarly, labeled PI(4,5)P2 products can also be visualized by autoradiography and identified by comparing them to unlabeled phospholipid standards. This can be done by wrapping the plate in plastic wrap and attaching autoradiography orientation markers. Overlay with X-ray film and place it in an X-ray film cassette. Expose the film at –80°C or room temperature overnight. Develop the film.

2. To quantify PI(4,5)P2 production, align the developed film with the TLC plate. Mark the position of PI(4,5)P2 on the plate with a pencil.
3. Using a razor blade, scrape the region containing PI(4,5)P2 on the TLC plate into a scintillation vial containing 5 mL Cytoscint* scintillation cocktail (Thermo Fisher).
4. Read the cpm of ^{32}P of scraped samples using a liquid scintillation counter.
5. Additionally, read the cpm of ^{32}P using 1 μL of the reaction mixture from step 3 in Section 4.1.
6. Since the chemical reaction for PI(4,5)P2 generation is PI(4)P + ATP → PI(4,5)P2 + ADP, we can use the following equation to calculate PI(4,5)P2 production (pmol/min/μg PIPKI protein):

$$\mathrm{PI(4,5)P2\ production} = \frac{\text{cpm of scraped spot from the TLC plate} \times [\text{ATP}](\text{pmol})}{\text{cpm of total reation x reaction time}(\text{min}) \times \text{PIPKI protein}(\mu\text{g})}$$

In our condition, [ATP] is 2500 pmol, reaction time is 10 min, the cpm of total reaction should be cpm from 1 μL of the reaction mixture in step 5 multiplied by the total volume of 40 μL. Therefore, the equation can be rewritten as:

$$\mathrm{PI(4,5)P2\ production} = \frac{\text{cpm of scraped spot from the TLC plate} \times 2500\ \text{pmol}}{\text{cpm of total reation} \times 10\ \text{min} \times \text{PIPKI protein}(\mu\text{g})}$$

5. ASSESSMENT OF ACTIN CYTOSKELETAL REORGANIZATION INDUCED BY PIPKIγ OVEREXPRESSION

One of the most well-studied functions of all PIPKIs and their catalytic product, PI(4,5)P2, is the regulation of actin cytoskeletal reorganization, which is essential for cell motility, membrane trafficking, and pathogen infection (Tan et al., 2015; van den Bout & Divecha, 2009). Infection of intracellular pathogens, such as *Listeria*, *Vaccinia*, and *Shigella* (Frischknecht et al., 1999), or stimulation of platelet-derived growth factor and pervanadate (Rozelle et al., 2000), induces drastic changes in actin cytoskeleton that can be classified into two morphologically distinct structures in phalloidin-stained cells: actin comets (phalloidin-stained F-actin structure with a head and a tail) and foci (F-actin-labeled round spots). It is believed that pathogens use actin comets and foci along with the associated vesicles for intracellular propulsion (Frischknecht et al., 1999; Gouin, Welch, & Cossart, 2005). The formation of actin comets and foci is dependent on PI(4,5)P2- and PI(4,5)P2-regulated actin regulatory proteins (Rivera, Vasilescu, Papayannopoulos, Lim, & Mayer, 2009; Roach et al., 2012). In the absence of infection or external stimulation, overexpression of any of the three PIPKI family members alone is sufficient to induce the formation of actin comets and foci (Martin, 1998; Roach et al., 2012), supporting the critical role of PIPKIs in this process. Here, we present a protocol for visualizing actin comets and foci in PIPKIγ-transfected mammalian cells by fluorescently labeled phalloidin staining of F-actin. One caution is that overexpression of PIPKI can lead to cell death or toxicity, which may make the explanation of experimental results difficult (Martin, 1998; Roach et al., 2012). To achieve reproducible and reliable results, actin structure of cells should be imaged within 24 h after transfection.

1. Plate cells, e.g., Cos7, in Dulbecco's modified Eagle medium containing 10% fetal bovine serum, 100 U/mL penicillin, and 100 mg/mL streptomycin, on glass coverslips in a 12-well plate 1 day before experiment.
2. Add 0.5 μg plasmid DNA carrying GFP-PIPKIγ wild-type, PA-binding defective mutants, or other mutants, to a sterile 1.5-mL Eppendorf tube along with 50 μL of 150 m*M* NaCl and 4 μL of 25 m*M* deacylated polyethylenimine (PEI) (average molecular weight, 25,000 Da) [prepared according as described (Thomas et al., 2005)]. Vortex briefly. Incubate the DNA/PEI mixture at room temperature for 10–30 min.
3. Add the mixture to cells drop by drop. Rock the plate gently to evenly distribute the DNA/PEI complex.
4. (Optional) To study the regulation of PIPKI wild-type protein by the enzymatic activity of PLDs and DGKs, cells can be treated with vehicle control DMSO, pan-PLD inhibitor FIPI (5-fluoro-2-indolyl-deschlorohalopemide) (Cat # F5807, Sigma-Aldrich, St. Louis, MO), pan-PLD inhibitor (Cat # VU0155056, Avanti Polar Lipids), PLD1 inhibitor (Cat VU0359595, Avanti Polar Lipids), PLD2 inhibitor (Cat # VU0285655-1, Avanti Polar Lipids), or DGK inhibitor (Cat # R59022, Calbiochem, La Jolla, CA). Alternatively, GFP-PIPKIγ can be transfected into cells in which individual PA-generating enzyme has been knocked down by small interfering RNAs or small hairpin RNAs.

5. 16–24 h after transfection, fix cells with 2% paraformaldehyde at room temperature for 10 min.
6. Wash cells with PBS, three times each with a 5-min interval.
7. Permeabilize cells with 0.1% Triton X-100 in PBS for 10 min.
8. Wash cells with PBS once.
9. Stain cells with rhodamine-conjugated phalloidin (Cat # R415, Thermo Fisher) (1:200 in PBS) for 1 h in the dark at room temperature.
10. Wash cells with PBS, three times each with a 5-min interval between each wash.
11. Pipette 3–5 μL of antifade mounting media [4% *n*-propyl gallate (w/v) in 90% glycerol (v/v) and 10% PBS] onto a microscope slide.
12. Using a forceps, gently pick-up the coverslip, use a kimwipe to remove excess buffer. To mount coverslips onto a microscope slide, place coverslips cell-side down at a 45-degree angle and slowly lower coverslips toward the surface of the slide. Aspirate any excess mounting media and allow it to air dry for a couple of minutes before sealing the edges of the coverslips with clear nail polish.
13. Observe slides under an epifluorescent microscope or confocal microscope. Identify cells transfected with GFP-PIPKIγ by green fluorescence. Examine the membrane localization of GFP signals, and the number of actin comets and foci in cells expressing PIPKIγ wild-type and mutant proteins.

6. SUMMARY

Over the last few decades, it has become well accepted that lipids, including PI(4,5)P2 and PA described in this chapter, play critical roles in signaling transduction (Lemmon, 2008). The cloning and study of the PI(4,5)2 and PA-metabolizing enzymes, especially PIPKIs and PLDs, have revealed important cellular and physiological roles of these two important lipids (Frohman, 2015; Tan et al., 2015). Using molecular and cell biology techniques, including some described in this chapter, and mouse models, it is generally accepted that the local production of PI(4,5)P2 and PA mediates their diverse functions (Frohman, 2015; Tan et al., 2015). However, it remains technically challenging to dissect the specific role of locally produced PI(4,5)P2 and PA and to identify specific enzymes that are responsible for their generation at various subcellular locations. The current PI(4,5)P2 and PA biosensors mainly detect the production of these two lipids on the plasma membrane (Lu, Tay, He, & Du, 2016; Tan et al., 2015; Zhang et al.,

2014). More sensitive approaches that can effectively monitor intracellular PI(4,5)P2 and PA are urgently needed. In addition, due to the lack of specific antibodies for immunofluorescent staining, our knowledge about the localization of PIPKIs and PLDs is mainly based on overexpression studies that often nonspecifically change the subcellular localizations of these enzymes. With the rapid progress in gene-editing technology, it has become feasible to fuse an epitope tag or introduce specific mutations to the genomic locus of each member of PIPKs and PLDs using the clustered regularly interspaced short palindromic repeats (CRISPRs)/Cas9 technique (Doudna & Charpentier, 2014). These engineered proteins are expected to express at the similar levels of the endogenous proteins, thus will certainly help us gain better understanding of PI(4,5)P2 and PA signaling and their contribution to cellular functions, physiology, and pathology.

ACKNOWLEDGMENTS

This study was supported by a grant from National Institutes of Health (NIH) (HL119478) and grants from the Cancer Prevention and Research Institute of Texas (CPRIT) (RP130425 and RP160775) to G.D.

REFERENCES

Buser, C. A., & McLaughlin, S. (1998). Ultracentrifugation technique for measuring the binding of peptides and proteins to sucrose-loaded phospholipid vesicles. *Methods in Molecular Biology*, *84*, 267–281.

Cockcroft, S. (2001). Signalling roles of mammalian phospholipase D1 and D2. *Cellular and Molecular Life Sciences*, *58*(11), 1674–1687.

Di Paolo, G., & De Camilli, P. (2006). Phosphoinositides in cell regulation and membrane dynamics. *Nature*, *443*(7112), 651–657.

Doudna, J. A., & Charpentier, E. (2014). Genome editing. The new frontier of genome engineering with CRISPR-Cas9. *Science*, *346*(6213), 1258096.

Doughman, R. L., Firestone, A. J., & Anderson, R. A. (2003). Phosphatidylinositol phosphate kinases put PI4,5P(2) in its place. *Journal of Membrane Biology*, *194*(2), 77–89.

Du, G., Altshuller, Y. M., Kim, Y., Han, J. M., Ryu, S. H., Morris, A. J., et al. (2000). Dual requirement for rho and protein kinase C in direct activation of phospholipase D1 through G protein-coupled receptor signaling. *Molecular Biology of the Cell*, *11*(12), 4359–4368.

Frischknecht, F., Moreau, V., Rottger, S., Gonfloni, S., Reckmann, I., Superti-Furga, G., et al. (1999). Actin-based motility of vaccinia virus mimics receptor tyrosine kinase signalling. *Nature*, *401*(6756), 926–929.

Frohman, M. A. (2015). The phospholipase D superfamily as therapeutic targets. *Trends in Pharmacological Sciences*, *36*(3), 137–144.

Gouin, E., Welch, M. D., & Cossart, P. (2005). Actin-based motility of intracellular pathogens. *Current Opinion in Microbiology*, *8*(1), 35–45.

Ishihara, H., Shibasaki, Y., Kizuki, N., Wada, T., Yazaki, Y., Asano, T., et al. (1998). Type I phosphatidylinositol-4-phosphate 5-kinases. Cloning of the third isoform and deletion/substitution analysis of members of this novel lipid kinase family. *Journal of Biological Chemistry*, *273*(15), 8741–8748.

Jang, J. H., Lee, C. S., Hwang, D., & Ryu, S. H. (2012). Understanding of the roles of phospholipase D and phosphatidic acid through their binding partners. *Progress in Lipid Research, 51*(2), 71–81.

Jenkins, G. H., Fisette, P. L., & Anderson, R. A. (1994). Type I phosphatidylinositol 4-phosphate 5-kinase isoforms are specifically stimulated by phosphatidic acid. *Journal of Biological Chemistry, 269*(15), 11547–11554.

Jenkins, G. M., & Frohman, M. A. (2005). Phospholipase D: A lipid centric review. *Cellular and Molecular Life Sciences, 62*(19–20), 2305–2316.

Lemmon, M. A. (2008). Membrane recognition by phospholipid-binding domains. *Nature Reviews Molecular Cell Biology, 9*(2), 99–111.

Loijens, J. C., & Anderson, R. A. (1996). Type I phosphatidylinositol-4-phosphate 5-kinases are distinct members of this novel lipid kinase family. *Journal of Biological Chemistry, 271*(51), 32937–32943.

Lu, M., Tay, L. W., He, J., & Du, G. (2016). Monitoring phosphatidic acid signaling in breast cancer cells using genetically encoded biosensors. *Methods in Molecular Biology, 1406*, 225–237.

Martin, T. F. (1998). Phosphoinositide lipids as signaling molecules: Common themes for signal transduction, cytoskeletal regulation, and membrane trafficking. *Annual Review of Cell and Developmental Biology, 14*, 231–264.

Parker, G. J., Loijens, J. C., & Anderson, R. A. (1998). Detection of phosphatidylinositol-4-phosphate 5-kinase activity using thin-layer chromatography. *Methods in Molecular Biology, 105*, 127–139.

Rivera, G. M., Vasilescu, D., Papayannopoulos, V., Lim, W. A., & Mayer, B. J. (2009). A reciprocal interdependence between Nck and PI(4,5)P(2) promotes localized N-WASp-mediated actin polymerization in living cells. *Molecular Cell, 36*(3), 525–535.

Roach, A. N., Wang, Z., Wu, P., Zhang, F., Chan, R. B., Yonekubo, Y., et al. (2012). Phosphatidic acid regulation of PIPKI is critical for actin cytoskeletal reorganization. *Journal of Lipid Research, 53*(12), 2598–2609.

Rozelle, A. L., Machesky, L. M., Yamamoto, M., Driessens, M. H., Insall, R. H., Roth, M. G., et al. (2000). Phosphatidylinositol 4,5-bisphosphate induces actin-based movement of raft-enriched vesicles through WASP-Arp2/3. *Current Biology, 10*(6), 311–320.

Rusten, T. E., & Stenmark, H. (2006). Analyzing phosphoinositides and their interacting proteins. *Nature Methods, 3*(4), 251–258.

Sciorra, V. A., Rudge, S. A., Prestwich, G. D., Frohman, M. A., Engebrecht, J., & Morris, A. J. (1999). Identification of a phosphoinositide binding motif that mediates activation of mammalian and yeast phospholipase D isoenzymes. *The EMBO Journal, 18*, 5911–5921.

Shulga, Y. V., Topham, M. K., & Epand, R. M. (2011). Regulation and functions of diacylglycerol kinases. *Chemical Reviews, 111*(10), 6186–6208.

Stace, C. L., & Ktistakis, N. T. (2006). Phosphatidic acid- and phosphatidylserine-binding proteins. *Biochimica et Biophysica Acta, 1761*(8), 913–926.

Sun, Y., Thapa, N., Hedman, A. C., & Anderson, R. A. (2013). Phosphatidylinositol 4,5-bisphosphate: Targeted production and signaling. *Bioessays, 35*(6), 513–522.

Tan, X., Thapa, N., Choi, S., & Anderson, R. A. (2015). Emerging roles of PtdIns(4,5)P2—Beyond the plasma membrane. *Journal of Cell Science, 128*(22), 4047–4056.

Thomas, M., Lu, J. J., Ge, Q., Zhang, C., Chen, J., & Klibanov, A. M. (2005). Full deacylation of polyethylenimine dramatically boosts its gene delivery efficiency and specificity to mouse lung. *Proceedings of the National Academy of Sciences of the United States of America, 102*(16), 5679–5684.

van den Bout, I., & Divecha, N. (2009). PIP5K-driven PtdIns(4,5)P2 synthesis: Regulation and cellular functions. *Journal of Cell Science, 122*(Pt. 21), 3837–3850.

Viaud, J., Mansour, R., Antkowiak, A., Mujalli, A., Valet, C., Chicanne, G., et al. (2016). Phosphoinositides: Important lipids in the coordination of cell dynamics. *Biochimie*, *125*, 250–258.
Zhang, Y., & Du, G. (2009). Phosphatidic acid signaling regulation of Ras superfamily of small guanosine triphosphatases. *Biochimica et Biophysica Acta*, *1791*(9), 850–855.
Zhang, F., Wang, Z., Lu, M., Yonekubo, Y., Liang, X., Zhang, Y., et al. (2014). Temporal production of the signaling lipid phosphatidic acid by phospholipase D2 determines the output of extracellular signal-regulated kinase signaling in cancer cells. *Molecular and Cellular Biology*, *34*(1), 84–95.
Zhao, H., & Lappalainen, P. (2012). A simple guide to biochemical approaches for analyzing protein–lipid interactions. *Molecular Biology of the Cell*, *23*(15), 2823–2830.

AUTHOR INDEX

Note: Page numbers followed by "*f*" indicate figures, and "*t*" indicate tables.

C

D

E

G

H

K

L

M

N

O

P

Q

R

S

T

U

X

Y

Z

SUBJECT INDEX

Note: Page numbers followed by "*f*" indicate figures, and "*t*" indicate tables.

D

H

I

J

K

L

N

O

P

T

Kolibri 360 GmbH
Ferdinand-Jühlke-Straße 7
99095 Erfurt, Deutschland
produktsicherheit@kolibri360.de